Intelligent Semantic Analysis
FOR HEALTHCARE

Intelligent Semantic Analysis FOR HEALTHCARE

Edited by

SONALI VYAS
Professor, JK Lakshmipat University, Jaipur, Rajasthan, India

SAMIYA KHAN
Assistant Professor, Computer Science, University of Southampton, Delhi, India

PRIYADARSAN PATRA
Vice-Chancellor, NIST University, Odisha, India

KORHAN CENGIZ
Associate Professor, Biruni University, Istanbul, Turkey

ELSEVIER

ACADEMIC PRESS
An imprint of Elsevier

Academic Press is an imprint of Elsevier
125 London Wall, London EC2Y 5AS, United Kingdom
50 Hampshire Street, 5th Floor, Cambridge, MA 02139, United States

ISBN: 978-0-443-33512-9

For information on all Academic Press publications visit our website at https://www.elsevier.com/books-and-journals

Publisher: Mara Conner
Acquisitions Editor: Saniya Puri
Editorial Project Manager: Thangminlun Kipgen
Production Project Manager: Maria Shalini
Cover Designer: Arumugam Kothandan

Typeset by TNQ Tech

Contents

Contributors

Nidhi Agarwal
Faculty of Social Science and Humanities, Lincoln University College, Petaling Jaya, Malaysia

Azamat Ali
Lovely Professional University, Phagwara, Punjab, India

Emerson Galang Aliswag
National University, Manila, Philippines

N. Ambika
Jain College Vasavi Campus, Department of Computer Science, Bangalore, Karnataka, India

Ulfat Andrabi
Mittal School of Business, Lovely Professional University, Phagwara, Punjab, India

Aaliya Ashraf
School of Business, Chandigarh University, Chandigarh, Punjab, India

Manju Bala
Chandigarh Business School of Administration, Landran, India

V. Bhanumathi
Department of Electronics and Communication Engineering, College of Engineering Guindy Campus, Anna University, Chennai, Tamil Nadu, India

Tarandeep Kaur Bhatia
Amity School of Engineering and Technology, Amity University, Mohali, Punjab, India

Priyanka Chhibber
Mittal School of Business, Lovely Professional University, Phagwara, Punjab, India

Geetika Dikshit
Maulana Azad National Institute of Technology, Bhopal, Madhya Pradesh, India

Shivendra Dubey
Department of Artificial Intelligence and Machine Learning, Manipal University, Jaipur, Rajasthan, India

Selena Firmin
Federation University, Ballarat, VIC, Australia

Shaurya Gupta
School of Computer Science, University of Petroleum and Energy Studies, Dehradun, Uttarakhand, India

Amelie Gyrard
Machine-to-Machine Measurement (M3), Paris, France

Tanvi Hungund
California State University, Fullerton, CA, United States

Md Warif Islam Mahim
Department of Bachelor of Business Administration, School of Business and Economics, United International University, Dhaka, Bangladesh

Rajiv Iyer
Department of CSE, Amity School of Engineering and Technology, Amity University Maharashtra, Mumbai, Maharashtra, India

Atanu Jana
Department of Electronics and Communication Engineering, NIST University, Berhampur, Odisha, India

Ahmed Imran Kabir
Department of Bachelor of Business Administration, School of Business and Economics, United International University, Dhaka, Bangladesh

Shreyasi Kadam
Department of Pharmaceutics, Genesis Institute of Pharmacy, Kolhapur, Maharashtra, India

Ouali Khawla
Faculty of Economics, Commerce and Management Sciences Badji Mokhtar, Annaba, Algeria

Shobhit Kukreti
Carnegie Mellon University, Pennsylvania, PA, United States

Pradnya S. Kulkarni
Dr. Vishwanath Karad MIT World Peace University, Pune, Maharashtra, India

Rohit Kumar
ABV-IIITM, Gwalior, Madhya Pradesh, India

Puneet Kumar
Kalinga University, Naya Raipur, Chhattisgarh, India

Bouguerra Mabrouk
Faculty of Economics, Commerce and Management Sciences Badji Mokhtar, Annaba, Algeria

Sandipan Mallik
Department of Electronics and Communication Engineering, NIST University, Berhampur, Odisha, India

Vedprakash Maralapalle
L. S. Raheja School of Architecture, School of Architecture, Mumbai, Maharashtra, India

Dewan Ariyan Mehdi
Department of Bachelor of Business Administration, School of Business and Economics, United International University, Dhaka, Bangladesh

Amaresh Kumar Mohanty
Department of Computer Science and Engineering, NIST University, Berhampur, Odisha, India

Nigar K. Mujawar
Department of Pharmaceutics, ASPM College of Pharmacy, Sindhudurg, Maharashtra, India

Sonali Nikam
Department of Pharmaceutical Chemistry, D. Y. Patil Education Society, D. Y. Patil College of Pharmacy, Kolhapur, Maharashtra, India

U. Padmavathi
Shiv Nadar University, Chennai, Tamil Nadu, India

Bandhan Panda
Department of Computer Science and Engineering, NIST University, Berhampur, Odisha, India

Snehal Panda
Department of Electronics and Communication Engineering, NIST University, Berhampur, Odisha, India

Pranali Pangam
Department of Pharmaceutical Chemistry, Dr. N. J. Paulbudhe College of Pharmacy, Ahmednagar, Maharashtra, India

K. Parkavi
School of Computer Science and Engineering, Vellore Institute of Technology, Vellore, Tamil Nadu, India

Swapnali Patil
Department of Pharmaceutics, Vijayrao Naik College of Pharmacy, Shirwal, Maharashtra, India

Priyadarsan Patra
Department of Computer Science and Engineering, NIST University, Berhampur, Odisha, India

Satya Narayana Patro
Department of Radiology, University of Arkansas for Medical Sciences (UAMS), Little Rock, AR, United States

A. Pavithra
Department of Electronics and Communication Engineering, College of Engineering Guindy Campus, Anna University, Chennai, Tamil Nadu, India

Vincent Pham
Eqimonie, Paris, France

G Sekhar Reddy
Department of Computer Science and Engineering, NIST University, Berhampur, Odisha, India

Grari Sabrina
Faculty of Economics, Commerce and Management Sciences Badji Mokhtar, Annaba, Algeria

Pournima Sankpal
Department of Pharmaceutical Chemistry, Bharati Vidyapeeth College of Pharmacy, Palus, Sangli, Maharashtra, India

A. Saranya
Department of CSE, Amity School of Engineering and Technology, Amity University Maharashtra, Mumbai, Maharashtra, India

Rahul Shrivastava
SRM Institute of Science and Technology, Tiruchirappalli, Tamil Nadu, India

N. Sudarssan
Department of Artificial Intelligence and Data Science, Sri Eshwar College of Engineering, Coimbatore, Tamil Nadu, India

Ashwini R. Suryawanshi
Sant Gajanan Maharaj College of Pharmacy, Kolhapur, Maharashtra, India

Kunjabihari Swain
Department of Electrical Engineering, NIST University, Berhampur, Odisha, India

Yalouli Tarek
Faculty of Economics, Commerce and Management Sciences Badji Mokhtar, Annaba, Algeria

Neethu Rose Thomas
School of Advanced Sciences, Vellore Institute of Technology, Vellore, Tamil Nadu, India

Shiva Tushir
Department of Pharmacy, Panipat Institute of Engineering and Technology, Samalkha, India

Navita Vashista
Department of Management Studies, Panipat Institute of Engineering and Technology, Samalkha, India

Goguldeep Veluswamy
Department of Radiology, University of Arkansas for Medical Sciences (UAMS), Little Rock, AR, United States

Sonali Vyas
IET, JK Lakshmipat University, Jaipur, Rajasthan, India; Department of Computer Science, JK Lakshmipat University, Jaipur, Rajasthan, India

Preface

The rapid convergence of artificial intelligence, semantic technologies, and cyber–-physical systems is transforming healthcare at an unprecedented pace. What was once a fragmented ecosystem of data silos and reactive practices is evolving into an interconnected, context-aware paradigm. This volume, *Intelligent Semantic Analysis for Healthcare*, arises from a shared recognition that the true power of this transformation lies not merely in the abundance of data, but in its meaning—its semantics.

In today's clinical environment, data are increasingly multimodal, spanning electronic health records, medical imaging, and real-time wearable sensor streams. Yet, without semantic coherence, these vast resources remain underutilized. Our guiding principle is that semantic intelligence—enabled through knowledge representation, ontologies, large language models (LLMs), and interoperable frameworks—forms the essential backbone of next-generation healthcare.

To provide a comprehensive understanding of this complex domain, we have structured this book into a layered journey:

- **Part I** lays the foundational frameworks, examining advanced data analytics, sustainability in smart healthcare, and cyber–physical system architectures.
- **Part II** explores the physical layer, focusing on IoT, wearables, and the Internet of Medical Things (IoMT), which serves as the primary data acquisition interface.
- **Part III** the semantic core of the book, delves into knowledge graphs, large language models, and hybrid semantic–statistical systems that enable explainable and trustworthy decision-making.
- **Part IV** presents domain-specific clinical applications, illustrating how semantic integration translates into actionable intelligence in radiology, oncology, and diagnostic innovation.
- **Part V** addresses the critical issues of trust, including privacy, security, and resilience, which are indispensable for real-world deployment.
- **Part VI** expands the discussion to sustainability, policy frameworks, and emerging paradigms, such as the metaverse, offering a forward-looking perspective on intelligent healthcare ecosystems.

A distinctive strength of this volume lies in its interdisciplinary nature. Contributions spanning computer science, biomedical engineering, and clinical research underscore the cross-domain collaboration required to address today's complex healthcare challenges. The inclusion of chapters on large language models, hybrid architectures, and emerging agentic workflows is especially timely, reflecting how these technologies are reshaping the way clinical knowledge is generated, interpreted, and applied.

As editors, we are deeply grateful to all the contributing authors for their scholarly rigor and commitment. Their work reflects a shared vision—to move from data-driven healthcare to meaning-driven healthcare, where insights are not only accurate, but also interpretable, ethical, and truly actionable.

We hope that this volume will serve both as a valuable reference and as a catalyst—-stimulating new research directions, fostering interdisciplinary collaboration, and contributing to the realization of intelligent, equitable, and sustainable healthcare systems worldwide.

Editors
Prof. Sonali Vyas
Professor, JK Lakshmipat University, Jaipur, Rajasthan, India
Dr. Samiya Khan
Assistant Professor, Computer Science, University of Southampton, Delhi
Prof. Priyadarsan Patra
Vice-Chancellor, NIST University, Odisha, India
Prof. Korhan Cengiz
Associate Professor, Prince Mohammad Bin Fahd University, Saudi Arabia

PART I

Foundational frameworks of intelligent healthcare

CHAPTER 1

Fundamentals of data analysis for healthcare informatics

Neethu Rose Thomas[1] and K. Parkavi[2]
[1]School of Advanced Sciences, Vellore Institute of Technology, Vellore, Tamil Nadu, India; [2]School of Computer Science and Engineering, Vellore Institute of Technology, Vellore, Tamil Nadu, India

1. Introduction

The history of data analysis in healthcare dates to the early 20th century when statistical methods were first applied to medical research and public health. Early efforts focused on epidemiology and biostatistics, providing insights into disease patterns and treatment outcomes (Hamer et al.; Doll and Hill). The introduction of computers in the mid-twentieth century revolutionized data analysis, allowing for more sophisticated techniques and the handling of larger datasets. In the 1980 and 1990s, the rise of electronic health records (EHRs) (Dick et al., 1997) began to digitize patient data, making it more accessible for analysis. This period also saw the development of basic data mining (Cios and Moore) and machine learning techniques (Coenen, 2004; Saria et al., 2018). However, these early methods often struggled with the complexity and variability of healthcare data. The 21st century has witnessed exponential growth in healthcare data (Raghupathi and Raghupathi, 2014) due to advancements in medical imaging (Jha and Topol, 2016) genomics (Esteva et al., 2019; Dash et al., 2019), wearable devices, and mobile health applications (Vijayan et al., 2021; Reddy and Aggarwal). This explosion of data has driven the need for more advanced and intelligent data analysis techniques capable of uncovering deeper insights and supporting clinical decision-making. More than 600 medications require dose adjustments for several levels to treat renal failure, a task that is difficult for human prescribers to accomplish without aid but can be done accurately by computers. Multiple studies have shown that computer-based decision support can improve physician performance and, in certain cases, patient outcomes (Bates and Gawande, 2003; Belle et al., 2015).

Healthcare is a challenging system with many players, including pharmaceutical corporations, hospitals, and healthcare decision-makers. This sector is similarly governed by stringent laws and regulations. Now a days there is a shift toward the traditional doctor-patient relationship. The doctor becomes a collaborator, and the patient participates in the therapeutic process (Raghupathi and Raghupathi, 2014). Healthcare is no longer limited to the treatment of patients. Decision-makers should prioritize promoting healthy attitudes and preventing diseases that are avoidable (Gupta and Rathore,

Intelligent Semantic Analysis for Healthcare
ISBN 978-0-443-33512-9,
https://doi.org/10.1016/B978-0-443-33512-9.00023-8

2013). This became obvious and essential, particularly during the Covid-19 pandemic (Obermeyer and Emanuel, 2016).

Need for Intelligent Data Analysis (IDA) Techniques in Healthcare

1. Handling Complex and High-Dimensional Data: IDA techniques, such as deep learning and machine learning, can effectively manage and interpret diverse data types.
2. Improving Diagnostic Accuracy: IDA approaches can help detect subtle patterns and correlations in medical data that human practitioners may overlook, resulting in more precise diagnoses and improved patient outcomes.
3. Personalized Medicine: IDA techniques can process and analyze a vast amount of data, enabling personalized treatment plans that improve efficacy and reduce adverse effects.
4. Enhancing Predictive Analytics: IDA techniques improve the accuracy of predictions, allowing for proactive interventions and better resource allocation.
5. Operational Efficiency: IDA can streamline processes, manage resources efficiently, and enhance patient care coordination to optimize operations and reduce costs.

The application of machine learning in healthcare changed the world of data analysis. Big data analytics can process and evaluate data from EHRs, wearable devices, and other sources, providing actionable insights for clinicians and researchers (Dash et al., 2019; Reddy and Aggarwal). The use of AI and machine learning enhances diagnostic, patient monitoring, and treatment planning (Rajkomar et al., 2019; Palanisamy and Thirunavukarasu, 2019). Deep learning models, including convolutional neural networks (CNNs) and recurrent neural networks (RNNs), have made significant advances in image identification, disease prediction, and customized treatment (Esteva et al., 2019). Natural language processing (NLP) draws meaningful information from unstructured textual data, such as clinical notes, research papers, and patient comments (Reddy and Aggarwal). The efforts to enhance interoperability and data sharing among healthcare systems are underway, enabling more comprehensive data analysis and collaboration across institutions.

Combining data from several modalities, including imaging, genomics, and clinical records, will be crucial for comprehensive patient analysis. As AI and ML models become more complicated, explainability becomes increasingly important to ensure that healthcare providers understand how decisions are made. Ensuring the ethical use of data and protecting patient privacy will remain a top priority. The integration of IoT devices in healthcare will generate continuous streams of data from range of sources, including smart home systems and wearable devices. Continued advancements in intelligent data analysis will enable more precise and tailored treatment approaches based on a thorough grasp of each patient's unique characteristics.

2. Fundamentals of data analysis

Data analysis is a fundamental process that covers examining, cleaning, transforming, and modeling data to uncover useful information, support decision-making, and draw conclusions. For decades, traditional data analysis methodologies based on statistical concepts and well-established analytical procedures served as the foundation for analyzing healthcare data. Traditional approaches have limits in terms of scalability, adaptability, and handling complex, high-dimensional, and unstructured data.

As the velocity, volume, and variety of healthcare data continue to expand, traditional methods often lack the flexibility to adapt quickly, necessitating extensive reconfiguration and redevelopment of analytical frameworks. The complexity of high-dimensional data, such as genomic information, poses significant challenges for traditional methods, which may suffer from the "curse of dimensionality." Traditional models may not leverage the full predictive power of complex datasets, leading to suboptimal forecasting performance. Traditional methods may either overfit or underfit the data, reducing the generalizability and robustness of predictive models. Maintaining the privacy and security of sensitive patient data is difficult (Belle et al., 2015), necessitating robust encryption and de-identification procedures. Conventional statistical techniques are not well-suited for analyzing free-text data, which requires natural language processing (NLP) and advanced text mining techniques. Traditional approaches, which typically rely on batch processing, are not conducive to real-time analytics.

2.1 Data analysis techniques

Data analysis encompasses a range of techniques, each serving distinct purposes and providing unique insights that convert raw data into valuable information, that may be utilized for decision-making, policy formulation, and scientific research.

- Descriptive analysis involves summarizing and visualizing data using measures such as mean, median, and standard deviation. For example, calculating the average length of stay for patients with specific conditions can help manage hospital resources efficiently.
- Inferential analysis uses regression analysis and hypothesis testing to make predictions about a population, such as determining the efficacy of a new treatment during clinical trials.
- Predictive analysis uses historical data to forecast future events (forecast patient outcome) like predicting the likelihood of readmission, which helps in proactive care planning, commonly through machine learning algorithms and time series analysis.
- Prescriptive analysis suggests possible actions based on the data, often incorporating optimization and simulation techniques to recommend the best course of action like optimizing surgery schedules to reduce waiting times.
- Exploratory data analysis (EDA) aims to discover patterns, spot anomalies, and test hypotheses with minimal assumptions, typically through data visualization and

clustering. It helps identify unexpected patterns and correlations within large datasets, aiding in the discovery of potential risk factors for diseases.

2.2 Data types

Data structures serve as the framework for organizing, managing, and storing data efficiently. They enable effective data manipulation and retrieval, which are critical for performing various analytical tasks (Belle et al., 2015; Gupta and Rathore, 2013). Data can be split into two categories: unstructured and structured data. Effective use of data structures ensures that data is organized in a way that optimizes computational efficiency, facilitating complex analyses and the generation of meaningful insights (Palanisamy and Thirunavukarasu, 2019).

2.2.1 Structured data

Structured data is information that is highly organized into predefined formats, typically stored in databases and spreadsheets with clearly defined fields and schemas. This type of data is easily searchable and analyzable using traditional data analysis techniques and tools. Like electronic health records (EHRs), tracking patient outcomes, identifying trends in public health, and optimizing hospital operations. These standardized data formats facilitate the integration and sharing of information across different healthcare systems and providers, enhancing the continuity and quality of patient care. Healthcare organizations may use structured data to make data-driven decisions, improve clinical workflows, and maintain regulatory compliance, all of which lead to better patient outcomes and more effective healthcare delivery.

2.2.2 Unstructured data

Unstructured data is data which does not correspond to a precise format, making it harder to organize and analyze. Unstructured data includes free-text clinical notes, medical photographs, audio recordings, patient feedback, social media posts, and emails. Natural language processing, machine learning, and image analysis (Wang et al., 2018) are a few advanced technologies used to extract meaningful insights from unstructured data. These tools can transform unstructured data into actionable information, enabling more comprehensive patient care, early detection, and personalized means of treatment. As the volume of unstructured data continues to grow, effectively harnessing this information becomes increasingly important for advancing healthcare outcomes and improving operational efficiencies.

2.2.3 Semi-structured data

Semi-structured data is a mixture of structured and unstructured data, comprising components of both (Piwek et al., 2016; Vest and Gamm, 2010). This type of data is not stored in traditional databases but still includes organizational properties, such as tags or markers that provide some degree of structure. Like XML files, JSON documents, and HL7 messages, which are commonly used for exchanging information between different healthcare systems. Electronic health records (EHRs) often incorporate semi-structured data, such as clinical narratives that are tagged with metadata. The flexibility of semi-structured data makes it suitable for integrating diverse data sources and supporting interoperability across various healthcare applications. Semi-structured data requires special tools and procedures due to its hybrid nature. By effectively utilizing semi-structured data, healthcare organizations can improve data sharing, enhance clinical decision-making, and streamline administrative processes.

3. Introduction to intelligent data analysis

Intelligent data analysis (IDA) techniques represent a significant advancement over traditional methods, particularly in the healthcare domain. IDA is poised to revolutionize healthcare by providing deeper insights, more accurate predictions, and improved decision-making capabilities. IDA is the utilization of sophisticated algorithms and computational techniques to analyze complex datasets. It integrates concepts from artificial intelligence, machine learning, data mining and statistical modeling (Wang et al., 2018) to automate and enhance the data analysis process. The goal of IDA is to extract meaningful relationships, patterns, and knowledge from data, enabling accurate predictions and informed decisions.

3.1 Characteristics of intelligent data analysis techniques

IDA is inherently adaptive, that allows us to handle complicated and high-dimensional data, like genomic sequences and medical imaging. Secondly, IDA is highly flexible, able to integrate and analyze wide ranges of data types, encompassing structured, unstructured, and semi-structured data. Thirdly, IDA emphasizes automation and efficiency, utilizing advanced algorithms to automate the preprocessing, analysis, and interpretation of data, thereby reducing human error and increasing processing speed. Lastly, IDA focuses on providing actionable insights, leveraging predictive analytics and pattern recognition to support decision-making in real-time.

3.2 Advantages of employing intelligent techniques over traditional methods

- Intelligent approaches provide higher scalability and efficiency.
- Intelligent approaches excel in integrating and analyzing diverse data sets.
- Intelligent approaches can smoothly incorporate data from many sources and formats.
- Intelligent techniques are particularly adept at analyzing unstructured data.
- Intelligent techniques can recognize complex patterns and correlations in data.
- Intelligent techniques can detect anomalies and rare events in data.

- Intelligent techniques provide valuable decision support to clinicians.
- Intelligent techniques streamline administrative and clinical workflows, reducing redundancy and optimizing resource allocation.
- Intelligent techniques are capable of continuous learning and improvement, adapting to new data.
- Intelligent systems can incorporate feedback from clinicians and patients to refine their models and improve performance.
- Intelligent techniques incorporate robust data privacy measures

Intelligent Data Analysis (IDA) in healthcare helps in semantic understanding of complicated medical data by combining application of machine learning for predictive inference, deep learning architectures (e.g., Transformers, CNNs, RNNs) for multimodal feature extraction, Natural Language Processing for clinical text interpretation, Data Mining for association discovery, and Statistical Analysis for validation. This integrated approach enables automated knowledge discovery, intelligent clinical reasoning, and precision medicine initiatives, which increases diagnostic accuracy, therapeutic efficacy, and healthcare system efficiency. Unfortunately, implementing IDA in healthcare presents unique challenges, including data privacy concerns, algorithmic bias mitigation, interpretability requirements for clinical trust, and integration with legacy healthcare IT systems. All of these must be considered for successful deployment in healthcare.

4. Machine learning

Machine learning (ML) is the heart of intelligent data analysis, offering powerful tools for uncovering patterns, making predictions, and generating insights from complex healthcare datasets. It entails training models on vast datasets to recognize patterns, create decisions, and predict outcomes. ML techniques are roughly categorized into three categories: supervised learning, unsupervised learning, and reinforcement learning. ML has transformative potential, from predicting patient outcomes and diagnosing diseases to personalizing treatment plans and managing healthcare operations efficiently. ML models can analyze complex datasets, including medical images, genomic sequences, and electronic health records, offering insights that are often beyond the reach of traditional statistical methods.

4.1 Supervised learning algorithms

Supervised learning entails training an algorithm with labeled data, which means that the input data is associated with the appropriate output. The model teaches how to map inputs to outputs in order to predict the outcome of new, previously unknown inputs. This method comprises a training phase in which the model iteratively updates its parameters to lessen the difference between predicted and actual outputs. Supervised learning is commonly divided into two types: classification and regression. In healthcare, supervised learning is often used for disease prediction, image analysis for medical diagnostics, and individualized treatment recommendations.

4.1.1 Decision trees

Decision trees is a flexible and user-friendly machine learning technique that can be utilized for both classification and regression. They operate by recursively partitioning the dataset into subgroups depending on the value of the input attributes, producing a decision tree (tree-like model). Every internal node represents a feature, each branch has a decision rule, and each leaf node is a result. This hierarchical structure makes decision trees simpler to understand, as going from root to leaf illustrates the decision-making process. Decision trees are particularly valuable for clinical decision support, where they can help predict patient outcomes, diagnose diseases, and recommend treatments based on patient data. Their capacity to manage both numerical and categorical data, plus their resilience to missing values, increases their usefulness in complex medical datasets. However, decision trees, particularly deep trees, are subject to overfitting and may not perform well with new data.

4.1.2 Random forests

Random Forests are an advanced decision tree-based ensemble learning technique for classification and regression applications. During training, they create a huge number of decision trees and output the mode (classification) or mean (regression) prediction. This method, developed by Leo Breiman and Adele Cutler, leverages the power of bagging (bootstrap aggregating) to improve model accuracy and robustness. A random selection of features is chosen for each tree, encouraging variety among the trees and lowering the likelihood of overfitting. This characteristic makes Random Forests particularly effective in handling large datasets with complex interactions and noise. In healthcare, it is used to forecast patient outcomes, identify risk factors, and categorize medical conditions. Despite their strengths, Random Forests can be computationally costly, necessitating tweaking of hyperparameters such as the number of trees and maximum depth to optimize performance.

4.1.3 Support vector machines (SVMs)

Support Vector Machines is a complex and adaptable technique utilized for regression, classification, and outlier detection. SVMs, which were designed by Vladimir Vapnik and colleagues in the 1990s, work by determining the ideal hyperplane for separating data into different classes. SVMs are particularly effective in high-dimensional environments and are well-suited to problems with more dimensions than samples. It helps in disease diagnosis, medical image analysis, and bioinformatics. Kernel functions improve their ability to deal with complex, non-linear interactions by transforming the input space into a higher-dimensional space that allows for linear separation. Despite its advantages, SVMs are computationally demanding and sensitive to kernel and regularization parameter selection, necessitating careful tweaking and cross-validation.

4.1.4 k-nearest neighbors (k-NN)

The k-Nearest Neighbors (k-NN) algorithm is a simple, yet powerful, instance-based learning technique used for regression and classification tasks. It operates on the principle that identical data points tend to be close to each other in a feature space (Hellerstein, 2008). To categorize a new point of data, the algorithm selects the "k" nearest elements from the training set and assigns the most common class to them. For regression, the result is the average of the nearest neighbors' values (Petrova-Antonova and Tancheva, 2020). The value of "k" has a considerable impact on the performance of the k-NN method, with lower values rendering the model sensitive to noise and higher values smoothing out the predictions.

k-NN has been applied to a variety of tasks including disease prediction, patient risk assessment, and pattern recognition in medical imaging. Its non-parametric nature means that it makes no presumptions regarding the underlying data distribution, which makes it adaptable to diverse types of data. Nevertheless, k-NN's performance may suffer with data that is highly dimensional due to the "curse of dimensionality," in which the concept of distance becomes less meaningful. Moreover, it can be computationally intensive, as it requires calculating the distance to every training example for each prediction.

4.2 Unsupervised learning algorithms

Unsupervised learning techniques are intended to detect patterns and relationships in datasets without prelabeled outcomes. Unsupervised learning works with input data alone, aiming to uncover the underlying structure. Methods for unsupervised learning include clustering methods like k-means and hierarchical clustering, as well as dimensionality reduction techniques like Independent Component Analysis (ICA) and Principal Component Analysis (PCA). Unsupervised learning has numerous uses, most notably in the study of huge and complicated datasets. Unsupervised learning aids in anomaly detection, crucial for identifying rare diseases and unusual patient responses to treatments.

4.2.1 K-means clustering

K-means clustering partitions of a dataset into discrete, non-overlapping groups or clusters. The algorithm aims to minimize the variance within each cluster and maximize the variance between clusters. It begins by selecting "k" initial centroids, either randomly or through specific methods like the k-means++ algorithm. Each data point gets assigned to the closest centroid using a distance measure, usually Euclidean distance. The centroids are recalculated as the meaning of all points in each cluster, and the assignment process is repeated iteratively until convergence, when the cluster assignments no longer change significantly. It is used for patient segmentation, disease outbreak detection, and image analysis. For instance, it can identify subgroups of patients with similar characteristics or health conditions, enabling tailored interventions and treatment plans. In epidemiology, k-means helps detect clusters of disease occurrences, assisting in understanding and controlling outbreaks. Additionally, in medical imaging, it aids in segmenting different anatomical structures and identifying abnormalities, improving diagnostic accuracy. k-means clustering has limitations, such as sensitivity to the initial choice of centroids and difficulties establishing the appropriate number of clusters.

4.2.2 Hierarchical clustering

Hierarchical clustering is used for creating a cluster hierarchy through a series of nested partitions. This method can be either agglomerative ("bottom-up") or divisive ("top-down"). In agglomerative clustering, each point of data is assigned to its own cluster, and pairs of clusters are iteratively merged based on their similarity until all points are in the same cluster. Divisive clustering, conversely, starts with one cluster containing all data points and splits it into smaller clusters. The process continues until each point of data is in its own cluster. The results are often visualized using a dendrogram, which displays hierarchy and can be cut at a desired level to retrieve a specific number of clusters.

In healthcare, hierarchical clustering is valuable for discovering nested groupings within data, such as different subtypes of diseases or patient segments based on genetic and clinical similarities. It is particularly useful in genomic studies for constructing phylogenetic trees that depict evolutionary relationships among species or individuals. Hierarchical clustering also aids in the analysis of medical imaging data, helping to identify tissue types or anomalies. Hierarchical clustering does not require a predetermined number of clusters, which allows for greater freedom in exploratory data analysis. However, it is computationally costly with large datasets and is sensitive to noise and outliers, which can have a substantial impact on the resulting cluster structure.

4.2.3 Principal component analysis (PCA)

Principal Component Analysis (PCA) is a widely used dimensionality-reducing technique in data analysis that converts high-dimensional data to a lower-dimensional space while retaining as much variability as feasible. PCA achieves this by identifying the principal components, or directions in the data which account for the most variance. These principal components are orthogonal to each other and ranked by the amount of variance they explain. The first principal component captures the largest variance, the second captures the next greatest amount, and so on. This transformation is achieved through linear combinations of the original variables, effectively simplifying the dataset while retaining its essential structure.

In healthcare, PCA is particularly valuable for dealing with complex datasets, such as those involving genomic data (can reduce the dimensionality of gene expression data), medical imaging (compressing and enhancing images by focusing on the most significant features), and patient records (clustering patients with similar health profiles). Despite its strengths, PCA assumes linearity and can be less effective with highly non-linear data structures.

4.2.4 Independent component analysis (ICA)

Independent Component Analysis (ICA) is a sophisticated computation approach that divides a multivariate signal into additive, independent components. It is a type of blind source separation, which means it attempts to decompose a signal into underlying sources without much information about the signal's origin. ICA aims to make the resulting components statistically independent and non-Gaussian. This makes ICA particularly useful for situations where the goal is to identify and separate individual signals from a mixture, especially when these signals are not necessarily orthogonal.

ICA has been effectively used in a variety of domains, including medical imaging, electroencephalography (to isolate artifacts such as eye blinks and muscle activity from brain signal recordings), and functional magnetic resonance imaging (helps identify and separate different brain activation patterns related to distinct cognitive processes). Moreover, it can be used in the analysis of genomic data to identify underlying biological signals that contribute to gene expression profiles. One of the strengths of ICA is its ability to reveal hidden factors that are not apparent with other dimensionality reduction techniques. However, it is sensitive to the choice of algorithm and requires careful interpretation of the independent components, which might sometimes be challenging to associate with specific physiological or biological phenomena.

4.3 Semi-supervised learning

Semi-supervised learning covers the gap among supervised and unsupervised learning by using both unlabeled and labeled data to train. By leveraging the structure and patterns within the unlabeled data, semi-supervised learning can improve the model's accuracy and generalization compared to purely supervised methods. This technique is commonly employed in healthcare for problems that include disease classification and medical image analysis, where labeled data is limited but unlabeled data is abundant. For instance, semi-supervised learning can enhance diagnostic models by learning from a few annotated medical images along with a larger set of unannotated images, resulting in better disease detection and classification.

Collectively, these techniques empower healthcare professionals to derive actionable insights from vast amounts of data, ultimately improving patient care, advancing medical research, and optimizing healthcare delivery systems.

5. Deep learning

Deep learning is a form of machine learning that uses neural networks with numerous layers to represent complex patterns in data. Each layer in a deep learning model converts input data into increasingly abstract representations, allowing the machine to learn complex characteristics and relationships. This method has transformed areas such as image and natural language processing, speech recognition, and self-learning systems. Convolutional Neural Networks (CNNs) are particularly useful in medical imaging, allowing for tasks such as tumor recognition and organ delineation. Recurrent Neural Networks (RNNs) excel at dealing with sequential data, making them valuable for forecasting medical outcomes and analyzing time-series data from wearable devices. Despite its transformative promise, deep learning faces several hurdles, including the need for big, labeled datasets, significant processing resources, and the interpretability of complicated models.

5.1 Artificial neural networks (ANNs)

Artificial Neural Networks (ANNs) are computer models inspired by the neural networks seen in the human brain that are used to recognize patterns and solve complicated issues. They are composed of layers of neurons or interconnected nodes, each of which performs simple computations to process and transmit input data through the network. An ANN's basic structure comprises of an input layer that collects input data, either one or several hidden layers that can execute calculations and feature alterations, and an output layer that creates the ultimate prediction or classification. Weighted connections are used to connect each neuron in one layer to neurons in the next. The neurons use activation functions on the weighted sum of their inputs, which introduces nonlinearity and allows the network to learn and encode complicated relationships. ANNs are trained using techniques such as backpropagation, which modifies the weights of the connections to reduce the difference between projected and actual outputs. This learning process enables artificial neural networks to adapt to new input and improve their performance over time. ANNs are widely utilized in a variety of applications, including image and audio recognition, natural language processing, and medical diagnosis of cancer, diabetes, and cardiovascular problems, as well as illness detection, localization, and prediction. However, they demand significant computer resources and big datasets for efficient training, and their "black box" nature can be difficult to interpret.

5.1.1 Activation function

Activation functions are critical components of neural networks that introduce nonlinearity into the model, enabling it to learn and represent complex patterns in the data. These functions determine the output of a neuron by applying a mathematical transformation to the weighted sum of its inputs. Common activation functions include the Sigmoid, Tanh, and Rectified Linear Unit (ReLU). The Sigmoid function maps input values to a range between 0 and 1, making it useful for probability-based outputs, but it can suffer from vanishing gradients during backpropagation.

$$\text{Sigmoid}, \sigma(x) = \frac{1}{1 + e^{-x}} \tag{1.1}$$

The Tanh function maps inputs to a range between −1 and 1, offering stronger gradients than Sigmoid but still facing vanishing gradient issues.

$$\tanh(x) = \frac{2}{1 + e^{-2x}} - 1 \tag{1.2}$$

ReLU, which outputs the input directly if it is positive and zero otherwise, has become popular due to its efficiency and ability to mitigate the vanishing gradient problem, though it can suffer from "dying ReLU" where neurons stop activating.

$$\text{ReLU}(x) = \max(0, x) \tag{1.3}$$

Variants like Leaky ReLU and Parametric ReLU address this issue. Choosing the appropriate activation function is crucial for the performance and training of neural networks, impacting their ability to model complex data accurately.

5.1.2 Training process

The training process of deep learning involves iterative learning, where a neural network adjusts its parameters to minimize the difference between predicted and actual outputs. This process begins with initializing the weights and biases of the network. Training data, typically split into batches, is then fed into the network. For each batch, the network processes the input through its layers, computing weighted sums and applying activation functions to generate predictions. The loss function (to measure the error), such as Mean Squared Error (L1 loss the Median Absolute Error (MAE), and L2 loss the Root Mean Squared Error (RMSE)) for regression or Cross-Entropy for classification, quantifies the prediction error.

To optimize the network, the backpropagation algorithm computes gradients of the loss function with respect to each weight using the chain rule. These gradients indicate how to adjust the weights to reduce the loss. An optimization algorithm, like Stochastic Gradient Descent (SGD) or Adam, updates the weights accordingly. This cycle of forward pass, loss computation, backpropagation, and weight updating repeats for many epochs until the model converges or reaches a predefined number of iterations.

5.1.3 Overfitting and regularization

Overfitting occurs when the model learns noise and details from the training data, resulting in poor generalization to new data. Regularization techniques such as dropout

(Randomly drops neurons during training to prevent co-adaptation), weight decay, and batch normalization are often employed to prevent overfitting and improve generalization. Hyperparameters like learning rate, batch size, and network architecture are tuned to enhance performance. Validation data helps monitor the model's performance and avoid overfitting by providing an unbiased evaluation. The final model is evaluated on test data to ensure its effectiveness in real-world applications.

5.2 Deep learning architectures

5.2.1 Convolutional neural networks (CNNs)

Convolutional Neural Networks (CNNs) are deep learning algorithms that excel at processing and analyzing visual data. CNNs, inspired by the biological visual cortex, learn spatial hierarchies of information from input images via a series of convolutional, pooling, and fully connected layers. Convolutional layers apply filters to the input image, then use convolution processes to extract local features like edges, textures, and forms. These characteristics are subsequently down sampled using pooling layers, which minimize spatial dimensions and computation while maintaining the most important information, before being classified using fully connected layers. CNNs have the advantage of capturing spatial and temporal connections in images using shared weights and biases across the network, making them more efficient in terms of parameter utilization than fully linked networks. This feature allows CNNs to perform well in tasks such as image recognition, object detection, and semantic segmentation. After several layers of convolution and pooling, the retrieved features are sent into fully connected layers for classification or regression (Sarker, 2021).

CNNs have revolutionized various applications, including medical image analysis, where they assist in detecting anomalies in radiographs and MRIs, and pathology slide analysis. Despite their effectiveness, CNNs require large datasets and significant computational power for training. Techniques such as transfer learning, where pre-trained models on large datasets are fine-tuned for specific tasks, have made CNNs more accessible for a broader range of applications.

5.2.2 Recurrent neural networks (RNNs)

Recurrent Neural Networks (RNNs) (Cho et al., 2014) are a type of neural network that detects patterns in sequences of data such as time series, audio, and text. Unlike standard neural networks, RNNs feature connections that create directed cycles, allowing them to retain a 'memory' of prior inputs. This memory enables RNNs to capture temporal dependencies and context throughout time, making them especially useful for jobs where the sequence of inputs matters. In an RNN, each neuron in the hidden layer receives input from both the previous layer and its own previous state. This feedback loop allows information to be stored, making RNNs useful for applications such as language modeling, audio recognition, and machine translation. Traditional RNNs, on the other hand, have difficulty dealing with long-term dependencies due to issues such as vanishing and expanding gradients during backpropagation across time (Sarker, 2021).

To address these constraints, Hochreiter and Schmidhuber proposed enhanced variations of Long Short-Term Memory (LSTM) (Hochreiter and Schmidhuber), while Cho et al. introduced Gated Recurrent Unit (GRU) networks in 2014 (Cho et al., 2014). LSTMs (which can be stacked to create deep architectures, allowing the learning of even more complex patterns and hierarchies in sequential data) and GRUs (for the gating mechanisms to selectively update the hidden state of the network at each time) include gating mechanisms that regulate the flow of information, allowing them to capture long-range dependencies more effectively. These models have been frequently used in NLP applications like sentiment analysis (Zhang et al., 2018), text generation, and named entity recognition (Li et al., 2020).

RNNs have also been applied in time series analysis of patient vitals, predicting disease progression, and analyzing sequential data from EHRs. Despite their advantages, RNNs can be computationally intensive, and training them can be challenging. Recent developments, such as the Transformer model, have started to replace RNNs in many applications due to their ability to handle long-range dependencies more efficiently.

5.2.3 Generative adversarial networks (GANs)

Generative Adversarial Networks (GANs) are a type of deep learning model created by Ian Goodfellow and colleagues in 2014. GANs are made up of two neural networks: a generator and a discriminator, which are trained at the same time using adversarial competition. The generator generates synthetic data samples that resemble real data, while the discriminator assesses these samples and seeks to separate them from genuine data. The generator seeks to provide increasingly realistic data (fake data samples) to deceive the discriminator, whereas the discriminator strives to improve its ability to detect fake data (Pereira et al., 2024). This adversarial process is repeated until the generator delivers high-quality synthetic data that the discriminator cannot distinguish from actual data.

5.2.4 Autoencoders

Autoencoders are a form of artificial neural network used for unsupervised learning and efficient coding of input data. They are made up of two basic parts: an encoder, which compresses the input into a latent-space representation, and a decoder, which reconstructs the input from that representation. The purpose of an autoencoder is to learn a compact, meaningful representation of the data while reducing the reconstruction error between the original input and the reconstructed output (Sarker, 2021). Autoencoders are used to reduce dimensionality, denoise images, detect anomalies, and extract features. It is used to compress and denoise medical images, reducing storage requirements and improving image quality. They can also assist in identifying anomalies in patient data, which may indicate potential health issues.

5.2.5 Transformer networks

Vaswani et al. introduced transformer networks, a sort of deep learning model, in 2017. Transformers, unlike standard neural networks, handle input data via self-attention techniques rather than convolutional or recurrent structures. This self-attention mechanism enables transformers to balance the relevance of various aspects of the incoming data, allowing them to capture long-term dependencies and relationships more effectively than earlier models. One of the primary benefits of transformers is their capacity to analyze whole data sequences at the same time, resulting in quicker training times and higher performance on complex tasks. However, they necessitate extensive processing resources and big volumes of training data, which can be limited in some cases.

Transformers have changed NLP by allowing for more efficient and powerful models. They are the foundation of cutting-edge models like BERT, GPT, and T5. These models have demonstrated exceptional performance in tasks such as sentiment analysis, machine translation, and text production. Transformer networks are being leveraged to analyze and interpret large volumes of clinical text data, such as EHRs, medical literature, and patient notes. Additionally, transformers are being explored for applications in genomics, medical imaging, and drug discovery.

6. Data mining and pattern recognition

Data mining is the process of extracting patterns, correlations, and insights from huge datasets using machine learning, statistical, and computational techniques. It consists of several major steps: data cleansing, data selection, data integration, data transformation, knowledge presentation, and pattern detection. The purpose of data mining is to gather useful information from raw data and convert it to a comprehensible structure for future use. This process can identify trends, predict outcomes, and generate actionable insights.

In healthcare, data mining is essential for boosting medical research, patient care, and optimizing resource allocation. For example, it is useful to detect forged insurance claims, predict disease outbreaks, and identify effective treatment plans based on historical data. Techniques such as association rule mining can reveal relationships between different medical conditions, while clustering methods can segment patient populations for targeted interventions. Predictive modeling can predict patient outcomes, which is useful in preventive treatment and individualized therapy. Despite its benefits, data mining confronts several obstacles, including maintaining data privacy, handling unstructured data, and integrating data from several sources.

6.1 Data sources

Data collection in healthcare involves gathering information from a variety of sources to support clinical decision-making, research, and operational efficiency (De Lusignan et al., 2011). Key sources include World wide web, EHRs (Wang et al., 2018), medical imaging systems, wearable devices and mobile health applications (Piwek et al., 2016), surveys and patient-reported outcomes and administrative databases capture billing, insurance claims, and resource utilization data, essential for financial and operational analysis. Additionally, health information exchanges (HIEs) (Vest and Gamm, 2010) facilitate the sharing of data across different healthcare organizations, enhancing care coordination.

6.2 Data preprocessing

6.2.1 Data cleaning

Data cleaning or preprocessing assures that the data utilized for analysis is complete, accurate, and ready for meaningful insights. Healthcare data often comes with challenges such as missing values, inconsistencies, and errors due to manual entry, different data formats, and various sources. These issues can have an enormous effect on the quality and consistency of data analysis outcomes. The primary issues in healthcare data cleaning and preprocessing are handling missing values, removing duplicate records, data privacy and security, heterogeneity of data sources, ensuring data integrity. Data cleaning (Hellerstein, 2008) is one of the important parts of machine learning. Python, OpenRefine (Petrova-Antonova and Tancheva, 2020; Li et al.), Trifacta Wrangler, IBM Infosphere Quality Stage, Cleenex (Pereira et al., 2024), TIBCO clarity are some of the data cleaning tools.

6.2.2 Data integration

Data integration is the process of merging data from several sources to create a cohesive view, allowing for more extensive analysis and decision-making. This procedure is critical since healthcare data is frequently scattered across multiple systems, including EHRs, medical imaging systems, laboratory information systems, and patient-generated data via mobile apps and wearable devices. Effective data integration enables the consolidation of disparate information, providing clinicians with a holistic view of a patient's health. It also provides advanced analytics, resulting in more accurate treatment planning, diagnosis, and population health management. However, it creates issues of data privacy, interoperability, and security.

6.3 Data transformation

Data transformers are critical components of data processing pipelines, designed to change and optimize raw data into an appropriate format for analysis and modeling. These transformations can involve a variety of procedures like normalization, standardizing, encoding categorical variables, and feature scaling. These strategies increase the performance of machine learning algorithms as well as model correctness by modifying data. Effective data transformation eliminates biases, reduces variability, and ensures that data adheres to the assumptions of diverse analytical methodologies. This phase is critical for preparing data for further analyses and guaranteeing solid, dependable results in intelligent data analysis systems. Pandas, TenserFlow Transforms, OpenRefine, scikit-learn are some of the methods used for data transformation.

6.4 Pattern recognition

Pattern recognition systems automatically discover and classify patterns and regularities in input data, such as shapes, textures, or sequences. It is the use of algorithms and statistical approaches to recognize and classify data based on previously learned patterns. Pattern recognition's primary benefit is its capacity to process and interpret vast amounts of complicated data properly and fast. This can lead to earlier diagnosis and intervention, more individualized treatment regimens, and better patient results. The challenges include assuring the accuracy and reliability of recognition systems, dealing with variability in medical data, and resolving ethical concerns about data security and privacy.

It is used in medical imaging to detect anomalies such as tumor in X-rays, MRIs, and CT scans. In genomics, pattern recognition helps identify genetic markers associated with specific diseases. It also aids in analyzing electrocardiograms (ECGs) to detect irregular heartbeats and in monitoring vital signs from wearable devices to recognize patterns indicative of health issues.

7. Natural language processing (NLP)

Natural Language Processing (NLP) is a form of artificial intelligence that enables machines to read, interpret, and synthesize human language in meaningful and usable ways. It is critical in evaluating and extracting insights from massive amounts of unstructured textual data contained in electronic health records (EHRs), clinical notes, research literature, and patient-generated content such as social media posts or patient forums. Tokenization and part-of-speech tagging are examples of basic NLP techniques. More advanced applications include sentiment analysis, named entity recognition (NER), and language translation. By applying NLP, healthcare providers can automate clinical documentation, extract key information for research purposes, identify patterns in patient symptoms or treatment outcomes, and even monitor public health trends. However, challenges in healthcare NLP include handling medical terminology variations, ensuring accuracy in understanding context-specific language nuances, and maintaining patient data privacy and confidentiality. Advances in machine learning and deep learning have greatly increased NLP models' capabilities, enabling more accurate and efficient analysis of healthcare text data.

7.1 Named entity recognition (NER)

Named Entity Recognition (NER) serves as a crucial technique in the field of Natural Language Processing (NLP), focusing on the identification and classification of important entities such as symptoms, diseases, medications, treatments, laboratory results and anatomical terms found within unstructured clinical text, which encompasses electronic health records (EHRs), discharge summaries, and radiology reports. Within the context of biomedicine, the process of named entity recognition, often referred to as Biomedical Named Entity Recognition (BioNER) or clinical NER, facilitates the automatic structuring of unstructured text data. This organization enables its application in big data analysis, supports clinical decision-making, aids in research endeavors, and contributes to the creation of disease registries. Recent advancements in the performance and reliability of named entity recognition in healthcare settings have been achieved through the application of deep learning and transformer-based models. A recent study demonstrated that a medical NER system achieved high precision and recall from extensive EHR notes, enabling the extraction of information on a large scale at the institutional level (Macri et al., 2023).

Innovative frameworks also address challenges that are specific to biomedical language, including nested entities, non-contiguous entity spans, and specialized terminology. This enhances the dependability and precision of entity recognition in clinical environments (Ji et al., 2023).

7.2 Medical information extraction

Medical Information Extraction (IE) is an automated way to get structured, clinically relevant data from unstructured text sources like case reports, clinical notes, EHR narratives, and discharge summaries. This automated approach plays a crucial role in transforming raw information into usable data for healthcare applications and this holds significant importance in the field of healthcare informatics, as substantial amounts of crucial patient information are in free-text format, which restricts its use for decision-making, analysis, and research purposes. Recent developments in Natural Language Processing (NLP), especially with transformer-based architectures and models tailored for specific domains, have greatly improved the feasibility and precision of information extraction. A recent study from 2024 showcased a highly effective system using large language models to accurately extract clinical features from medical records, achieving a low null-rate, all while working on standard consumer-grade hardware (Wang et al., 2024). In a similar vein, systems built on transformers have proven effective in extracting medication details and contextual information such as dosage adjustments from clinical notes, attaining impressive F_1-scores in benchmark assessments (Chen et al., 2023). Additionally, the integration of NER with relation-extraction modules enables the organization of categories and their connections such as disease-symptom, treatment-outcome, or drug-adverse reaction associations from unstructured clinical text (da Silva et al., 2023). By employing these techniques, medical information extraction facilitates the conversion of narrative clinical data into organized datasets that aid in various subsequent tasks, including population-level analytics, registry creation, cohort identification, and clinical decision-making support.

7.3 Medical text classification

Medical text classification refers to the automatic categorization of unstructured clinical text such as discharge summaries, physician notes, radiology reports, and EHR narratives into predefined categories (e.g., disease diagnosis, treatment type, risk group, or ICD codes). This process is vital because clinical text often contains rich diagnostic and treatment information that does not appear in structured fields. With the arrival of deep learning and domain-adapted language models, classification performance has improved substantially. For instance, a comparative study of deep learning models on unstructured medical notes showed that Transformer-based encoders and pretrained models such as Bidirectional Encoder Representations from Transformers (BERT) variants, significantly outperform older architectures (RNNs, CNNs) across various disease classification tasks, especially in class imbalance condition (Lu et al., 2022). More recently, hybrid and knowledge-enhanced models further push the boundaries: for example, a model integrating medical knowledge graphs with a BERT backbone Knowledge Graph Enhanced Multi-Type Text BERT (KG-MTT-BERT) was shown to produce state-of-the-art results for multi-type medical text classification, outperforming baseline deep-learning approaches (He et al., 2022). Through these advances, medical text classification is becoming a dependable tool for automated triage, disease surveillance, cohort identification, and population-level analytics reducing clinician burden while enabling scalable and consistent analysis of large volumes of clinical text.

8. Ethical considerations in intelligent data analysis

Privacy concerns in healthcare data are paramount due to the sensitive nature of medical information. The Health Insurance Portability and Accountability Act (HIPAA) in the United States, for instance, sets strict guidelines for the protection of personal health information (PHI) to prevent unauthorized access and breaches. However, with the growing digitization of health records and the use of advanced data analytics, maintaining privacy has become increasingly challenging. Data breaches can lead to significant harm, including identity theft, discrimination, and financial loss. Ensuring data anonymization and implementing robust cybersecurity measures are critical strategies to mitigate these risks. Ethical considerations also demand transparency about how patient data is used and shared, reinforcing the importance of informed consent and trust between patients and healthcare providers.

8.1 Guidelines for ethical practices in data analysis

Ensuring ethical practices in data analysis projects requires adherence to a comprehensive set of guidelines that prioritize transparency, fairness, and responsibility. First, obtaining informed consent from individuals whose data will be analyzed is crucial, ensuring they understand how their information will be used and the potential impacts. Data privacy and confidentiality must be properly enforced by establishing strong security measures and following regulations such as GDPR and HIPAA. Secondly, it is vital to address and mitigate biases in data and algorithms to prevent unfair treatment of any group. This involves using diverse datasets and incorporating fairness-aware algorithms to detect and correct biases. Ethical practices also include ensuring transparency in data analysis methods and results. This means clearly documenting and communicating the processes and algorithms used, making them understandable to stakeholders and allowing for independent verification.

It is also vital to develop data governance frameworks that specify roles, duties, and responsibility for data administration and analysis. Audits and evaluations should be undertaken on a regular basis to guarantee that ethical standards and legislation are met. Additionally, researchers and analysts must be aware of and navigate the ethical implications of their work, considering the societal impacts and striving to use data analysis for the public good. Lastly, fostering a culture of ethical awareness and continuous education among data professionals is vital to keeping up with evolving standards and technologies. This includes training programs and resources that emphasize the importance of ethics in data analysis.

9. Future trends and directions

Future trends in intelligent data analysis (IDA) are set to revolutionize various industries, particularly healthcare. One significant trend is the integration of multi-modal data, combining information from medical imaging, genomics, EHRs, and wearable devices to create a comprehensive patient profile. Advances in explainable AI (XAI) are crucial for ensuring transparency and trust in AI-driven choices, as they address the black-box character of many machine learning algorithms. Transfer learning is another emerging trend, enabling models trained on large datasets to be adapted to specific healthcare applications with minimal additional training. Real-time analytics powered by edge computing and the IoT will facilitate continuous monitoring and immediate intervention, improving patient outcomes. Moreover, the focus on secure privacy techniques like federated learning will grow, ensuring patient data security while enabling collaborative analysis across institutions. As these trends evolve, they will significantly enhance the capabilities of IDA, leading to more personalized, effective, and efficient healthcare solutions.

10. Conclusion

The rapid advancements in intelligent data analysis (IDA) techniques are transforming the landscape of healthcare. These innovative methods, including machine learning, natural language processing (NLP), deep learning and advanced data mining offer unprecedented capabilities for analyzing complex, high-dimensional data. They address many limitations of traditional data analysis by providing scalable, efficient, and adaptable solutions that can handle diverse and voluminous healthcare datasets.

The integration of IDA into healthcare systems facilitates comprehensive patient analysis by merging data across various modalities such as genomics, imaging, and electronic health records (EHRs). This multifaceted approach enhances the accuracy of diagnostics, enables personalized treatment plans, and supports precision medicine initiatives. Furthermore, IDA techniques improve operational efficiency by shortening workflows and optimizing allocation of resources, resulting in improved patient outcomes and cost savings for healthcare providers.

However, the implementation of IDA in healthcare also raises important ethical and legal considerations. Ensuring data privacy, addressing algorithmic bias, maintaining transparency and explainability, and adhering to regulatory compliance are critical to the responsible use of these technologies. Ongoing research is focused on developing explainable AI models and robust frameworks to safeguard patient data and build trust among users.

Looking forward, the future of IDA in healthcare is promising. Emerging trends such as real-time data processing, integration of IoT devices, and advancements in AI algorithms will continue to drive innovation. As these technologies evolve, they will further enhance the ability to provide precise, individualized patient care and support ongoing medical research.

In conclusion, intelligent data analysis techniques are revolutionizing healthcare by enabling more sophisticated, efficient, and ethical data management and analysis. These advancements promise to boost patient care, drive medical research forward, and optimize healthcare delivery systems, heralding a new era of healthcare innovation.

References

Bates, D.W., Gawande, A.A., 2003. Improving Safety with Information Technology. Available: www.nejm.org.

Belle, A., Thiagarajan, R., Soroushmehr, S.M.R., Navidi, F., Beard, D.A., Najarian, K., 2015. Big data analytics in healthcare. Biomed Res. Int. 2015. https://doi.org/10.1155/2015/370194.

Chen, A., Yu, Z., Yang, X., Guo, Y., Bian, J., Wu, Y., 2023. Contextualized medication information extraction using Transformer-based deep learning architectures. J. Biomed. Inf. 142 (Jun). https://doi.org/10.1016/j.jbi.2023.104370.

Cho, K., et al., Sep. 2014. Learning Phrase Representations using RNN Encoder-Decoder for Statistical Machine Translation. Available: http://arxiv.org/abs/1406.1078.

Cios K. J. and Moore G. W., Uniqueness of Medical Data Mining Keywords: Medical Data Mining; Unique Features of Medical Data Mining and Knowledge Discovery; Ethical; Security and Legal Aspects of Medical Data Mining.

Coenen, F., 2004. Printed in the United Kingdom data mining: past, present and future. Knowl. Eng. Rev. 00, 0–1. https://doi.org/10.1017/S0000000000000000.

da Silva, D.P., da Rosa Fröhlich, W., de Mello, B.H., Vieira, R., Rigo, S.J., 2023. Exploring named entity recognition and relation extraction for ontology and medical records integration. Inform. Med. Unlocked 43 (Jan). https://doi.org/10.1016/j.imu.2023.101381.

Dash, S., Shakyawar, S.K., Sharma, M., Kaushik, S., Dec. 2019. Big data in healthcare: management, analysis and future prospects. J. Big Data 6 (1). https://doi.org/10.1186/s40537-019-0217-0.

De Lusignan, S., et al., 2011. Reporting Observational Studies of the Use of Information Technology in the Clinical Consultation A Position Statement from the IMIA Primary Health Care Informatics Working Group (IMIA PCI WG).

Dick, R.S., Steen, E.B., Detmer, D.E., 1997. Computer-Based Patient Record : An Essential Technology for Health Care, Revised Edition. National Academies Press.

Doll R. and Hill A. B., Smoking and Carci Noma of Tie Lung Preliminary Report.

Esteva, A., et al., Jan. 01, 2019. A Guide to Deep Learning in Healthcare. Nature Publishing Group. https://doi.org/10.1038/s41591-018-0316-z.

Gupta, V., Rathore, N., 2013. Deriving Business Intelligence from Unstructured Data. Available: http://www.irphouse.com/ijict.htm.

Hamer, W. H., Major Greenwood, B., and Udny Yule, G., %ection of Eptbemiologp anb 0tate Llebtcine. The Statistics of Anti-typhoid and Anti-cholera Inoculations, and the Interpretation of Such Statistics in General.

He, Y., Wang, C., Zhang, S., Li, N., Li, Z., Zeng, Z., Oct. 2022. KG-MTT-BERT: Knowledge Graph Enhanced BERT for Multi-Type Medical Text Classification. Available: http://arxiv.org/abs/2210.03970.

Hellerstein, J.M., 2008. Quantitative Data Cleaning for Large Databases. Available: http://db.cs.berkeley.edu/jmh.

Hochreiter S., J. Schmidhuber. Long Short- Term Memory.

Jha, S., Topol, E.J., Dec. 13, 2016. Adapting to Artificial Intelligence: Radiologists and Pathologists as Information Specialists. American Medical Association. https://doi.org/10.1001/jama.2016.17438.

Ji, L., Yan, D., Cheng, Z., Song, Y., Oct. 2023. Improving unified named entity recognition by incorporating mention relevance. Neural Comput. Appl. 35 (30), 22223–22234. https://doi.org/10.1007/s00521-023-08820-6.

Li, L., Lud€ascher, B., and Zhang, Q., Towards More Transparent, Reproducible, and Reusable Data Cleaning with OpenRefine.

Li, J., Sun, A., Han, J., Li, C., Mar. 2020. A Survey on Deep Learning for Named Entity Recognition. https://doi.org/10.1109/TKDE.2020.2981314.

Lu, H., Ehwerhemuepha, L., Rakovski, C., Dec. 2022. A comparative study on deep learning models for text classification of unstructured medical notes with various levels of class imbalance. BMC Med. Res. Methodol. 22 (1). https://doi.org/10.1186/s12874-022-01665-y.

Macri, C.Z., et al., Nov. 2023. A case study in applying artificial intelligence-based named entity recognition to develop an automated ophthalmic disease registry. Graefe's Arch. Clin. Exp. Ophthalmol. 261 (11), 3335–3344. https://doi.org/10.1007/s00417-023-06190-2.

Obermeyer, Z., Emanuel, E.J., Sep. 2016. Predicting the future — big data, machine learning, and clinical medicine. N. Engl. J. Med. 375 (13), 1216–1219. https://doi.org/10.1056/nejmp1606181.

Palanisamy, V., Thirunavukarasu, R., Oct. 01, 2019. Implications of Big Data Analytics in Developing Healthcare Frameworks – A Review. King Saud bin Abdulaziz University. https://doi.org/10.1016/j.jksuci.2017.12.007.

Pereira, J.L.M., Fonseca, M.J., Fonseca, M.J., Lopes, A., 2024. Cleenex: support for user involvement during an iterative data cleaning process. ACM J. Data Inform. Quality 16 (1), 26. https://doi.org/10.54499/UIDB/50021/2020.

Petrova-Antonova, D., Tancheva, R., 2020. Data cleaning: a case study with openrefine and trifacta wrangler. In: Communications in Computer and Information Science. Springer Science and Business Media Deutschland GmbH, pp. 32–40. https://doi.org/10.1007/978-3-030-58793-2_3.

Piwek, L., Ellis, D.A., Andrews, S., Joinson, A., Feb. 2016. The rise of consumer health wearables: promises and barriers. PLoS Med. 13 (2). https://doi.org/10.1371/journal.pmed.1001953.

Raghupathi, W., Raghupathi, V., 2014. Big data analytics in healthcare: promise and potential. Health Inf. Sci. Syst. 2, 3. https://doi.org/10.1186/2047-2501-2-3.

Rajkomar, A., Dean, J., Kohane, I., Apr. 2019. Machine learning in medicine. N. Engl. J. Med. 380 (14), 1347–1358. https://doi.org/10.1056/nejmra1814259.

Reddy, C. K., and Aggarwal, C. C., HealtHcare Data analytics.

Saria, S., Butte, A., Sheikh, A., 2018. Better medicine through machine learning: what's real, and what's artificial? PLoS Med. 15 (12). https://doi.org/10.1371/journal.pmed.1002721.

Sarker, I.H., Nov. 01, 2021. Deep Learning: A Comprehensive Overview on Techniques, Taxonomy, Applications and Research Directions. Springer. https://doi.org/10.1007/s42979-021-00815-1.

Vest, J.R., Gamm, L.D., May 2010. Health information exchange: persistent challenges and new strategies. J. Am. Med. Inform. Assoc. 17 (3), 288–294. https://doi.org/10.1136/jamia.2010.003673.

Vijayan, V., Connolly, J., Condell, J., McKelvey, N., Gardiner, P., Aug. 02, 2021. Review of Wearable Devices and Data Collection Considerations for Connected Health. MDPI AG. https://doi.org/10.3390/s21165589.

Wang, Y., Kung, L.A., Byrd, T.A., Jan. 2018. Big data analytics: understanding its capabilities and potential benefits for healthcare organizations. Technol. Forecast. Soc. Change 126, 3–13. https://doi.org/10.1016/j.techfore.2015.12.019.

Wang, L., Ma, Y., Bi, W., Lv, H., Li, Y., Mar. 2024. An entity extraction pipeline for medical text records using large language models: analytical study. J. Med. Internet Res. 26, e54580. https://doi.org/10.2196/54580.

Zhang, L., Wang, S., Liu, B., Jul. 2018. Deep learning for sentiment analysis: a survey. Wiley Interdiscip. Rev.: Data Min. Knowl. Discov. 8 (4). https://doi.org/10.1002/widm.1253.

CHAPTER 2

From efficiency to sustainability: The evolution of smart healthcare systems

Nigar K. Mujawar[1] and Ashwini R. Suryawanshi[2]
[1]Department of Pharmaceutics, ASPM College of Pharmacy, Sindhudurg, Maharashtra, India; [2]Sant Gajanan Maharaj College of Pharmacy, Kolhapur, Maharashtra, India

1. Introduction

With an emphasis on social, economic, and environmental sustainability, sustainable smart healthcare systems combine cutting-edge technologies to represent a paradigm shift in the delivery of healthcare (Voulvoulis et al., 2022). These technologies combine data analytics, artificial intelligence (AI), and the Internet of Things (IoT) to maximize resource efficiency, enhance operational effectiveness, and optimize patient care. Healthcare professionals may track medical equipment, monitor patients remotely, and expedite administrative procedures by utilizing connected gadgets and sensors. In addition to improving patient outcomes and lowering healthcare costs and environmental impact, this real-time data facilitates proactive healthcare interventions and individualized treatment programs. The focus on sustainability ensures that these systems serve long-term ecological stewardship, community well-being, and present healthcare needs (Barnes and Zvarikova, 2021). Because IoT makes it possible to continuously monitor patient health parameters like heart rate, blood pressure, and glucose levels, it is essential to the development of sustainable smart healthcare systems. Healthcare providers may get real-time data from connected devices, which makes customized care plans and early intervention possible. AI systems examine enormous volumes of medical data to find trends, forecast patient outcomes, and improve treatment plans. Machine learning models help make healthcare decisions more accurate, improve diagnostic accuracy, and propel advances in precision medicine. Data analytics methods, such as data visualization and predictive analytics, convert unprocessed data into meaningful insights. These observations guide quality improvement programs in healthcare facilities, enhance resource allocation, and inform population health management strategies (Ghazal et al., 2021).

Optimizing patient outcomes through accessible, affordable, and individualized treatment is the main goal of sustainable smart healthcare systems. These systems seek to decrease healthcare inequities, improve clinical results, and increase patient happiness by incorporating cutting-edge technologies. Another crucial objective is efficient resource management, which guarantees the prudent use of energy, infrastructure,

Intelligent Semantic Analysis for Healthcare
ISBN 978-0-443-33512-9,
https://doi.org/10.1016/B978-0-443-33512-9.00015-9

and medical resources. Sustainable methods reduce carbon emissions and support ecologically friendly healthcare operations. In order to provide healthcare services to marginalized populations through telemedicine, remote diagnostics, and mobile health applications, scalability and accessibility are also crucial goals. Iterative feedback loops promote innovation and adaptation to changing healthcare requirements and technological breakthroughs, hence driving continuous improvement (Ahmad et al., 2022). In order to address the problems of the current healthcare landscape and ensure readiness for future needs, sustainable smart healthcare systems combine technology innovation with sustainable practices. This approach to healthcare delivery is comprehensive. By increasing productivity, boosting patient care, and encouraging environmental responsibility throughout the healthcare industry, these systems have the potential to completely change the way healthcare is delivered (Marques et al., 2019).

2. Benefits of sustainable smart healthcare systems

Numerous advantages of sustainable smart healthcare systems include improved accessibility to services, reduced environmental impact, and a transformation in healthcare delivery.

2.1 More efficiency in the delivery of healthcare

The use of IoT, AI, and data analytics simplifies procedures like treatment planning, diagnosis, and patient monitoring. Healthcare professionals may make well-informed decisions more quickly because to real-time data gathering and analysis, which lowers wait times and increases patient throughput. Predictive analytics and automated workflows improve scheduling, resource allocation, and inventory control, which eventually raises operational efficiency in healthcare facilities (Dwivedi et al., 2022).

2.2 Improved accessibility to medical services

By utilizing telemedicine, remote monitoring, and mobile health apps, sustainable smart healthcare systems go beyond conventional limits to provide medical services. Patients do not need to physically attend healthcare institutions in order to receive ongoing care, tests, and specialized consultations in rural or underserved locations. By improving early diagnosis, intervention, and treatment of medical disorders, this increased accessibility improves health outcomes and lessens disparities in healthcare access (Mishra and Singh, 2023a).

2.3 Lessened environmental impact and resource consumption

Healthcare facilities with sustainable practices integrated into their operations such as energy-efficient structures, renewable energy sources, and conscientious waste management have a less environmental impact. Smart gadgets with IoT capabilities optimize energy usage and track environmental parameters, guaranteeing resource efficiency and lowering greenhouse gas emissions and waste production. Sustainable smart healthcare systems support environmental sustainability by integrating healthcare delivery with international conservation initiatives through the adoption of ecofriendly technologies and practices.

In addition to improving healthcare delivery's accessibility and efficiency, sustainable smart healthcare systems place a high priority on environmental responsibility. These systems open the door to a future in which healthcare is efficient and environmentally responsible, benefiting patients, healthcare providers, and society at large by embracing cutting-edge technologies and sustainable practices (Singh et al., 2019).

3. Sustainable smart healthcare systems components and technologies

The goal of sustainable smart healthcare systems is to reduce resource consumption and environmental impact while improving the quality, effectiveness, and accessibility of healthcare services. These systems use data driven strategies and cutting edge technologies to build a more patient-centered, long lasting healthcare environment. Sustainable smart healthcare systems components and technologies shown in Table 2.1.

Sustainable smart healthcare systems can deliver effective, economical, and ecologically conscious medical care by combining these elements. These systems use state-of-the-art technology to improve patient care, expedite processes, and lessen the environmental impact of healthcare activities (Tripathy et al., 2022).

4. Environmental Sustainability Initiatives

Environmental sustainability initiative in sustainable smart healthcare systems is shown in Fig. 2.1.

4.1 Green building design in healthcare facilities

4.1.1 Energy efficiency

Sustainable healthcare facility design aims to reduce energy usage through a variety of ways. This involves using high-performance insulation, energy-efficient windows, and modern HVAC systems during the construction process. Furthermore, using LED lighting and smart lighting systems that change based on occupancy and natural light availability helps to reduce energy use. The adoption of energy-efficient medical equipment and building systems is also essential for reducing overall energy usage. These design and operational principles together help to create a more sustainable and cost-effective healthcare environment (Silva et al., 2023a).

Table 2.1 Sustainable smart healthcare systems components and technologies.

Sr. No.	Component and technologies	Description	Examples
1	Integration of IoT-enabled devices	Leveraging devices that record and transmit health data in real time.	There are smart medical devices, wearable health monitors, and remote patient monitoring tools (Goyal et al., 2021).
2	Wearable health monitors	Devices that keep an eye on vital signs and other health data, like fitness trackers and smartwatches.	Fitness trackers and smartwatches (Adeghe et al., 2024).
3	Remote patient monitoring	Tools that let doctors and nurses check on their patients' health from afar.	Remote monitoring tools for long-term conditions (Mucchi et al., 2021).
4	Smart medical devices	What it does is gather data and send it to healthcare providers.	Smart insulin pens, connected inhalers, and wireless blood pressure monitors (Raikar et al., 2023).
5	Utilization of electronic health records (EHRs)	Keeping and giving access to all kinds of medical information.	Health information systems that can talk to each other, centralized patient data stores, and data analytics tools (Reegu et al., 2023).
6	Centralized patient data	Extensive databases containing patient data.	Whole medical history integration in EHR systems (Serbanati, 2020).
7	Interoperable systems	Platforms that make it easy to transfer data between different healthcare environments.	Platforms for interoperable EHRs and health information exchanges (HIEs) (Reegu et al., 2023).
8	Data analytics tools	Software for trend analysis and decision support using EHR data.	Systems for clinical decision support and predictive analytics software (Lytras et al., 2020).
9	Role of AI-driven algorithms in healthcare optimization	AI tools to forecast results and optimize healthcare delivery.	Tools for operational optimization, algorithms for personalized medicine, and predictive analytics (Nadella et al., 2023).

Table 2.1 Sustainable smart healthcare systems components and technologies.—cont'd

Sr. No.	Component and technologies	Description	Examples
10	Predictive analytics	AI programs that forecast future health issues by analyzing patient data from the past.	Forecasting patient deterioration and disease outbreaks (Awotunde et al., 2022).
11	Personalized medicine algorithms	Medical treatment customization systems that make use of environmental, lifestyle, and genetic data.	Tools for genomic analysis and systems for recommending customized treatments (Andreu-Perez et al., 2015).
12	Operational optimization	AI solutions that improve healthcare operations' efficiency.	Tools for managing patient flow and optimizing staff scheduling (Pillai, 2023).

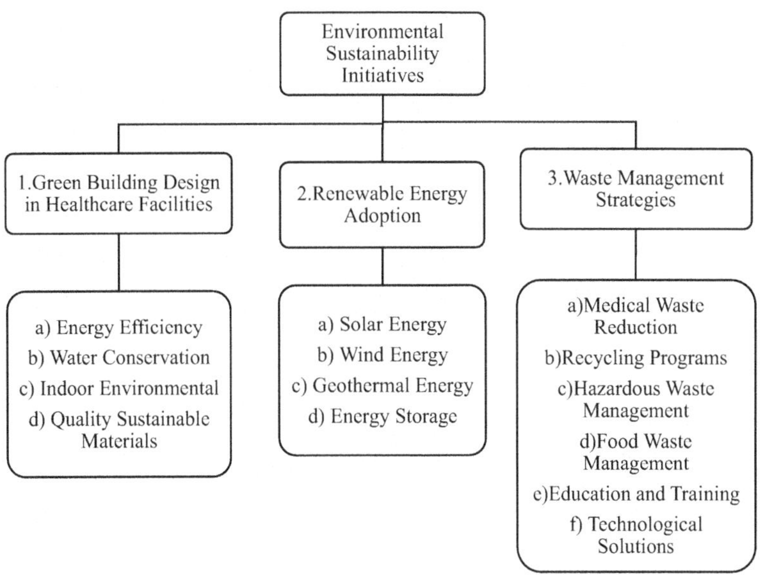

Figure 2.1 Environmental sustainability initiative in sustainable smart healthcare systems.

4.1.2 Water conservation

Effective water management in healthcare institutions requires various techniques. Installing low-flow fixtures and fittings, such as faucets, toilets, and showerheads, helps to save water. Implementing water recycling systems for non-potable use, such as irrigation and cooling, further reduces water consumption. Furthermore, using smart leak detection systems is critical for reducing water waste by locating and repairing leaks quickly. These measures improve water efficiency, contribute to the facility's overall sustainability goals, and save operational costs (Varma, 2022).

4.1.3 Indoor environmental quality

A variety of measures are used in healthcare institutions to maintain a healthy interior environment. High-efficiency air filters and modern ventilation systems are employed to maintain high air quality, while non-toxic building materials aid to reduce indoor pollutants. Maximizing natural light through careful building design improves patient recovery rates while significantly lowering energy use. Furthermore, using sound-absorbing materials reduces noise pollution, resulting in a quieter and more comfortable atmosphere for both patients and staff. These design and operational strategies help to improve indoor environmental quality and overall facility sustainability (Wang and Liu, 2024).

4.1.4 Sustainable materials

To reduce the environmental impact of healthcare facilities, sustainable building strategies include the use of recycled and local materials. Choosing durable materials increases the building's lifespan and decreases maintenance requirements, which contributes to long-term sustainability. Furthermore, pursuing certifications such as LEED (Leadership in energy and environmental design) assures adherence to established sustainability standards, demonstrating the facility's dedication to environmental responsibility. These measures collectively contribute to a more environmentally friendly and efficient healthcare environment (Vierra, 2016).

4.2 Renewable energy adoption

4.2.1 Solar energy

To boost renewable energy use in healthcare institutions, solar panels are erected on rooftops or other suitable sites to generate electricity, lowering dependency on traditional power sources. Solar water heaters, which use solar thermal systems, provide an effective way to heat water, further reducing reliance on traditional energy sources. These approaches help the facility achieve a more sustainable energy profile by lowering operational costs and reducing the environmental effect of energy consumption (Soares et al., 2018).

4.2.2 Wind energy

Small-scale wind turbines can be built on hospital grounds to complement energy needs, increasing the use of renewable energy in healthcare. Furthermore, connecting with off-site wind farms via power purchase agreements (PPAs) enables facilities to obtain clean energy. These measures lessen dependency on traditional power sources, promote sustainable energy practices, and help to cut operating expenses. Integrating wind energy solutions helps to improve the overall sustainability of healthcare facilities by offering a cleaner, renewable source of power (Brauch et al.).

4.2.3 Geothermal energy

Ground source heat pumps: Use geothermal systems for heating and cooling by tapping into the steady temperatures underground (Olabi et al., 2023).

4.2.4 Energy storage

To improve energy efficiency and sustainability, healthcare facilities can install battery storage systems to store excess renewable energy. This stored energy can be used during peak demand or power outages to ensure a consistent and uninterrupted energy supply. Furthermore, integration with smart grids enables more effective energy flow management and greater use of renewable energy. This combination of battery storage and smart grid integration promotes a more resilient and efficient energy system, decreasing dependency on traditional power sources while also contributing to the facility's overall sustainability goals (O'Dwyer et al., 2019).

4.3 Waste management strategies

4.3.1 Medical waste reduction

A number of critical procedures are required for effective waste management in healthcare facilities. Proper trash classification into categories such as general, recyclable, hazardous, and medical waste guarantees safe and efficient disposal. Using reusable medical instruments and devices helps to reduce waste from single-use items. Transitioning to EHRs reduces paper waste and contributes to sustainability. Collectively, these solutions improve waste management efforts, reduce environmental effect, and contribute to the overall goal of establishing an eco-friendlier healthcare environment (Lee and Lee, 2022).

4.3.2 Recycling programs

Effective waste management in healthcare facilities requires comprehensive recycling programs for paper, plastics, metals, and glass. These programs ensure that recyclable items are processed correctly and diverted from landfills. Furthermore, effective disposal and recycling of electronic trash (e-waste), such as obsolete medical equipment and computers, is critical for managing dangerous components and lowering environmental impact. Implementing these recycling methods reduces waste, promotes environmental sustainability, and assures the proper management of materials and equipment inside the facility (Razip et al., 2022).

4.3.3 Hazardous waste management

The safe management of hazardous waste in healthcare facilities entails guaranteeing its disposal through licensed services that follow regulatory criteria. This strategy ensures that hazardous materials are securely handled and disposed of. Furthermore, using onsite waste treatment technology like as autoclaves aids in sterilization and reduces the volume of medical waste. These technologies help facilities to handle garbage more effectively and responsibly, thereby lowering environmental impact and improving overall waste management practices. Combining approved disposal services with on-site treatment options improves safety and sustainability in healthcare waste management (World Health Organization, 2022).

4.3.4 Food waste management

Healthcare facilities can effectively minimize food waste using two main measures. First, implementing composting programs for organic food waste from cafeterias and kitchens helps turn this waste into valuable compost that may be utilized for gardening and landscaping, so minimizing landfill contributions. Second, collaborating with local organizations to give leftover food reduces food waste by providing meals to people in need, so promoting community welfare. These approaches not only decrease environmental effect, but they also encourage social responsibility and community involvement. Composting and donation programs work together to create a more sustainable and ethical food waste management system (Martin-Rios et al., 2018).

4.3.5 Education and training

Effective waste management in healthcare settings requires ongoing staff training and awareness efforts. Training programs should emphasize best practices for trash segregation, recycling, and reduction to ensure that employees handle waste responsibly and efficiently. Furthermore, implementing awareness campaigns educates patients and visitors about the facility's sustainability initiatives, encouraging an environmentally responsible culture. These programs work together to improve the facility's waste management efforts by ensuring that both personnel and the general public are aware of and engaged in sustainable practices, resulting in more successful waste reduction and recycling outcomes (Mahjoob et al., 2023).

4.3.6 Technology solutions

To improve waste management efficiency, healthcare facilities can use smart waste bins and data analytics. Smart garbage bins automatically sort and compact waste, increasing recycling efficiency and reducing contamination. Furthermore, utilizing data analytics to track trash generation patterns might assist uncover potential for reduction and optimization. Analyzing waste data allows facilities to make informed decisions about how to change procedures, increase segregation, and target waste reduction areas. These technologies work together to make waste management more effective and sustainable, optimizing resource utilization while minimizing environmental impact (Shukla and Hait, 2022).

5. Challenges in adoption of sustainable smart healthcare systems

5.1 Data security and privacy issues

5.1.1 Patient data protection

Healthcare systems must safeguard sensitive patient data, such as medical records, personal information, and financial details, to prevent breaches. Compliance with complicated and resource-intensive rules such as HIPAA in the United States, GDPR in Europe, and other regional data protection legislation is critical. The growing threat of cyberattacks, such as ransomware, poses major hazards to this data. To counteract these threats, strong cybersecurity measures like encryption, firewalls, and frequent security audits are required (Yigzaw et al., 2022).

5.1.2 Technical challenge

The development of IoT devices in healthcare, such as wearable monitoring and smart medical equipment, creates several access points for potential cyberattacks, making security critical. Furthermore, securely transferring patient data between various healthcare providers and systems while ensuring privacy is a huge difficulty. Implementing secure data exchange protocols and technologies, such as blockchain, can help reduce these risks (Wenhua et al., 2023).

5.1.3 Organizational challenges

To avoid inadvertent breaches and maintain compliance, it is critical that all healthcare professionals receive proper data security training. However, assigning adequate resources for data security measures might be difficult, especially for smaller healthcare facilities with tight budgets (Ma, 2021).

5.2 Interoperability of healthcare systems

5.2.1 Technical barriers

Incompatibility across systems used by healthcare facilities might create data silos. The absence of standardized protocols and data formats impedes seamless data interchange between various systems. While initiatives to standardize data formats like as HL7 and FHIR are progressing, they have not yet been widely adopted (Oliveira et al., 2023).

5.2.2 Integration challenges

Many healthcare institutions still use outdated systems that are difficult to interface with current technologies, resulting in costly and time-consuming upgrades and replacements. Furthermore, transferring data from legacy systems to new interoperable platforms is a complicated process that necessitates meticulous planning and execution to avoid data loss or corruption (Katuk et al., 2023).

5.2.3 Organizational barriers

Achieving interoperability in healthcare needs collaboration across stakeholders, such as healthcare providers, technology vendors, and regulatory agencies, which can be difficult to coordinate. Furthermore, healthcare personnel may be hesitant to adopt new interoperable systems because of the learning curve and potential interruption to existing workflows (de Mello et al., 2022).

5.3 Initial investment costs and ROI considerations

5.3.1 High initial investment

Sustainable smart healthcare systems require significant infrastructure investment, including energy-efficient buildings, renewable energy sources, and advanced medical equipment. Furthermore, obtaining and integrating new technologies like EHR systems, IoT devices, and cybersecurity solutions can be costly (Padmanabhan, 2017).

5.3.2 Financial constraints

Many healthcare facilities, particularly in underfunded areas, struggle to get the money required for long-term smart healthcare investments. Demonstrating a clear return on investment (ROI) for these projects can be difficult since while the long-term cost savings and efficiency advantages may be large, the early payback time might be lengthy (Horgan et al., 2020).

5.3.3 Operational challenges

Implementing new systems and technology in healthcare can disrupt daily operations, resulting in temporary inefficiencies and increased workloads for employees. Furthermore, the continual maintenance and improvements required for sustainable smart healthcare systems necessitate continuous investment, which might strain financial resources (Zhang et al., 2020).

6. Strategies for adoption and implementation of sustainable smart healthcare systems

6.1 Stakeholder collaboration and engagement

6.1.1 Identify key stakeholders

Healthcare practitioners, administrative personnel, IT professionals, and facility managers are among the internal stakeholders involved in the adoption of sustainable smart healthcare systems. External stakeholders include technology vendors, regulatory agencies, community organizations, and patients. Technology providers offer the tools and systems, while regulatory agencies guarantee that they meet legal requirements. Community organizations may support or gain from the system, while patients' experiences will be impacted. Engaging both internal and external stakeholders is critical for aligning objectives, addressing concerns, and assuring the successful installation and operation of new healthcare technologies (Negash and Sarmiento, 2023).

6.1.2 Building partnerships

Collaborative planning entails involving stakeholders early in the process to obtain feedback, align goals, and address issues, ensuring that new systems satisfy the needs of all parties. This early interaction aids in personalizing solutions to the unique needs of both internal and external stakeholders. Joint ventures with technology providers and industry specialists are also required. By collaborating with these professionals, healthcare institutions can benefit from their specialized knowledge and resources, allowing for a more seamless adoption and integration of new technologies. These tactics collectively improve the effectiveness of the adoption process, ensuring that systems meet the needs of users and stakeholders (Papadaki, 2017).

6.1.3 Communication, training

Continuous communication is essential for keeping stakeholders updated on progress, difficulties, and changes throughout the implementation process. Transparency guarantees that all parties are on the same page and can resolve any concerns as soon as they arise. Additionally, training programs are critical for educating employees to use new systems properly. Developing comprehensive training ensures that staff have the appropriate skills and information, allowing for a smooth transfer and minimizing disruptions. Together, these techniques facilitate the proper application and integration of new healthcare technologies, resulting in a successful and efficient adoption process (Gagnon et al., 2012).

6.1.4 Feedback mechanisms

Continuous feedback entails putting in place methods to collect ongoing feedback from users and stakeholders, which aids in identifying problems and making necessary changes. This ensures that the system can respond to user needs and operational challenges. Evaluation and improvement are equally crucial, necessitating regular reviews of the system's performance and modifications based on stakeholder input. This approach helps to improve functionality and happiness, ensuring that the system evolves to successfully fulfill the needs of its users. Together, these techniques promote a dynamic and responsive approach to system management, resulting in improved outcomes and user experiences (Kilbourne et al., 2014).

6.2 Infrastructure investment and technological readiness

6.2.1 Assessing infrastructure needs

Evaluate current infrastructure to determine if it can support new technologies and if changes are required. This examination guarantees that the current systems are capable of incorporating new solutions. Future-proofing focusses on planning for technology developments and scalability, ensuring that the infrastructure is viable and flexible in the long run. This method enables facilities to keep ahead of technology trends while also allowing for the inclusion of future advancements without repeated overhauls. Together, assessing present capabilities and planning for future-proofing guarantees that healthcare facilities may efficiently implement new technology while ensuring long-term operational efficiency and flexibility (Greenhalgh et al., 2017).

6.2.2 Financial planning

Budget allocation entails obtaining money for both the original investment and continuing maintenance costs of new technologies, including charges such as technology acquisition, installation, and training. Adequate finance is critical for effective implementation and operation. Cost-benefit analysis is critical for justifying these initiatives, indicating the possible ROI through efficiency and sustainability benefits. This study serves to legitimize the financial commitment by demonstrating how the advantages of the new system outweigh the costs, allowing for more informed decision-making and increasing confidence in the investment. Budget allocation and cost-benefit analysis are essential for assuring the financial viability and long-term success of new healthcare technologies (Eisman et al., 2020).

6.2.3 Technical readiness

System integration is critical for ensuring that new technologies integrate seamlessly with old systems, avoiding disruptions and increasing overall effectiveness. Proper integration enables seamless operation and better functionality throughout the healthcare facility. Cybersecurity measures are also vital, as they involve putting in place strong safeguards to protect data and systems from potential attacks and breaches. Strong cybersecurity measures are critical for ensuring data integrity and confidentiality, preventing unauthorized access, and protecting against cyber assaults. Effective system integration and rigorous cybersecurity measures work together to guarantee that new technologies improve operational efficiency while also ensuring a secure and reliable IT environment (Berlilana et al., 2021).

6.2.4 Pilot testing

Pilot tests are conducted in a controlled setting to evaluate the performance of new systems prior to a full-scale rollout. These tests aid in identifying potential faults and evaluating how effectively the system performs in real-world circumstances. The adjustment period follows, during which any issues detected during the pilot phase are addressed and adjustments are implemented. This phase guarantees that the system is improved and optimized for a smooth deployment. Trial runs and the adjustment period are essential for establishing that the new technology is effective, dependable, and ready for wider usage, lowering the likelihood of disruptions and improving overall implementation success (Herranz et al., 2024).

6.3 Regulatory and policy support

6.3.1 *Navigating regulations*

Compliance with regulations and standards, such as HIPAA in the US or GDPR in Europe, ensures that new systems fulfill legal and ethical norms. This devotion is critical for avoiding legal complications and guaranteeing the ethical treatment of sensitive data. Documentation is essential in this process since it allows for the creation of accurate records that demonstrate compliance. Proper documentation facilitates audits and reviews by demonstrating compliance with regulatory standards. Addressing compliance standards and keeping thorough documentation are both critical to ensure that new systems are legally compliant and efficiently managed (Corte-Real et al., 2024).

6.3.2 *Policy development*

Internal policies are critical for managing data privacy, security, and the use of new technology in a healthcare setting. Developing and enforcing these policies helps guarantee that all employees follow best practices and regulatory obligations. Guidelines and procedures are also crucial, as they give an organized approach to system implementation, operation, and maintenance. Establishing clear rules helps to streamline procedures, guaranteeing consistent and efficient usage of new systems while reducing risk. Internal policies, along with well defined standards and processes, form a solid framework for managing new technologies, assuring their effective and secure use within the facility (Dion and Evans, 2024).

6.3.3 *Engaging with policymakers*

Advocacy entails working with lawmakers and regulatory organizations to advance legislation and win funds for long term healthcare projects. This proactive approach influences the formulation of rules that promote innovation and sustainability in healthcare. Feedback and influence are also important since providing input on new regulations ensures that healthcare institutions' needs and concerns are acknowledged. Participating in the policy-making process helps establish laws that support effective and sustainable practices. Advocacy and feedback work together to ensure that healthcare systems have favorable policies and regulations in place, which improves their ability to develop and sustain new solutions (Omaghomi et al., 2024).

6.3.4 Adapting to change

Regulatory updates entail remaining up to date on regulatory changes and updating systems and practices to ensure continuous compliance. This proactive strategy enables healthcare institutions to react to new legislative needs while adhering to industry norms. Policy changes are also significant since they require examining and updating internal policies on a regular basis to ensure compliance with changing rules and best practices. Regular changes keep policies current and effective, promoting compliance and operational excellence. Together, regulatory changes and policy adjustments are critical for ensuring a compliant and adaptive healthcare system that fulfills both legal and operational standards (Raghavan et al., 2021).

7. Future prospects and potential of sustainable smart healthcare systems

7.1 Transformative impact on healthcare delivery

Sustainable smart healthcare systems have the potential to radically improve healthcare delivery by combining sophisticated technologies and ecofriendly practices. These systems improve patient care by enabling real-time monitoring, personalized treatment plans, and easier access to medical information. Healthcare institutions may provide more efficient, accurate, and responsive care by incorporating technologies like IoT devices, AI powered diagnostics, and telemedicine. Furthermore, sustainable methods lessen the environmental impact of healthcare operations, making it more ecologically responsible and resource efficient (Katal).

7.2 Long-term sustainable goals

Sustainable smart healthcare systems contribute to long-term sustainability goals by reducing energy consumption, minimizing waste, and boosting resource efficiency. Healthcare institutions can reduce their environmental impact by implementing energy-efficient infrastructure, renewable energy sources, and effective waste management methods. Long-term aims include obtaining LEED certifications, improving the lifecycle of facilities and equipment, and establishing a sustainable culture within the healthcare industry. These activities help to create a more sustainable future while maintaining excellent levels of patient care (Silva et al., 2023).

7.3 Innovation and growth of smart healthcare technologies

The future of smart healthcare is defined by continual innovation and expansion. Emerging technologies, including enhanced AI, blockchain for safe data exchange, and next-generation medical devices, are propelling healthcare systems forward. Smart healthcare technology innovations promise to increase diagnostic accuracy, operational efficiency, and patient outcomes. As research and development in this subject progresses, new solutions will emerge, altering healthcare delivery and extending the opportunities for personalized and efficient treatment. Embracing these developments will be critical to being at the forefront of healthcare technology and delivering continuous improvements in patient care and operational efficiency (Mishra and Singh, 2023b).

8. Case studies in smart healthcare systems

Two case studies in smart health care system discussed in Table 2.2.

Table 2.2 Case studies in smart health care system.

Sr. No.	Case study	Background	Implementation	Outcomes
1	Reducing readmissions with remote patient monitoring	A medical center and a technology business partnered to reduce readmissions among patients with chronic heart failure (CHF).	RPM devices were provided for vital sign monitoring. Data was integrated into the electronic health record system. Healthcare personnel monitored the data and intervened as necessary.	25% lower rates of readmission. Heightened involvement of the patient. Considerable financial savings (Gjeka et al., 2021).
2	Enhancing patient results with AI and telehealth	Medical center got government funding and collaborated with a major tech company to develop an AI-driven telemedicine platform.	Created a platform for AI-powered health advice and virtual consultations. Artificial intelligence examined patient data to generate tailored suggestions. Worked along with legislators and medical specialists.	A 30 percent gain in patient results. Improved access to specialized care for those who live far away. Less paperwork for healthcare practitioners to handle (Kvedar et al., 2014).

9. Conclusion

Sustainable smart healthcare systems provide various advantages, including better patient outcomes, more efficiency, and a lower environmental impact. These systems combine sophisticated technologies like IoT, AI, and telemedicine to provide high quality, personalized care while reducing resource consumption and waste. Despite the obvious benefits, adoption hurdles remain, including high initial costs, technology integration issues, and opposition to change within healthcare organizations. To solve these challenges, strategic approaches are necessary. These include government incentives, infrastructural investments, healthcare professional training programs, and stakeholder collaboration. As we look ahead, the integration of sustainable practices and smart technologies has the potential to transform healthcare, making it more robust, efficient, and environmentally friendly. The vision for the future includes a healthcare system that not only serves the needs of patients and providers but also fits with global sustainability goals, resulting in a healthier planet and people.

References

Adeghe, E.P., Okolo, C.A., Ojeyinka, O.T., 2024. A review of wearable technology in healthcare: monitoring patient health and enhancing outcomes. OARJ Multidiscip. Stud. 7 (01), 142–148.

Ahmad, K.A., Khujamatov, H., Akhmedov, N., Bajuri, M.Y., Ahmad, M.N., Ahmadian, A., May 1, 2022. Emerging trends and evolutions for smart city healthcare systems. Sustain. Cities Soc. 80, 103695.

Andreu-Perez, J., Leff, D.R., Ip, H.M., Yang, G.Z., April 13, 2015. From wearable sensors to smart implants—toward pervasive and personalized healthcare. IEEE (Inst. Electr. Electron. Eng.) Trans. Biomed. Eng. 62 (12), 2750–2762.

Awotunde, J.B., Folorunso, S.O., Ajagbe, S.A., Garg, J., Ajamu, G.J., February 2022. AiIoMT: IoMT-based system-enabled artificial intelligence for enhanced smart healthcare systems. Machine Learning for Critical Internet of Medical Things: Applications and Use Cases 3, 229–254.

Barnes, R., Zvarikova, K., 2021. Artificial intelligence-enabled wearable medical devices, clinical and diagnostic decision support systems, and Internet of Things-based healthcare applications in COVID-19 prevention, screening, and treatment. Am. J. Med. Res. 8 (2), 9–22.

Berlilana, N.T., Ruangkanjanases, A., Hariguna, T., Sarmini, December 13, 2021. Organization benefit as an outcome of organizational security adoption: the role of cyber security readiness and technology readiness. Sustainability 13 (24), 13761.

Brauch MD, Toledano P, Mehranvar L, Iliopoulos T, Sasmal S, Aydos M. Scaling Investment in Renewable Energy Generation to Achieve Sustainable Development Goals 7 (Affordable and Clean Energy) and 13 (Climate Action) and the paris agreement: Roadblocks and Drivers.

Corte-Real, A., Nunes, T., da Cunha, P.R., February 16, 2024. Reflections about blockchain in health data sharing: navigating a disruptive technology. Int. J. Environ. Res. Publ. Health 21 (2), 230.

de Mello, B.H., Rigo, S.J., da Costa, C.A., da Rosa Righi, R., Donida, B., Bez, M.R., Schunke, L.C., March 2022. Semantic interoperability in health records standards: a systematic literature review. Health Technol. 12 (2), 255–272.

Dion, H., Evans, M., February 8, 2024. Strategic frameworks for sustainability and corporate governance in healthcare facilities; approaches to energy-efficient hospital management. Benchmark Int. J. 31 (2), 353–390.

Dwivedi, R., Mehrotra, D., Chandra, S., March 1, 2022. Potential of Internet of Medical Things (IoMT) applications in building a smart healthcare system: a systematic review. J. Oral Biol. Craniofac. Res. 12 (2), 302–318.

Eisman, A.B., Kilbourne, A.M., Dopp, A.R., Saldana, L., Eisenberg, D., January 1, 2020. Economic evaluation in implementation science: making the business case for implementation strategies. Psychiatry Res. 283, 112433.

Gagnon, M.P., Desmartis, M., Labrecque, M., Car, J., Pagliari, C., Pluye, P., Frémont, P., Gagnon, J., Tremblay, N., Légaré, F., February 2012. Systematic review of factors influencing the adoption of information and communication technologies by healthcare professionals. J. Med. Syst. 36, 241–277.

Ghazal, T.M., Hasan, M.K., Alshurideh, M.T., Alzoubi, H.M., Ahmad, M., Akbar, S.S., Al Kurdi, B., Akour, I.A., August 23, 2021. IoT for smart cities: machine learning approaches in smart healthcare—a review. Future Internet 13 (8), 218.

Gjeka, R., Patel, K., Reddy, C., Zetsche, N., August 2021. Patient engagement with digital disease management and readmission rates: the case of congestive heart failure. Health Inf. J. 27 (3), 14604582211030959.

Goyal, S., Sharma, N., Bhushan, B., Shankar, A., Sagayam, M., 2021. IoT enabled technology in secured healthcare: applications, challenges and future directions. In: Cognitive Internet of Medical Things for Smart Healthcare: Services and Applications, pp. 25–48.

Greenhalgh, T., Wherton, J., Papoutsi, C., Lynch, J., Hughes, G., Hinder, S., Fahy, N., Procter, R., Shaw, S., November 1, 2017. Beyond adoption: a new framework for theorizing and evaluating non-adoption, abandonment, and challenges to the scale-up, spread, and sustainability of health and care technologies. J. Med. Internet Res. 19 (11), e8775.

Herranz, C., Gómez, A., Hernández, C., González-Colom, R., Contel, J.C., Cano, I., Piera-Jiménez, J., Roca, J., April 2024. Toward sustainable adoption of integrated care for prevention of unplanned hospitalizations: a qualitative analysis. Int. J. Integrated Care 24 (2).

Horgan, D., Borisch, B., Richer, E., Bernini, C., Kalra, D., Lawler, M., Ciliberto, G., Van Poppel, H., Paradiso, A., Riegman, P., Triberti, S., May 27, 2020. Propelling health care into the twenties. Biomed. Hub 5 (2), 1–53.

Katal N. AI-driven healthcare services and infrastructure in smart cities. In Smart Cities (pp. 150-170). CRC Press.

Katuk, N., Abdullah, W.A., Sugiharto, T., Ahmad, I., November 9, 2023. Smart technology: ecosystem, impacts, challenges and the path forward. Inform. Sys. Smart City 3 (1).

Kilbourne, A.M., Almirall, D., Eisenberg, D., Waxmonsky, J., Goodrich, D.E., Fortney, J.C., Kirchner, J. E., Solberg, L.I., Main, D., Bauer, M.S., Kyle, J., December 2014. Protocol: adaptive Implementation of Effective Programs Trial (ADEPT): cluster randomized SMART trial comparing a standard versus enhanced implementation strategy to improve outcomes of a mood disorders program. Implement. Sci. 9, 1–4.

Kvedar, J., Coye, M.J., Everett, W., February 1, 2014. Connected health: a review of technologies and strategies to improve patient care with telemedicine and telehealth. Health Aff. 33 (2), 194–199.

Lee, S.M., Lee, D., November 10, 2022. Effective medical waste management for sustainable green healthcare. Int. J. Environ. Res. Publ. Health 19 (22), 14820.

Lytras, M.D., Papadopoulou, P., Sarirete, A., January 2020. Smart Healthcare: emerging technologies, best practices, and sustainable policies. In: Innovation in Health Informatics, pp. 3–8.

Ma, C., November 1, 2021. Smart city and cyber-security; technologies used, leading challenges and future recommendations. Energy Rep. 7, 7999–8012.

Mahjoob, A., Alfadhli, Y., Omachonu, V., May 10, 2023. Healthcare waste and sustainability: implications for a circular economy. Sustainability 15 (10), 7788.

Marques, G., Pitarma, R., M. Garcia, N., Pombo, N., September 24, 2019. Internet of things architectures, technologies, applications, challenges, and future directions for enhanced living environments and healthcare systems: a review. Electronics 8 (10), 1081.

Martin-Rios, C., Demen-Meier, C., Gössling, S., Cornuz, C., September 1, 2018. Food waste management innovations in the foodservice industry. Waste Manag. 79, 196–206.

Mishra, P., Singh, G., 2023a. Smart healthcare in sustainable smart cities. In: Sustainable Smart Cities: Enabling Technologies, Energy Trends and Potential Applications. Springer International Publishing, Cham, pp. 195–219.

Mishra, P., Singh, G., 2023b. Internet of medical things healthcare for sustainable smart cities: current status and future prospects. Appl. Sci. 13 (15), 8869.

Mucchi, L., Jayousi, S., Gant, A., Paoletti, E., Zoppi, P., July 13, 2021. Tele-monitoring system for chronic diseases management: requirements and architecture. Int. J. Environ. Res. Publ. Health 18 (14), 7459.

Nadella, G.S., Satish, S., Meduri, K., Meduri, S.S., September 8, 2023. A systematic literature review of advancements, challenges and future directions of AI and ML in healthcare. Int. J. Mach. Learn. Sustain. Develop. 5 (3), 115–130.

Negash, Y.T., Sarmiento, L.S., February 1, 2023. Smart product-service systems in the healthcare industry: intelligent connected products and stakeholder communication drive digital health service adoption. Heliyon 9 (2).

Olabi, A.G., Mahmoud, M., Obaideen, K., Sayed, E.T., Ramadan, M., Abdelkareem, M.A., June 1, 2023. Ground source heat pumps: recent progress, applications, challenges, barriers, and role in achieving sustainable development goals based on bibliometric analysis. Therm. Sci. Eng. Prog. 41, 101851.

Oliveira, N.R., Santos, Y.D., Mendes, A.C., Barbosa, G.N., Oliveira, M.T., Valle, R., Medeiros, D.S., Mattos, D.M., December 29, 2023. Storage standards and solutions, data storage, sharing, and structuring in digital health: a Brazilian case study. Information 15 (1), 20.

Omaghomi, T.T., Elufioye, O.A., Onwumere, C., Arowoogun, J.O., Odilibe, I.P., Owolabi, O.R., 2024. General healthcare policy and its influence on management practices: a review. World J. Adv. Res. Rev. 21 (2), 441–450.

O'Dwyer, E., Pan, I., Acha, S., Shah, N., March 1, 2019. Smart energy systems for sustainable smart cities: current developments, trends and future directions. Appl. Energy 237, 581–597.

Padmanabhan, P., November 17, 2017. The Big Unlock: Harnessing Data and Growing Digital Health Businesses in a Value-Based Care Era. Archway Publishing.

Papadaki, M., May 29, 2017. Adaptation through collaboration: developing novel platforms to advance the delivery of advanced therapies to patients. Front. Med. 4, 56.

Pillai, A.S., July 3, 2023. AI-Enabled hospital management systems for modern healthcare: an analysis of system components and interdependencies. J. Adv. Analyt. Healthcare Manag. 7 (1), 212–228.

Raghavan, A., Demircioglu, M.A., Taeihagh, A., January 2021. Public health innovation through cloud adoption: a comparative analysis of drivers and barriers in Japan, South Korea, and Singapore. Int. J. Environ. Res. Publ. Health 18 (1), 334.

Raikar, A.S., Kumar, P., Raikar, G.V., Somnache, S.N., June 27, 2023. Advances and challenges in IoT-based smart drug delivery systems: a comprehensive review. Appl. Sys. Innov. 6 (4), 62.

Razip, M.M., Savita, K.S., Kalid, K.S., Ahmad, M.N., Zaffar, M., Rahim, E.E., Baleanu, D., Ahmadian, A., September 1, 2022. The development of sustainable IoT E-waste management guideline for households. Chemosphere 303, 134767.

Reegu, F.A., Abas, H., Gulzar, Y., Xin, Q., Alwan, A.A., Jabbari, A., Sonkamble, R.G., Dziyauddin, R.A., April 7, 2023. Blockchain-based framework for interoperable electronic health records for an improved healthcare system. Sustainability 15 (8), 6337.

Serbanati, L.D., January 1, 2020. Health digital state and Smart EHR systems. Inform. Med. Unlocked 21, 100494.

Shukla, S., Hait, S., January 2022. Smart waste management practices in smart cities: current trends and future perspectives. In: Advanced Organic Waste Management. Elsevier, pp. 407–424.

Silva, B.V., Holm-Nielsen, J.B., Sadrizadeh, S., Teles, M.P., Kiani-Moghaddam, M., Arabkoohsar, A., 2023. Sustainable, green, or smart? Pathways for energy-efficient healthcare buildings. Sustain. Cities Soc. 26, 105013.

Singh, P., Dixit, V., Kaur, J., December 2019. Green healthcare for smart cities. In: Green and Smart Technologies for Smart Cities. CRC Press, pp. 91–130.

Soares, N., Martins, A.G., Carvalho, A.L., Caldeira, C., Du, C., Castanheira, É., Rodrigues, E., Oliveira, G., Pereira, G.I., Bastos, J., Ferreira, J.P., November 1, 2018. The challenging paradigm of interrelated energy systems towards a more sustainable future. Renew. Sustain. Energy Rev. 95, 171–193.

Tripathy, S.S., Imoize, A.L., Rath, M., Tripathy, N., Bebortta, S., Lee, C.C., Chen, T.Y., Ojo, S., Isabona, J., Pani, S.K., December 31, 2022. A novel edge-computing-based framework for an intelligent smart healthcare system in smart cities. Sustainability 15 (1), 735.

Varma, V.G., January 1, 2022. Water-efficient technologies for sustainable development. In: Current Directions in Water Scarcity Research, 6. Elsevier, pp. 101–128.
Vierra, S., September 2016. Green Building Standards and Certification Systems. National Institute of Building Sciences, Washington, DC.
Voulvoulis, N., Giakoumis, T., Hunt, C., Kioupi, V., Petrou, N., Souliotis, I., Vaghela, C.J., July 1, 2022. Systems thinking as a paradigm shift for sustainability transformation. Glob. Environ. Change 75, 102544.
Wang, Y., Liu, L., April 29, 2024. Research on sustainable green building space design model integrating IoT technology. PLoS One 19 (4), e0298982.
Wenhua, Z., Qamar, F., Abdali, T.A., Hassan, R., Jafri, S.T., Nguyen, Q.N., January 20, 2023. Blockchain technology: security issues, healthcare applications, challenges and future trends. Electronics 12 (3), 546.
World Health Organization, February 1, 2022. Global Analysis of Healthcare Waste in the Context of COVID-19: Status, Impacts and Recommendations. World Health Organization.
Yigzaw, K.Y., Olabarriaga, S.D., Michalas, A., Marco-Ruiz, L., Hillen, C., Verginadis, Y., De Oliveira, M. T., Krefting, D., Penzel, T., Bowden, J., Bellika, J.G., January 2022. Health data security and privacy: challenges and solutions for the future. Roadmap Success. Digit. Health Ecosys. 1, 335–362.
Zhang, L., Ren, J., Yuan, H., Yang, Z., Wang, W., Guo, M., Cheng, G., Zhou, L., Tao, S., Zhang, L., Cui, H., October 24, 2020. Evaluation of smart healthcare systems and novel uv-oriented solution for integration, resilience, inclusiveness and sustainability. In: 5th International Conference on Universal Village (UV). IEEE, pp. 1–28.

CHAPTER 3

Architectural analysis of cyber-physical systems for healthcare: A review

V. Bhanumathi[1], A. Pavithra[1] and N. Sudarssan[2]
[1]Department of Electronics and Communication Engineering, College of Engineering Guindy Campus, Anna University, Chennai, Tamil Nadu, India; [2]Department of Artificial Intelligence and Data Science, Sri Eshwar College of Engineering, Coimbatore, Tamil Nadu, India

1. Introduction

A smart healthcare Cyber-Physical System (CPS) is a unique cyber-physical system, which combines embedded software control devices, networking capabilities, and complex physiological dynamics of patients in the modern medical field. In the process of communication, device, and information system interaction of smart healthcare cyber-physical systems, medical cyber-physical data are generated digitally, stored electronically, and accessed remotely by medical staff or patients. Cyber Physical System describing a broad range of complex, multi-disciplinary, physically-aware next-generation engineered system that integrates embedded computing technologies into the physical world.

CPSs are systems that integrate various physical processes, networking, and computation. Advances in proficient technologies and sciences have led to the application of CPS in numerous domains, including aerospace, automotive, industry, and medicine. In basic terms, it is a multidisciplinary strategy that facilitates the requirement for physical input and output by providing an interlink between the internet and its clients. Nonetheless, in the current decade (the smart era), the medical area has grown significantly in importance, and advancements and inventions in this field are taking place a lot gradually. In general, it have significant applications in the fields of robotics, energy, automation, health care, smart buildings, transportation, and physical security, among others. So, while designing CPS architecture for healthcare applications, special attentions need to be paid to ensure data security. There are also a number of other important issues to consider, for example, the requirement to store and manage the huge volume of data collected from thousands of medical sensors. Therefore, database management systems should be efficient and reliable. As medical data can provide useful insight into actions (treatments) necessary to save a patient's life, all data should be readily available and accessible to authorized medical personnel anytime from anywhere.

In addition, healthcare applications require huge computing resources for intelligent decision making based on the massive patient data. However, the networks of wireless sensors collecting the patient data are severely constrained in energy, processing, and

Intelligent Semantic Analysis for Healthcare
ISBN 978-0-443-33512-9
https://doi.org/10.1016/B978-0-443-33512-9.00005-6

storage capacity. The significant advancements in Wireless Body Area Networks (WBANs), medical sensors, and Cloud Computing have made CPS a strong contender and an attractive choice for healthcare applications, especially in-hospital and for in home patient care (Milenković et al., 2006). This is due to its ability to combine active users' input (such as smart feedback systems and digital patient records) with passive users' input (such as sensors) to enhance decision-making and overall control in healthcare. As depicted in Fig. 3.1, CPS consists of cyber and physical worlds where the physical world communicates with the cyber world through embedded systems, sensors, and actuators. The sensors serve as data collection devices, continuously observing and collecting data transmitted to the cyber system. A closed loop system is then created by processing and analyzing these collected data in the cyber world through the cloud, for example, to serve as a stimulus for coordination and control signals sent to actuators.

Medical cyber-physical systems (MCPSs) are a type of cyber-physical system used in modern medicine to help prevent and identify COVID-19. Each MCPS has its own embedded control systems as well as independent network systems. The core foundation of MCPS covers both cyberspace (including network space) and physical space (including user space). The physical space serves as the physical foundation for MCPS. It contains all types of hard real-time health sensing devices, health diagnosis devices, and a user space made up of various users. It sends sensing information to cyberspace via sensing devices and receives control information from cyberspace to control physical equipment.

The cyberspace is the central component of MCPS, and it is in charge of processing, storing, and managing user and health information access security. As MCPS's neural center, the cyberspace receives sensing information from the physical space via network transmission systems, identifies, stores, analyzes, and processes it, and generates feedback control information that is sent to the physical space via network transmission systems. In contrast to the Internet of Things (IoT), CPS emphasizes the development and research of virtual applications in the physical world, builds a closed-loop enabling system based

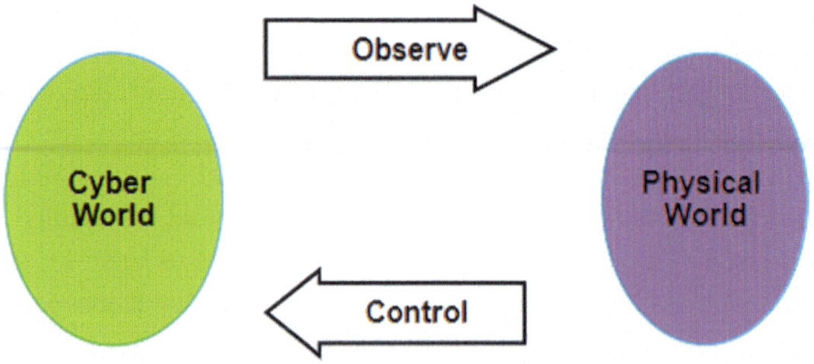

Figure 3.1 Overview of cyber-physical systems.

on state sensing, real-time analysis, scientific decision-making, and precise execution between cyber and physical space, and solves the problems of complexity and uncertainty.

MCPS builds the medical system into a medical union of cyberspace and physical space, Fig. 3.2 focuses on the interaction between the cyber system and the physical system in the medical environment, and updates the data and feedback data in real time by utilizing the dynamic data of the sensor network and monitoring system to move the calculation and monitoring from the independent system to the remote monitoring center. MCPS has the following characteristics: high integrity, limited physical resources, cross-domain transmission, multidomain networking, complicated space domain, dynamic reorganization, high automation, reliability, and CPS security. The medical field encompasses national health information, electronic medical records (EMR), patient care information, and operating room treatment information. Some of them are increasingly being managed by computer systems that include both hardware and software components, as well as high security and real-time capabilities. MCPS requires additional research and development in control and fusion systems, sensor and mobile networks, security, reliability and stability, embedded systems and abstract computing, model creation, security authentication, and data traceability.

2. Background

An overview of CPS in healthcare is given in this section. This discusses the key features and elements of CPS, as well as its advantages and the taxonomy that's utilized to characterize and group various healthcare modalities.

2.1 Autonomy level

Depending on the duties carried out by the individuals in the loop, cyber physical systems can be categorized as having a low to high degree of autonomy. The transition of

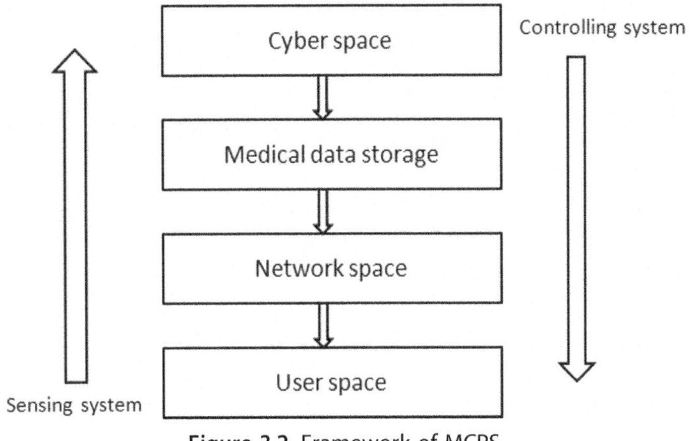

Figure 3.2 Framework of MCPS.

control from humans to machines and artificial dispersed networks is largely attributed to technologies like machine learning and artificial intelligence (Carreras Guzman et al., 2020). Two domains can be used to broadly classify many elements that are important in defining the autonomy degree in healthcare cyber physical systems (HCPS).

- **Equipment based:** The autonomy level is determined by variables such as the kind of medical gadget, the nature of the engagement, and the length of time spent with the patient. Medicare systems can be assisted by autonomous medical robots that serve food and medications to patients infected with the corona virus. A group of researchers from Hong Kong have developed a humanoid robot named "Grace" that can communicate with isolated coronavirus patients, monitor their body temperature and respond to touch thanks to a thermal camera implanted in their chest. To lessen the stress brought on by isolation, the robot will also engage socially with the corona virus sufferers (reuters).
- **Patient related:** The type of disease and the patient's risk level are examples of factors associated to the patient. High patient risk requires ongoing medical expert support in addition to the autonomous Medicare systems, whereas communicable diseases like the corona virus require a high level of autonomous Medicare cyber physical systems.

2.2 Security mechanisms

Numerous technologies are essential to the cyber physical system and can be categorized into specific domains. These domains include data collection through sensor technology and Internet of Things (IoT); managing large data and storing it in the cloud; using artificial intelligence and data analytics for decision making; control and coordination signals to smart machines and actuators; and so on. Due to the high degree of autonomy and criticality of healthcare applications, defense against cyber attacks must take a preventive approach (Dibaji et al., 2019). As a result, the security mechanism of the Healthcare Cyber Physical Systems must include the following:

- Recognizing cyber attacks against sensors.
- Sensitive cryptography for sending physiological information.
- Block chain technology and encryption techniques for safe cloud-based storage.
- Feedback signals to actuators that are encrypted.

2.3 Reliability

The healthcare industry places great importance on reliability, which is a crucial parameter for cyber-physical systems. The dependability of HCPS is illustrated in Fig. 3.3 Hardware components such as health sensors and actuators, software systems that compute a patient's health status, and communication networks that facilitate the transfer of patient data determine network reliability. For the initiation of cyber-physical systems to provide effective, dependable, and enhanced services, resource optimization and self-

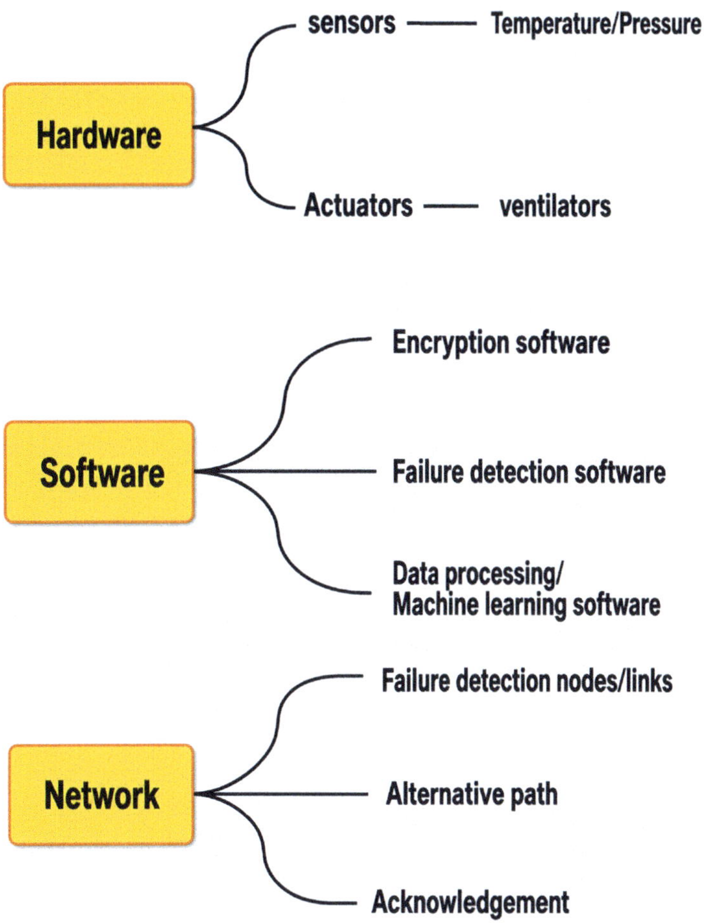

Figure 3.3 Reliability of HCPS.

adaptive behavior are necessary (Caesar et al., 2019). Cyber physical systems self-adaptive parts act according to the situation at hands on learning from historical data. Medical robots, for example, are intelligent devices that may self-organize in a changing environment to meet the demand for high-quality care.

3. Key features and elements of CPS in health care

To increase the efficacy, precision, and security of healthcare services, a cyber-physical system (CPS) in the healthcare industry combines computational and communication capabilities with physical processes. In the field of healthcare, a CPS's features are as follows:

3.1 Cyber-physical system in healthcare

The three Cs of CPS technologies are computation, communication, and control. It is distinguished by the close integration of hardware and software, which enables prolonged operation without the need for human intervention (Chen, 2017). In CPS, where sensors support physical activities and the cloud supports cyber activities, the integrated sensor-cloud architecture is essential. As seen in Fig. 3.4, the ecosystem of CPS consists of sensors, actuators, compute, storage, and communication. This assemblage of parts is known as a distributed control system. Through a number of advantages, including network integration, human system interaction, improved system performance, scalability, flexibility, optimization, and fast response time, CPS is a viable approach for connecting the physical and digital worlds (Haque et al., 2014). Healthcare systems now have a viable option for in-hospital and in–home patient care: CPS adapted to recent advancements in Cloud computing, medical sensors, and WBANs (Lee et al., 2012). These developments might enable CPS to keep an eye on and manage patient's illnesses in real time from a distance. Additionally, it would lessen the need for hospitalization of patients by increasing the care given to the elderly and those with impairments.

3.2 Taxonomy of CPS in healthcare

A taxonomy that explains and groups the cyber-physical systems techniques suggested for use in healthcare applications is necessary. We established a taxonomy for CPS in healthcare based on, which includes the following components:

- **Application:** CPS complexity varies depending on the application. It goes into one of two categories: regulated or aided. With the help of aided applications, people can keep an eye on their health without interfering with their daily activities. On the

Figure 3.4 CPS in healthcare: Building blocks.

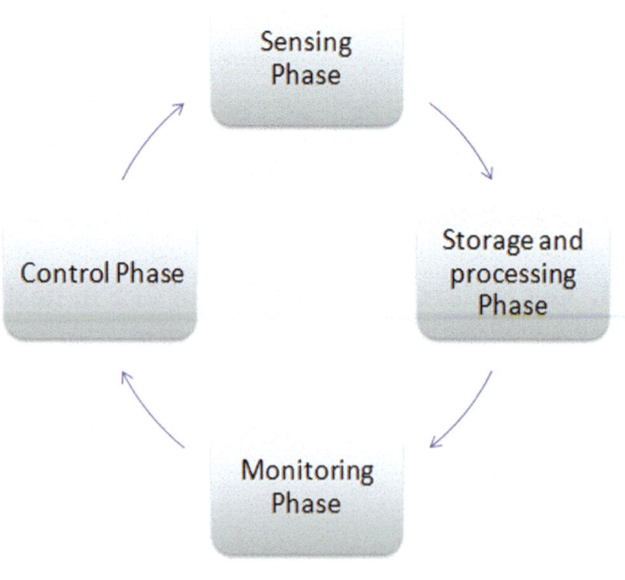

other hand, medical support needs to be easily accessible in regulated settings, requiring a high degree of close supervision and observation. Integrating these two now distinct facets of healthcare could create a single, expansive, sophisticated cyber-physical system with a number of benefits and drawbacks (Lee et al., 2012).

- **Infrastructure**: There are two possible server- or cloud-based architectures for CPS in the healthcare industry. While server-based infrastructure is suitable for small structures and needs individualized care, cloud-based infrastructure has become more and more popular in recent years due to its affordability, scalability and accessibility.
- **Data requirements:** Depending on the kind of biosensors used, data sizes may change for healthcare applications. Simple data like temperature or huge image data like magnetic resonance imaging (MRI) may be present.
- **Sensing:** Data from sensors serves as the input parameters for healthcare CPSs, hence sensing is regarded as the primary issue for healthcare applications. There are three primary branches in the sensing phase. The first category is sensor types, which might be diverse, homogeneous, or both. The second is the mode of data acquisition—passive or active—that the sensors employ. Third, whether the system is capable of sensing more than one characteristic or just one.
- **Data management:** In general, CPS takes into account the two primary methods of data storage: dispersed and warehousing (central).In contrast to warehousing, which stores data in a central database that can be queried, the distributed method views sensor devices as local databases and allows for distributed data processing under local administration. Decisions about data processing must be taken in light of the needs of the application and the resources at hand. It can be run in a cloud environment, at a centralized base station, or in a distributed fashion. For CPS to be used in healthcare applications, sensed sensor data must be accessed and processed quickly.
- **Security:** Ensuring data security should receive particular consideration when developing CPS architectures for healthcare applications. Encryption and privacy are the two primary components of security. In the healthcare industry, patient-doctor confidentiality and patient data privacy are requirements that must be strictly met. Healthcare applications that facilitate communication between professionals and users also need to be safeguarded. Depending on the application and security requirements, data encryption can be used to ensure security at both the user and network levels.
- **Control/Actuation:** Health information can offer insightful information that can be translated into the actions (treatments) required to preserve a patient's life. For the purpose of making informed choices and sending the data to the actuation components, authorized medical staff should be able to view the data at any time and from any location. Several application scenarios allow for the employment of either user-defined or automatic control/actuation techniques.

4. Reference architectural model for Cps implementation in healthcare applications

This section provides a reference architectural model specifically designed for the healthcare industry. It explains how different services, parts, and technologies can be combined to make it easier to extract useful insights from raw information.

4.1 Physical layer

The physical layer includes the processes for both data control and data acquisition, as seen in Fig. 3.5 Within this layer, wearable gadgets and sensors for monitoring and interacting with the human body are connected via WBANs, which become one of the primary elements in the data collecting process (Arefin et al., 2017). It consists of many kinds of nodes, such as non-invasive and invasive sensors that record crucial physiological indicators in real time. Nonetheless, the range of data collection can encompass diverse data sources outside WBANs, such as ambient sensors, electronic health records, and medical resources. The adaptability of data acquisition sources enables monitoring for various objectives, supporting various aspects of healthcare management and guaranteeing an all-encompassing approach to data collection. Big data that has been gathered is typically extensive and varied, encompassing formats that are semi-structured, unstructured, and organized. As a result, it adds complexity, necessitating the use of suitable data preparation methods in order to integrate and analyze data in later architectural model levels. Actuators are the physical layer's data control element. They provide feedback exchange within the network by interacting with end users in response to data they receive. They make it possible for physical acts to be carried out in response to calculated choices or preset guidelines.

4.2 Communication layer

To enable effective connection between the physical layer, which is where data is acquired, and the cyber layer, which is where data is processed and analyzed, this layer makes use of a wide range of wireless communication technologies, both short- and long-range. Technologies including Bluetooth Low Energy (BLE), Zigbee, UltraWide Band (UWB), Radio Frequency Identification (RFID), and Near Field Communication (NFC) are used for short-range communication. These technologies serve as bridges between the gathered big data and smart devices and the Fog/Edge layer, allowing reliable communication over comparatively small distances. Conversely, long-range wireless technologies, which have greater coverage, more bandwidth, and a longer reach, comprise WiFi, cellular networks (like 4G and 5G), and LPWANs (like LoRa and Sigfox). These technologies are essential for moving the collected and processed data to the cloud infrastructure, where it may be accessed, stored, and subjected to additional

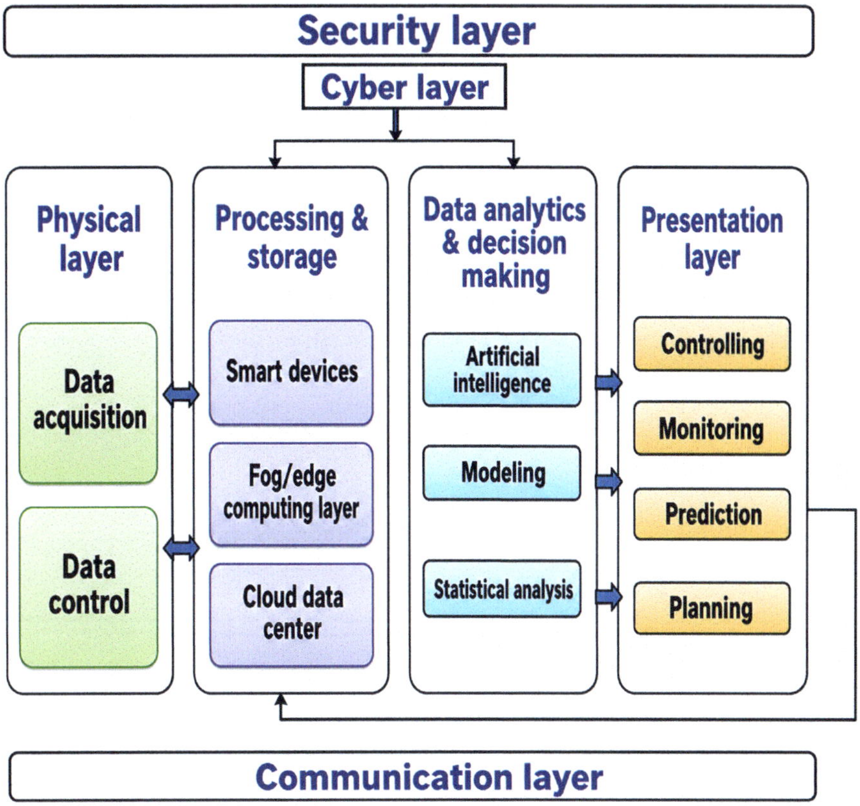

Figure 3.5 Comprehensive reference architectural model for CPSs implementation in healthcare applications.

analysis by a range of stakeholders, including administrators, academics, and healthcare professionals.

The choice of suitable wireless technology is critical for ensuring effective and dependable communication. When selecting the communication technology for a certain application, factors like data volume, latency requirements, energy efficiency, and sufficient coverage must be carefully taken into account. The communication layer in the architectural model enables the seamless integration and transfer of important data from the physical layer to higher levels by utilizing the capabilities of these technologies (Taleb et al., 2021). Table 3.1 offers a comprehensive summary of the features of the most popular wireless technologies, which can help choose the best technology for a given set of needs.

Table 3.1 Characteristics of mostly used wireless technologies (Khater et al., 2024).

Technology	Standard	Frequency band	Range (meters)	Data rate	Power consumption	Topology
Bluetooth	IEEE 802.15.1	2.4 GHz	Up to 100	3 Mbps	Moderate to high	P2P (Piconet), Star, Mcsh
Bluetooth low energy (BLE)	IEEE 802.15.1	2.4 GHz	<100	1 Mbps	Low	P2P (Piconet), Star, Mcsh
Zigbee	IEEE 802.15.4	2.4 GHz, 868 MHz, 915 MHz	10–100	250 kbps	Low	Star, Mesh, Cluster-tree
Wi-Fi	IEEE 802.11	2.4 GHz, 5 GHz	Up to 100	Up to several Gbps	Moderate to high	Star, Mesh
Ultra wide band (UWB)	IEEE 802.15.3	3.1–10.6 GHz	Up to 200	More than 100 Mb/s	Low	P2P, Mesh, Star

Khater, H.M., Sallabi, F., Serhani, M.A., Barka, E., Shuaib, K., Tariq, A., Khayat, M., 2024. Empowering healthcare with cyber-physical system—a systematic literature review. IEEE Access 12, 83952–83993. https://doi.org/10.1109/ACCESS.2024.3407376.

4.3 Cyber layer

4.3.1 Processing and storage

Smart devices act as intermediates between the physical and fog/edge layers, making up the core of this layer. These devices use short-range wireless communication technology to receive the data that the sensors send. In its capacity as a data gateway, it handles preliminary data processing duties such data aggregation and fundamental analysis before sending the information to the fog/edge layer. They can also apply preset rules or carry out basic calculations to get quick insights. However, smart devices are better suited for light data processing activities due to their constrained processing power and computing capabilities. Compared to smart devices, fog computing, which is situated close to the network's edge, offers more sophisticated data processing capabilities. They are capable of carrying out sophisticated data processing methods, such as those for data fusion, machine learning, and real-time analytics. With the use of these complex processing techniques, the gathered data may be analyzed more thoroughly, leading to more precise prediction-making, anomaly identification, and pattern recognition.

Additionally, it can help with data integration and correlation from many sources, improving overall comprehension of the gathered data and facilitating better informed decision-making. The processing and analysis processes can be completed closer to the data source by utilizing fog/edge computing, which lowers latency and speeds up decision-making. However, cloud storage systems have several advantages, including large computing capability, ample storage, and scalability to manage the demanding sophisticated data processing requirements. Big-data technologies like Hadoop, Spark, and Storm are frequently used in cloud-based infrastructures to facilitate the effective and concurrent processing of large-scale datasets. These technologies efficiently handle complex data processing tasks by utilizing the power of distributed computing. As an alternative, local databases may be used in accordance with the particular specifications and limitations of the healthcare application. Scalability, dependability, data protection, and regulatory compliance are necessary considerations when choosing the right data storage system. Healthcare companies can guarantee the safe and effective storage of processed data by making well-informed decisions.

4.3.2 Data analytics and decision making

As shown in Fig. 3.5, the main goal of this section is to use cutting-edge technology and methodologies to enhance decision-making processes and enable data analytics. AI comes out as one of the most popular methods among these strategies. AI, particularly ML algorithms like DL, is widely used to evaluate and understand the gathered data. These artificial intelligence methods use the data to find trends, spot abnormalities, and gain insightful knowledge that helps in decision-making. Expert systems can also be used to make use of domain-specific knowledge and offer intelligent decision assistance. They do this by analyzing patient data, applying rules and heuristics, and producing precise diagnoses and recommendations. The cyber layer is made possible by modeling techniques like data mining, optimization, and simulation, which allow for more in-depth analysis and useful scenarios or predictions. Optimization approaches discover the best answers based on predetermined objectives or limitations, while data mining techniques reveal hidden patterns and linkages. Through the use of virtual settings, simulation techniques allow for the assessment of the possible results and effects of various interventions or decisions.

Furthermore, statistical analysis occupies a pivotal role in the cyber layer. To investigate correlations between variables, test hypotheses, and reach relevant conclusions, statistical modeling, regression analysis, and hypothesis testing are utilized. In addition to facilitating data-driven decision-making and ensuring validity and reliability of the results, statistical analysis improves comprehension of observed trends. For processing, storing, analyzing, and decision-making in healthcare applications, the cyber layer serves as a strong foundation overall. Healthcare practitioners can obtain important insights, enhance patient monitoring and management, improve diagnosis and treatment plans, optimize resource allocation, and ultimately make evidence-based decisions for better healthcare outcomes through the integration of AI, modeling techniques, and statistical analysis. These methods can be used to smart devices, Fog/Edge, or the cloud, contingent upon the particular specifications and the healthcare application's architectural layout.

4.4 Presentation layer

Presenting analytical findings from the cyber layer—whether produced on the Edge or in the Cloud—is the purpose of this layer. A variety of approachable forms, such as graphs, charts, dashboards, and interactive visualizations, can be used to display these results. Users may easily comprehend trends, patterns, and anomalies thanks to these visualizations' clear and understandable depictions of the examined data. When it comes to healthcare applications, the presentation layer can facilitate the efficient flow of information to patients, giving them the ability to take charge of their own health. This may entail providing alerts for unusual health conditions, guidelines for changing one's lifestyle, reminders to take medication, and personalized advice. The presentation layer encourages informed decision-making and enhances patient engagement by providing the data in an easy-to-understand way. It also helps medical practitioners understand the results of their analyses for patients. By embracing the idea of the "human in the loop," it guarantees that medical professionals actively participate in decision-making procedures, allowing for accurate and thorough feedback as well as control.

Interactive interfaces that allow patients to examine their own health data, monitor their progress, and take an active role in managing their well-being can be made available by the presentation layer. This layer uses various device interfaces and services to meet different use cases and healthcare requirements. These comprise decision support tools and physician feedback systems, disease diagnostic and prognosis systems, fitness and general monitoring, and control systems for overseeing and directing healthcare operations.

4.5 Security layer

To shield the architecture from potential threats and weaknesses, security is a cross-layer service that is supplied to every layer. A variety of security methods and procedures can be used to accomplish this. Securing the individual components and the communication channels between them is the main goal in order to guarantee the security of the physical layer. Sensors, actuators, and other physical elements that have direct contact with the environment or the human body make up the majority of the physical layer. To guarantee that only authorized devices can connect and communicate within the physical layer, secure device authentication techniques should be put into place. Data integrity safeguards like digital signatures, message authentication codes (MAC), and checksums should be able to identify any tampering or illegal changes to the information gathered by the sensors.

Sensors, actuators, and other physical layer components should all use secure communication protocols to safeguard data transmission. Furthermore, to stop unwanted access or tampering with the physical components, physical security methods including installing physical barriers, securing equipment in enclosures resistant to tampering, or using anti-tamper mechanisms should be used. The physical layer can be better shielded from illegal access, data manipulation, and privacy violations by taking these security issues into account, assuring the system's overall security and integrity. To safeguard data while it is being transmitted and guarantee the legitimacy of communication endpoints, strong security measures must be implemented at the communication layer.

Data transmissions can be encrypted using encryption techniques, rendering the information unreadable by unauthorized parties. Sensitive data is shielded from interception and eavesdropping through the use of encryption algorithms and secure key exchange techniques. Verifying the identity of interacting entities can be done through secure authentication methods. To make sure that only authorized parties may access and trade data, methods like digital certificates, secure tokens, or biometric verification might be used. Secure authentication systems help stop unwanted access and data breaches by building trust between endpoints. To improve security, secure communication protocols like TLS (Transport Layer Security) or SSL (Secure Sockets Layer) can be used. By offering safe and secured communication routes, these protocols guard data integrity and stop it from being tampered with or altered while being transmitted.

Robust security measures are necessary to safeguard against potential cyber attacks in the cyber layer's data analytics and decision-making processes. Systems for detecting and stopping intrusions are essential for spotting and averting unwanted access attempts and questionable activity in the system. These programs track network activity, look for trends, and identify deviations that can point to a security breech. IDPS contributes to ensuring the integrity and confidentiality of data within the cyber layer by rapidly alerting administrators and taking relevant action. Security monitoring procedures are used in addition to IDPS to continuously check the system for vulnerabilities or security issues. This entails event analysis, real-time log monitoring, and security audits to find possible vulnerabilities and fix them early.

Security teams can minimize system impact and safeguard critical data processed within the cyber layer by keeping a tight eye on the system and promptly identifying and responding to security risks. In addition, processes for vulnerability management are put into place to find and fix any possible weaknesses in the system. This includes doing routine vulnerability scans, penetration tests, and security assessments to find flaws in software, configurations, or network infrastructure. Patches and updates can be implemented proactively to increase the cyber layer's security posture by resolving these vulnerabilities. In order to reduce the danger of application-level assaults, the presentation layer can also profit from secure coding techniques, user authentication methods, appropriate input validation, and routine device software updates and patches. Developers can lessen the possibility of exploitable vulnerabilities in the application code by adhering to recommended practices for coding. Robust user authentication methods, such biometric or multi-factor authentication, are crucial for guarding sensitive user data and preventing unwanted access. In order to stop malicious input from compromising the program, proper input validation is essential. Applications can reduce the risk of injection attacks and other security breaches by verifying and cleanup user input.

An overview of HCPS security is given in Table 3.2, which offers a thorough description of security considerations at various tiers.

5. Technologies in healthcare cyber physical systems

Cyber Physical Systems is an application space where numerous technologies are integrated for intelligent and effective operation of networked devices. In addition to serving as the foundation for communication in cyber-physical systems, the internet is also the most important technological advancement and a key enabler of other innovations such as blockchain, cloud computing, and IoT (Aceto et al., 2019). This section discusses the role that various technologies, including digital twins, IoT, big data, cloud computing, blockchain, AI, machine learning, and robots, play in the field of healthcare cyber physical systems (Fig. 3.6).

Table 3.2 Overview of HCPSs security (Khater et al., 2024).

Layer	Components	Security issues	Solutions
Physical layer	• Sensors • Actuators • EHRs • Hospital resources	• Individual components security • Network security	• Device authentication mechanisms • Checksums • Message authentication codes • Digital signature • Lightweight encryption • Access control mechanisms • Secure routing protocols • Anti-tamper mechanisms
Communication layer	• Wireless networks • Wired networks	• Data privacy during transmission • Authenticity of communication endpoints	• Secure key exchange mechanisms • Digital certificates • Biometric authentication • Secure tokens • Secure transmission protocols
Cyber layer	• Smart devices • Fog/Edge devices • Cloud • Intelligent devices • Data analytics and decision making processes	• Data confidentiality • Data integrity • Network vulnerabilities	• Encryption algorithms • Access controls • Regular backups • Intrusion detection and prevention systems • Monitoring of logs, event analysis • Security audits • Vulnerability scans • Penetration testing
Presentation layer	• Visualization tools • User interfaces	• Sensitive data protection • Authentication • Security breaches • Injection attacks • Cross-site scripting	• Secure coding practices • Multi-factor authentication • Proper input validation • End to end encryption

Khater, H.M., Sallabi, F., Serhani, M.A., Barka, E., Shuaib, K., Tariq, A., Khayat, M., 2024. Empowering healthcare with cyber-physical system—a Systematic literature review. IEEE Access 12, 83952–83993. https://doi.org/10.1109/ACCESS.2024.3407376.

5.1 Digital twin

A digital twin is a virtual representation of a physical object or process, as illustrated in Fig. 3.7. The digital twin is utilized in smart healthcare systems to generate virtual assets in cyberspace for the purpose of planning, controlling, and coordinating the use of digital resource information (Lu and Xu, 2018). The digital twin for SHCPS is essentially a simulation model for medical equipment and technologies, as well as for behavioral analysis of the patient treatment process, to help health care professionals research, evaluate, and forecast patients' health condition.

A three-tuple vector denoted by {resource id, allocation status, patient id} is used by the digital twin to keep a resource graph for different types of medical equipment. The patient's condition may affect the projected reallocation and the allocation status, which could be busy or idle. A graph with nodes representing health status at various time intervals and edges representing transitions based on changes in physiological parameters can be used to illustrate the patient treatment process. These graphs can be used to compute and classify a patient's health state using machine learning (Fig. 3.8).

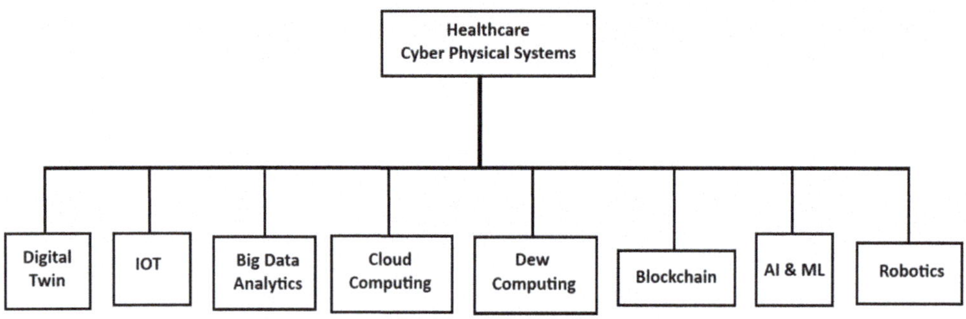

Figure 3.6 Technologies in healthcare cyber physical systems.

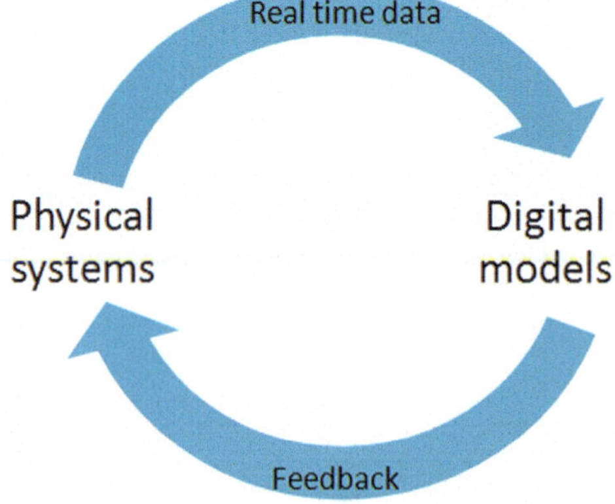

Figure 3.7 Digital twin model in HCPS.

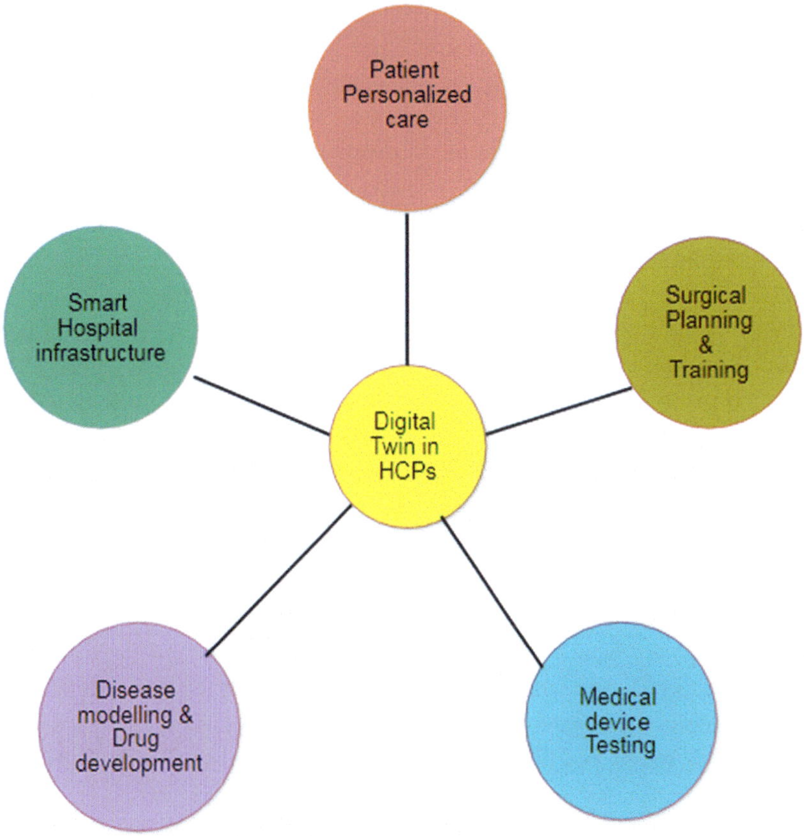

Figure 3.8 Digital twin domains in HCPS.

5.2 Internet of things (IoT)

New methods of machine-to-machine communication, such as Bluetooth, wireless networks, and other technologies including radio communication and near field communication, have been made possible by the Internet of Things (Aceto et al., 2019). High amounts of data are generated by sensor networks and sent to distant servers for analysis and control operations. This is made possible by the Internet of Things (IoT) at integration level smart healthcare cyber physical systems. Lightweight cryptographic algorithms and authentication are crucial for maintaining the integrity and secrecy of data and information flow because certain Medical IoT devices have limited processing power and memory. In the recent pandemic, medical IoT has assisted in remote patient monitoring (Ates et al., 2021; Nasajpour et al., 2020). Smart wearables including smart watches and bands, have been used to gather patients' blood pressure and heart rates, and smart thermometers have been used to detect body temperatures. These gadgets are linked to smartphones, which then transmit the information to cloud servers for medical analytics (Fig. 3.9).

5.3 Big data analytics

Machine-to-machine (M2M) communication among heterogeneous nodes and continuously collecting data from sensors in healthcare cyber physical systems with IoT of medical devices result in large data sets that may be semi-structured, unstructured, or structured and thus require processing, analysis, and storage in order to provide medical advice. Dealing with unstructured data, complexity, and validity issues presents a significant difficulty. The huge amounts of data collected from several sources in healthcare cyber-physical systems can have hidden patterns and linkages revealed by new computing paradigms with high processing power and big data technologies.

In order to quickly anticipate the symptoms, the epidemic has led to routine checks of physiological markers including as temperature, blood pressure, and heart rate. Human physiological parameters like temperature, blood pressure, and heart rate are among the various types of data generated in relation to the corona virus. Hospital data includes current patient intake, health status, and available facilities like beds and ventilators. City data includes the number of residents currently infected with the corona virus, the number of patients who are healthy, the number of corona deaths, and the number of residents vaccinated. Predictive analysis for future effects and a survey of spread are of interest to academics and analysts. Working with enormous datasets and finding these hidden patterns is made possible by deep learning models and high-end computing equipment.

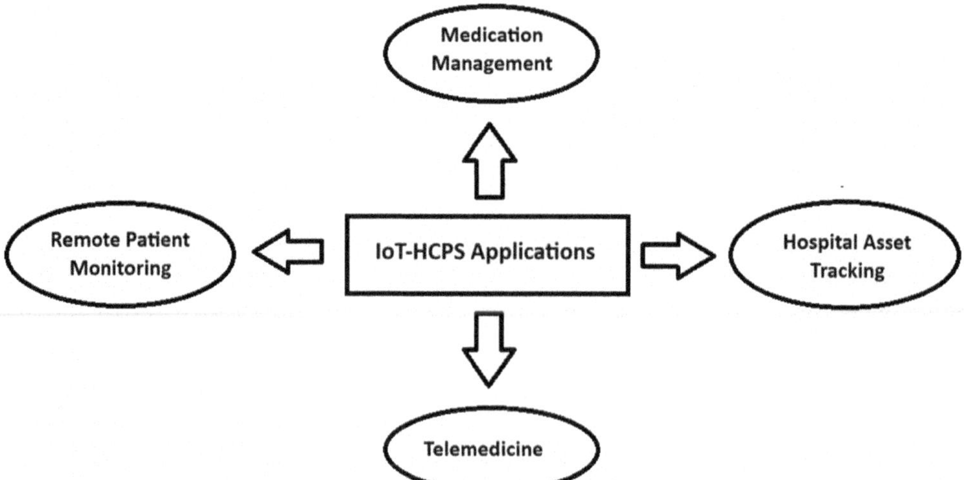

Figure 3.9 IoT-HCPS applications in HCPS.

Methodologies based on computational intelligence can also address the complexity and uncertainty of real-world data generated in real-time systems like healthcare CPS. These methodologies include artificial neural networks, which mimic the human brain and have many hidden layers with neurons to set learning parameters for pattern recognition, prediction, and classification of data sets; evolutionary algorithms, such as genetic programming and swarm intelligence, for natural selection; and fuzzy logic approach, which is based on approximation techniques and fuzzy rules for decision making and inferences (Iqbal et al., 2020).

5.4 Cloud computing

Patient medical records kept in separate locations that may be redundant or inconsistent depending on data management techniques are known as healthcare data silos. Cloud computing is a paradigm for computing where end users pay for infrastructure, resources, and services that are provided on a pay-per-use basis via cloud servers that give customers access to computational power and data storage. Within cyber-physical healthcare systems, Electronic Health Records can be securely maintained on a cloud server in an encrypted digital format. This allows for sharing and access by various entities, including banks, insurance companies, hospital administration, and patients. Ensuring the safety of patient information poses a challenge, since it is contingent upon the security framework implemented to protect the key generation center, which is responsible for maintaining the private keys of all legitimate users (Xu et al., 2019). Industry 4.0 standards have established new service models for cloud manufacturing, such as control as a service, machinery as a service, and industry automation as a service, in addition to the storage of encrypted data on cloud servers (Aceto et al., 2019). These novel cloud-based models can be used for real-world, demanding activities that need fog computing, closing the service to end users while facilitating effective, real-time information transfer and control between controllers and the healthcare level.

5.5 Dew computing

Through the reduction of transmission delay, the cloud architecture offers services that can be improved by the edge/fog computing paradigm, a distributed service architecture that increases the effectiveness of cyber-physical systems. By offering a new level of smart devices or smart interfaces closer in the network to IoT devices than edge devices, dew computing further minimizes latency and offers energy efficiency. In addition to controlling actuators and processing data from physical components, these smart systems also offer the technological advantages of scalability and reliability (Gushev, 2020). Through the provision of processing and analytical services in close proximity to the monitored region, the Dew computing layer proves to be highly beneficial for service provision in the healthcare arena (Manocha et al., 2021). For healthcare domains, a dew computing architecture with IoT devices, sensors, actuators, storage systems, network equipment, servers, and cloudlets at the fourth layer, called the edge server layer, which represents the edge or fog computing distributed service, and finally, cloud servers offering various infrastructure and software based services at the fifth layer, can be very time-efficient (Gushev, 2020). As a component of the healthcare cyber physical system of systems, the lightweight Dew computing paradigms applications on tablets and smartphones facilitate patient information sharing and interoperability, aiding in system collaboration.

5.6 Blockchain

Cloud servers' electronic health records can be securely and authentically accessed using a distributed database that is decentralized and distributed thanks to blockchain technology (Xu et al., 2019). It offers a decentralized platform for keeping tamper-proof records of events (Aceto et al., 2019) for a range of medical transactions, including those involving vaccines, devices, and patient-level occurrences. The blockchain is made up of a series of blocks, each containing several communications or transactions that are organized and hashed using a Merkle tree. With the exception of the first block, known as genesis, which does not contain a parent hash value, each block is uniquely recognized by a hash value and contains the hash of the previous block. A link like this with the parent block creates an unchangeable structure that is untouchable (Fig. 3.10).

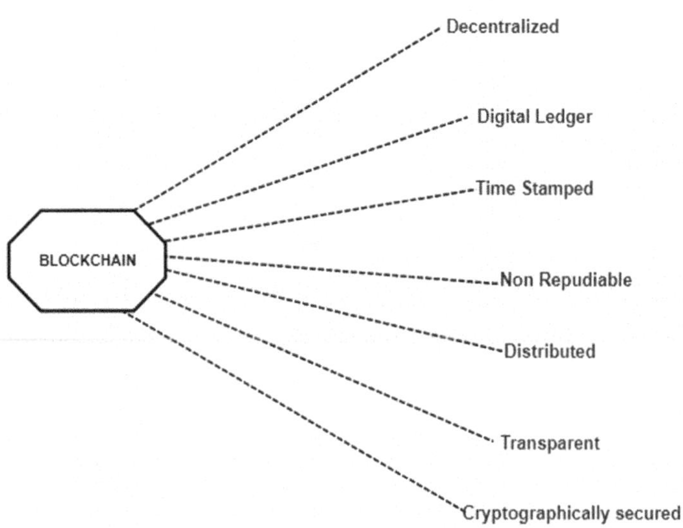

Figure 3.10 Blockchain characteristics.

5.7 Artificial intelligence and machine learning

Several facets of healthcare, such as medications, medical devices, patients, and illnesses, can benefit from the application of AI and machine learning. Together, researchers, academics, and professionals in the field of healthcare are attempting to glean insights and uncover patterns from massive patient data sets. These sizable datasets are also utilized to train machine learning models that assist medical professionals by classifying patients and assisting with automatic disease diagnosis.

Due to the current pandemic, research on corona virus drug discovery is urgently needed. Ventilators, which are now a life-saving tool for high-risk corona virus patients, are also in high demand. Studies on patients at high risk and the impact of disease on other body organs are also being conducted. Corona virus patients are classified as either low-risk or high-risk based on their level of danger. First, patients can be kept in isolation in their homes while various physiological parameters, such as their heart rate, blood pressure, and temperature, are sent to medical professionals for analysis. These professionals can then use machine learning techniques to predict the patients' symptoms and provide care via telemonitoring. Patients must be hospitalized for ongoing medical support, treatment, and observation during the following phase. By training the model using balanced datasets and selecting features using wrapper and filter-based techniques, AI and machine learning algorithms have been utilized to forecast the death rate of high-risk patients (Pourhomayoun and Shakibi, 2021).

5.8 Robots

Robots are autonomous devices that have been designed to carry out specific tasks precisely and accurately. These are artificial intelligence-based cognitive models that can frequently carry out pre-programmed tasks and continuously gather data from sensors about their surroundings. They can operate in complex contexts. In cyber-physical systems, robots play a wide range of roles. For example, medical robots help with surgery and patient care, industrial robots carry out production activities, and surveillance robots ensure security and safety.

Hospitals are overloaded with patients as a result of the recent pandemic's increasing coronavirus cases. Healthcare robots are becoming indispensable in patient care, offering anything from routine medication delivery to patient testing and assistance. Moxi and other robots carry out a number of tasks, such as picking up and dropping off patients, providing PPE supplies, and running COVID-19 testing (forbes). The Mitra robot supports medical personnel by taking patients' temperatures and facilitating video conferences between patients and family members (Nikolakis et al., 2019). Automation of production processes, such as the short development life cycle and high frequency of production of COVID-19 testing kits and ventilators, is made possible by industrial robots. Agile product development is the goal of industrial cyber-physical working environments with human-robot collaboration. This calls for context awareness in robots that can recognize human work zones and modify their area and speed accordingly (Nikolakis et al., 2019). Human behavior is changeable, unlike that of robots, thus it's necessary to constantly remind people to keep a safe distance when working together with fast-moving autonomous equipment. These reminders can take the form of audio-visual messaging. Therefore, in order to deliver control signals to robots and other autonomous machines and prevent mishaps caused by erroneous computations and motions, high precision algorithms are needed for depth estimation.

Robots for security can assist in monitoring compliance with COVID-19 guidelines, such as mask wear and social separation. They use multimodal technology to communicate through text, video, and audio files. They also take pictures, recognize objects, and provide reports.

6. Popular attacks on cyber physical systems

An attack with the malicious and destructive goal to exploit computer systems, architecture, or information systems is referred to as a cyber-attack. The attack typically takes the form of a person or group targeting other civil society organizations with the intention of obtaining the victim's private information, which is valuable and vulnerable. Cyber-physical systems (CPS) in healthcare, such as medical devices, hospital networks, and telemedicine platforms, are increasingly targeted by cyber-attacks. These attacks can have severe consequences, including compromising patient safety, privacy breaches, and disrupting healthcare services. Here are some common types of attacks on healthcare CPS:

6.1 Types of attackers

6.1.1 External threats

An external threat is defined as any vulnerability that can be used to obtain access to an environment from outside of it (Din and Paul, 2019). Some of the perpetual hazards are listed as follows in CPS:

Cyber Criminals: These are individuals or groups of people who utilize technology to commit cybercrimes in order to obtain personal data and information from organizations or businesses in order toprofit.

Hacktivists: They are a collection of persons that engage in malevolent acts for political or religious reasons, believing that they are battling political injustice.

State-Sponsored Attackers: They have defined aims that align with their country's political, commercial, or military interests. The government employs a number of talented hackers who find and exploit flaws before they are patched.

6.1.2 Insider risks

These are risks to an organization's security or data that originate from within, and they typically include former/current workers who work for the organization. These types of assaults are crucial to detect since they originate within the infrastructure (Din and Paul, 2019).

Some popular attacks on cyber physical systems are listed as:

Stuxnet: Stuxnet is a highly complex computer worm that enters and spreads on computers and PCs using several previously undisclosed Windows 0-day vulnerabilities. The primary goal of this Stuxnet was to produce real-world physical impacts. It used to target centrifuges in order to generate enriched uranium, which fuels nuclear devices, bombs, and reactors. When infecting a computer, it checks to see if it is connected to specified Siemens Programmable Logic Controllers (PLCs) (Mowla et al., 2018). PLCs are used to determine how computers interact with and control industrial gear, such as uranium centrifuges. The malicious worm then modifies the PLC's programming and algorithm, causing the centrifuges to spin too quickly and for too long. During this time, the PLCs send a false message to the computer controller claiming that everything is all right, masking and obscuring what is going on in the background. We may say that Stuxnet was developed by the spy agencies of the United States and Israel. This initiative, which was employed to create the worm, was dubbed "Operation Olympic Games" by President George W. Bush and maintained by President Barack Obama. The United States and Israel developed Stuxnet as a tool to derail and delay Iran's nuclear weapons program.

Distributed Denial of Service (DDoS): A malicious attempt to destroy regular traffic on a specific server, service, or network is known as distributed denial of service (DDoS). It involves flooding the target or surrounding infrastructure with fake data or an excessive amount of internet traffic. Their major method of efficacy is to leverage a large number of computer systems as attack traffic sources (Roehrs et al., 2019).A few examples of recent DDoS attacks are those from GitHub (2018), which was brought down for about 20 min only by an effective DDOS mitigation service that detected and immediately minimized the effect, and DYN (2016), which used a botnet with massive trickle-down effects to execute the then-largest DDoS attack because of malware called "Mirai."

Man in the Middle Attack: In order to obtain important and valuable data, the hacker may choose to intervene in the communication between the server and the client (Roehrs et al., 2017). The hack is carried out by changing the hacker's computer's IP address to the client's, tricking the server into thinking it is the real client and continuing to connect with them. As a cutting-edge defense against MITM (SSL Hijacking), Lenovo loaded Super Fish adware on its Windows computers in 2014. An email eavesdropping/email hijacking MITM attack in 2015 cost the British couple, the Luptons, £340,000. It was a catastrophic loss for them.

Phishing Attack: Phishing Attack: This technique involves delivering emails that seem to be from reliable sources, but in reality, the goal is to secretly insert dangerous malware—either as an attachment or from another believable source—into the target system. Sandboxing is a method of testing email content before it is utilized, and it can assist minimize this.

7. HCPS challenges

The implementation of CPS in healthcare systems brings particular difficulties that need to be carefully considered (Verma, 2022; Chen et al., 2022; Amin et al., 2020; Navaz et al., 2021). These difficulties may have an effect on data collection, processing, communication, and decision-making, among other facets of healthcare delivery. It is crucial to recognize and resolve these issues if CPS is to be successfully implemented and used in healthcare settings.

Some of the main issues surrounding the HCPSs are covered in the following.

1) **Sensor Accuracy and Reliability:** In order to facilitate accurate data gathering, healthcare sensors, including wearable technology and implanted sensors, must offer exact and consistent measurements. Sensor calibration, signal quality, environmental factors, and the effect of sensor placement on data accuracy are all areas that present challenges. To ensure accurate and dependable data collection, healthcare institutions need to carefully choose and calibrate their sensors.
2) **Power Management and Energy Efficiency:** In healthcare CPS applications, power management and energy efficiency are essential, particularly when wearable technology and implanted sensors are taken into account. These gadgets need to be made to minimize energy usage and guarantee accurate data collecting because they frequently run on small power supplies, such batteries. During data transmission, challenges include maximizing battery lifetime by device longevity.
3) **Medical Device Interoperability:** During the implementation of CPS, achieving interoperability between different medical devices is a challenging task. Multiple devices from various manufacturers that use different data formats and communication protocols are frequently a part of healthcare systems. It can be difficult to integrate various devices and make sure that data flow and interoperability go well. To facilitate smooth integration and cooperation between devices from many vendors, challenges include standardizing communication protocols, data formats and interoperability frameworks.
4) **Network Life and Reliability:** The infrastructure of the network should be able to sustain continuous operation devoid of frequent breakdowns or disruptions. Ensuring network scalability to handle an increase in the number of connected devices, preserving network stability and dependability to avoid data loss or delays, and putting in place fault-tolerant methods to efficiently manage network failures or congestion are among the challenges.
5) **Data Security and Privacy:** Sensitive and private patient data is handled by healthcare systems, and it needs to be shielded from attacks, data leaks, and

unwanted access. Putting strong security measures in place to protect data while it's in transit and at rest, like access limits, encryption, and authentication methods, presents challenges. Furthermore, maintaining adherence to data privacy laws and guidelines, such the General Data Protection Regulation (GDPR) in the European Union or the Health Insurance Portability and Accountability Act (HIPAA) in the United States, presents constant difficulties.

6) **Data Heterogeneity and Integration:** Medical imaging, wearable technology, electronic health records, and other sources all contribute significantly to the massive volumes of data generated by healthcare systems. Due to differences in data formats, structures, and semantics, integrating and harmonizing this heterogeneous data creates difficulties. This kind of heterogeneous data can be difficult to integrate and analyze, requiring complex feature extraction, preprocessing, and data integration approaches.
7) **Resource Management:** It might be difficult to manage network bandwidth, storage capacity, and computational resources in healthcare CPS. To manage the varied and dynamic demands of healthcare data processing and storage, resource allocation and utilization must be improved. Load balancing, resource provisioning, and resource optimization for maximum efficiency and economy pose obstacles.
8) **Quality of Processed Data:** In the processing and storage layer of healthcare CPS, ensuring the quality of processed data presents a major difficulty. The processing of data may include errors, inconsistencies, and inaccuracies as it moves through different stages. These may result from problems like algorithmic biases, data corruption, or mistakes in data transformation. To find and fix such problems, the implementation of strong data validation and quality control procedures presents a difficulty. It is necessary to use data quality assessments, such as data cleansing, outlier detection, and profiling, to find and fix anomalies or discrepancies in the processed data. Furthermore, the quality assurance process becomes more challenging when it comes to guaranteeing data completeness, integrity, and consistency across many sources and formats.
9) **Real-Time Response:** Healthcare CPS produces enormous volumes of data instantly. Timely processing and analysis of this data can be difficult, particularly when dealing with large-scale datasets. Healthcare CPS faces a significant issue in ensuring scalable and real-time analytics capabilities while upholding low latency and high dependability, since this directly affects timely decision-making, patient safety, and care delivery.
10) **Accurate Decision-Making and Data Analytics:** One of the most important challenges is ensuring that the methods utilized at the cyber layer for data analytics and decision-making processes are accurate. This entails confirming the validity of modeling approaches, assuring the soundness of statistical analysis, and evaluating the accuracy and dependability of AI models. Taking into account variables like

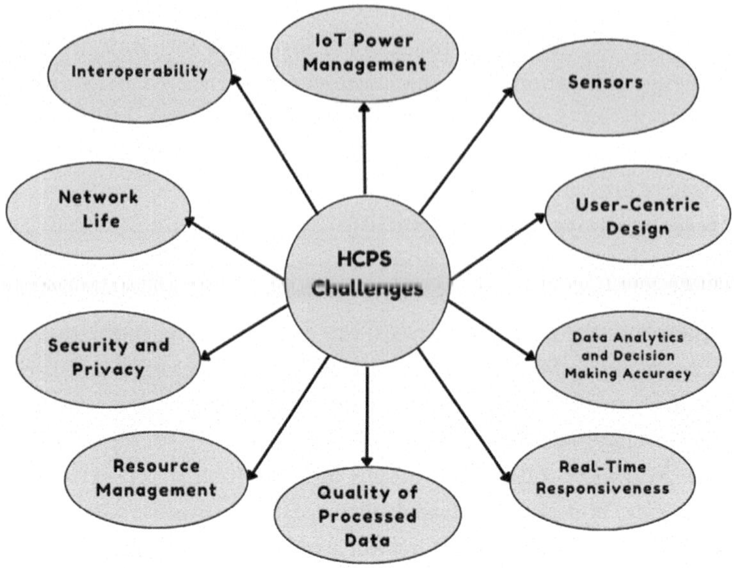

Figure 3.11 Challenges in the implementation of healthcare in CPS.

feature selection, algorithm selection, and data pretreatment, accurate model training is essential. Approaches for validation and verification are required in order to evaluate model performance and deal with possible biases. Similar issues for modeling methodologies include choosing suitable modeling approaches, fine-tuning parameters, and verifying models against real-world scenarios. Proper test selection, adherence to test assumptions, and proper result interpretation are necessary for statistical analysis.

11) **User-Centric Design:** It might be difficult to create visualizations that satisfy the requirements of different users, such as administrators, patients, and healthcare professionals. Effective decision-making requires that the visualizations be clear, easy to use, and give pertinent information to various user groups (Fig. 3.11).

References

Aceto, G., Persico, V., Pescape, A., 2019. A survey on information and communication technologies for industry 4.0: state-of-the-art, taxonomies, perspectives, and challenges. IEEE Commun. Surv. Tutor. 21 (4), 3467–3501. https://doi.org/10.1109/comst.2019.2938259.

Amin, S., Salahuddin, T., Bouras, A., 2020. Cyber physical systems and smart homes in healthcare: current state and challenges. In: IEEE International Conference on Informatics, IoT, and Enabling Technologies, ICIoT 2020. Institute of Electrical and Electronics Engineers Inc., Qatar, pp. 302–309. https://doi.org/10.1109/ICIoT48696.2020.9089638. http://ieeexplore.ieee.org/xpl/mostRecentIssue.jsp?punumber=9081868.

Arefin, M.T., Ali, M.H., Haque, A.K.M.F., 2017. Wireless body area network: an overview and various applications. J. Comput. Commun. 5 (7), 53–64.

Ates, H.C., Yetisen, A.K., Güder, F., Dincer, C., 2021. Wearable devices for the detection of COVID-19. Nat. Electron. 4 (1), 13–14. https://doi.org/10.1038/s41928-020-00533-1.

Caesar, B., Grigoleit, F., Unverdorben, S., 2019. (Self-)adaptiveness for manufacturing systems: challenges and approaches. SICS Softw. Inten. Cyber Phys. Sys. 34 (4), 191–200. https://doi.org/10.1007/s00450-019-00423-8.

Carreras Guzman, N.H., Wied, M., Kozine, I., Lundteigen, M.A., 2020. Conceptualizing the key features of cyber-physical systems in a multi-layered representation for safety and security analysis. Syst. Eng. 23 (2), 189–210. https://doi.org/10.1002/sys.21509. Available from: http://onlinelibrary.wiley.com/journal/10.1002/(ISSN)1520-6858.

Chen, H., 2017. Applications of cyber-physical system: a literature review. J. Ind. Integr. Manage. 2 (3), 1750012.

Chen, F., Tang, Y., Wang, C., Huang, J., Huang, C., Xie, D., Wang, T., Zhao, C., 2022. Medical cyber-physical systems: a solution to smart health and the state of the art. IEEE Trans. Comput. Social Syst. 9 (5), 1359–1386. https://doi.org/10.1109/TCSS.2021.3122807.

Dibaji, S.M., Pirani, M., Flamholz, D.B., Annaswamy, A.M., Johansson, K.H., Chakrabortty, A., 2019. A systems and control perspective of CPS security. Annu. Rev. Control 47, 394–411. https://doi.org/10.1016/j.arcontrol.2019.04.011.

Din, S., Paul, A., 2019. Smart health monitoring and management system: toward autonomous wearable sensing for Internet of Things using big data analytics. Future Gener. Comput. Syst. 91, 611–619. https://doi.org/10.1016/j.future.2017.12.059.

forbes. https://www.forbes.com/sites/saibala/2021/01/26/robots-have-become-an-essential-part-of-the-waragainst-covid-19/?sh=7aaf7bd25ef3. Accessed on July 10 2020.

Gushev, M., 2020. Dew computing architecture for cyber-physical systems and IoT. Internet Things 11, 100186. https://doi.org/10.1016/j.iot.2020.100186.

Haque, S.A., Aziz, S.M., Rahman, M., 2014. Review of cyber-physical system in healthcare. Int. J. Distributed Sens. Netw. 10 (4), 217415. https://doi.org/10.1155/2014/217415.

Iqbal, R., Doctor, F., More, B., Mahmud, S., Yousuf, U., 2020. Big data analytics and computational intelligence for cyber–physical systems: recent trends and state of the art applications. Future Gener. Comput. Syst. 105, 766–778. https://doi.org/10.1016/j.future.2017.10.021.

Khater, H.M., Sallabi, F., Serhani, M.A., Barka, E., Shuaib, K., Tariq, A., Khayat, M., 2024. Empowering healthcare with cyber-physical system - a systematic literature review. IEEE Access 12, 83952–83993. https://doi.org/10.1109/ACCESS.2024.3407376.

Lee, I., Sokolsky, O., Chen, S., Hatcliff, J., Jee, E., Kim, B., King, A., Mullen-Fortino, M., Park, S., Roederer, A., Venkatasubramanian, K.K., 2012. Challenges and research directions in medical cyber-physical systems. Proc. IEEE 100 (1), 75–90. https://doi.org/10.1109/JPROC.2011.2165270.

Lu, Y., Xu, X., 2018. Resource virtualization: a core technology for developing cyber-physical production systems. J. Manuf. Syst. 47, 128–140. https://doi.org/10.1016/j.jmsy.2018.05.003.

Manocha, A., Bhatia, M., Kumar, G., 2021. Dew computing-inspired health-meteorological factor analysis for early prediction of bronchial asthma. J. Netw. Comput. Appl. 179, 102995. https://doi.org/10.1016/j.jnca.2021.102995.

Milenković, A., Otto, C., Jovanov, E., 2006. Wireless sensor networks for personal health monitoring: issues and an implementation. Comput. Commun. 29 (13–14), 2521–2533. https://doi.org/10.1016/j.comcom.2006.02.011.

Mowla, N., Doh, I., Chae, K., 2018. Selective fuzzy ensemble learner for cognitive detection of bio-identifiable modality spoofing in MCPS. In: Proc. 20th Int. Conf. Adv. Commun. Technol. (ICACT), pp. 63–67.

Nasajpour, M., Pouriyeh, S., Parizi, R.M., Dorodchi, M., Valero, M., Arabnia, H.R., 2020. Internet of things for current COVID-19 and future pandemics: an exploratory study. J. Healthcare Inform. Res. 4 (4), 325–364. https://doi.org/10.1007/s41666-020-00080-6.

Navaz, A.N., Serhani, M.A., El Kassabi, H.T., Al-Qirim, N., Ismail, H., 2021. Trends, technologies, and key challenges in smart and connected healthcare. IEEE Access 9, 74044–74067. https://doi.org/10.1109/ACCESS.2021.3079217.

Nikolakis, N., Maratos, V., Makris, S., 2019. A Cyber Physical System (CPS) Approach for Safe Human-Robot Collaboration.

reuters. www.reuters.com/business/healthcare-pharmaceuticals/meet-grace-healthcare-robot-covid-19-created-2021-06-09/. Accessed on June 14 2021.

Pourhomayoun, M., Shakibi, M., 2021. Predicting mortality risk in patients with COVID-19 using machine learning to help medical decision-making. Smart Health 20, 100178. https://doi.org/10.1016/j.smhl.2020.100178.

Roehrs, A., Costa, C., Rosa, R., 2017. OmniPHR: a blockchain based interoperable architecture for personal health records. J. Biomed.Informat. 71, 70–81.

Roehrs, A., da Costa, C.A., da Rosa Righi, R., da Silva, V.F., Goldim, J.R., Schmidt, D.C., 2019. Analyzing the performance of a blockchain-based personal health record implementation. J. Biomed. Inf. 92. https://doi.org/10.1016/j.jbi.2019.103140. Available from: http://www.elsevier.com/inca/publications/store/6/2/2/8/5/7/index.htt.

Taleb, H., Nasser, A., Andrieux, G., Charara, N., Motta Cruz, E., 2021. Wireless technologies, medical applications and future challenges in WBAN: a survey. Wirel. Netw. 27 (8), 5271–5295. https://doi.org/10.1007/s11276-021-02780-2.

Verma, R., 2022. Smart city healthcare cyber physical system: characteristics, technologies and challenges. Wirel. Pers. Commun. 122 (2), 1413–1433. https://doi.org/10.1007/s11277-021-08955-6.

Xu, J., Xue, K., Li, S., Tian, H., Hong, J., Hong, P., Yu, N., 2019. Healthchain: a blockchain-based privacy preserving scheme for large-scale health data. IEEE Internet Things J. 6 (5), 8770–8781. https://doi.org/10.1109/jiot.2019.2923525.

Further reading

Celdrán, A.H., Pérez, M.G., Clemente, F.J., Pérez, G.M., 2018. Sustainable securing of medical cyber-physical systems for the healthcare of the future. Sustainable Comput. Inf. Syst. 19, 138–146.

Chen, F., Tang, Y., Wang, C., Huang, J., Huang, C., Xie, D., Wang, T., Zhao, C., 2021. Medical cyber--physical systems: a solution to smart health and the state of the art. IEEE Trans. Comput. Social Syst. 9 (5), 1359–1386.

Mohammed, M.A., Lakhan, A., Zebari, D.A., Abd Ghani, M.K., Marhoon, H.A., Abdulkareem, K.H., Nedoma, J., Martinek, R., 2024. Securing healthcare data in industrial cyber-physical systems using combining deep learning and blockchain technology. Eng. Appl. Artif. Intell. 129, 107612.

Nair, M.M., Tyagi, A.K., Goyal, R., 2019. Medical cyber physical systems and its issues. Procedia Comput. Sci. 165, 647–655.

Ramasamy, L.K., Khan, F., Shah, M., Prasad, B.V., Iwendi, C., Biamba, C., 2022. Secure smart wearable computing through artificial intelligence-enabled internet of things and cyber-physical systems for health monitoring. Sensors 22 (3), 1076.

Wang J, Abid H, Lee S, Shu L, Xia F. A Secured Health Care Application Architecture for Cyber-Physical Systems. arXiv preprint arXiv:1201.0213. December 31, 2011.

PART II

The physical layer: IoT, wearables, and cyber-physical systems

CHAPTER 4

Transforming healthcare with intelligent data analysis: Harnessing wearable technology, generative AI, and cloud computing

Tanvi Hungund[1] and Shobhit Kukreti[2]
[1]California State University, Fullerton, CA, United States; [2]Carnegie Mellon University, Pennsylvania, PA, United States

1. Introduction

Intelligent data analysis techniques are playing an increasingly critical role in healthcare, driven significantly by the proliferation of wearable technology. These devices represent a transformative shift in health monitoring, enabling personalized care through real-time data analysis and fundamentally altering the landscape of health management.

With the rise in chronic diseases such as heart disease, diabetes, and cancer, coupled with an aging global population and diverse lifestyles, there is a growing imperative for remote and continuous health monitoring solutions. Wearable sensors, such as smartwatches, are capable of tracking an extensive array of health metrics including physical activity levels, sleep patterns, blood pressure, heart rate, skin temperature, oxygen saturation, and various biochemical markers like lactate, cortisol, glucose, electrolytes, and pH levels, in addition to environmental factors.

This chapter delves into the profound impact of wearable technology on health monitoring analytics, transcending traditional applications in athletic contexts. It explores the current challenges faced by healthcare systems in adopting and integrating wearable devices, while also proposing innovative solutions to enhance their efficacy and reliability. A central focus lies on the transformative potential of advanced semantic analytics, which facilitate the extraction of actionable insights from the complex datasets generated by wearable devices.

By deciphering patterns and correlations within these datasets, healthcare providers can make informed decisions that lead to more precise diagnoses, personalized treatment plans, and proactive health management strategies.

Furthermore, the chapter examines the pivotal roles of cloud computing and Artificial Intelligence/Machine Learning (AI/ML) technologies in harnessing the vast amounts of data collected from wearable devices. Cloud computing offers scalable and secure infrastructure for storing, processing, and analyzing healthcare data in real-time, enabling seamless integration across healthcare systems and facilitating remote

Intelligent Semantic Analysis for Healthcare
ISBN 978-0-443-33512-9,
https://doi.org/10.1016/B978-0-443-33512-9.00011-1

access to patient information. AI/ML algorithms enhance the analytical capabilities of wearable technology by automating data interpretation, predicting health outcomes, and optimizing clinical decision-making processes. This integration of cloud computing and AI/ML not only enhances the efficiency of healthcare delivery but also drives innovation in personalized medicine, ultimately improving patient outcomes and healthcare quality on a global scale.

The convergence of wearable technology with intelligent data analysis techniques represents a paradigm shift in healthcare, empowering patients and healthcare providers alike with actionable insights and personalized care solutions. By addressing challenges and leveraging advancements in semantic analytics, cloud computing, and AI/ML technologies, healthcare systems can maximize the potential of wearable devices to deliver proactive, patient-centric care that meets the evolving needs of modern healthcare practices.

2. Wearable technology in healthcare

Wearable devices have evolved from niche tools for athletes to indispensable assets in healthcare, fostering advancements in physical well-being monitoring and personalized health management. These devices track vital signs, monitor activity levels, and collect data that can revolutionize healthcare delivery beyond conventional methods. By leveraging machine learning (ML) and Artificial Intelligence (AI) algorithms, the data gathered from wearables offer insights into an individual's health status, enabling early detection of health issues and facilitating personalized healthcare interventions. This capability not only enhances proactive healthcare but also empowers individuals to make informed decisions about their well-being, potentially mitigating the onset and progression of chronic diseases. A pivotal advantage of AI-driven wearable health technology lies in its emphasis on preventive care, aligning with strategies to detect health issues early. According to the California Department of Managed Healthcare, preventive care includes interventions that help prevent health problems or identify them before they escalate into serious conditions. Wearable devices play a crucial role in promoting healthy behaviors by offering incentives, reminders, and feedback on activities like physical exercise, hydration, and nutritional intake. By fostering healthier lifestyles, wearables contribute to reducing healthcare costs associated with hospitalizations and emergency visits, thus alleviating financial burdens on individuals and healthcare systems alike. Moreover, wearable technology facilitates remote patient monitoring, enabling healthcare providers to deliver personalized care plans based on real-time data insights. AI algorithms further enhance these capabilities by analyzing trends in an individual's diet, exercise habits, and overall health metrics, thereby tailoring recommendations that aim to optimize health outcomes and prevent the onset of chronic conditions. This integrated approach not only enhances the quality of care but also empowers individuals to take proactive steps toward maintaining their health and well-being.

The integration of wearable devices with AI and ML technologies represents a transformative shift in healthcare, offering unprecedented opportunities for proactive health management and personalized interventions. By leveraging real-time data analytics and promoting healthy behaviors, wearable technology stands at the forefront of preventive healthcare strategies, potentially reshaping how individuals engage with their health and interact with healthcare providers in the future. As research and development continue to expand in this field, the potential for wearables to improve health outcomes and reduce healthcare costs remains promising, highlighting their critical role in the future of healthcare delivery.

3. Semantic analysis techniques in healthcare

By definition per Wikipedia, Semantic analysis is the process of extracting meaning from text and semantic analysis creates the art of possible by making sense of data from varied sources. Hence, Semantic analysis techniques are at the center of extracting insights from diverse health data sources, including wearable devices and sensors, electronic health records (EHRs), and clinical databases. Applying the semantic analytics techniques, even a layman can get details in an easy-to-understand textual output and recommendations on actions they need to take to improve their overall health. For instance, wearable ECG monitors continuously monitor heart rhythms, alerting users and healthcare providers to anomalies for timely intervention (Lam et al., 2019). In chronic conditions like diabetes, wearables monitor glucose levels, enabling real-time adjustments in treatment plans for improved outcomes. Moreover, semantic analysis of genomic data integrates genetic and clinical information, enhancing personalized treatment strategies and predicting therapy responses accurately.

Additionally, semantic analysis in healthcare with wearable devices and sensors provides major advantages by transforming raw healthcare data into actionable insights that enhance patient care and clinical decision-making. This approach enables early and proactive detection and prevention of health issues through continuous monitoring, allowing for timely interventions to prevent complications or exacerbations of chronic conditions. It supports personalized medicine by tailoring treatments based on individual health trends and genetic factors, improving treatment efficacy and patient outcomes. Remote monitoring capabilities facilitate proactive healthcare management outside traditional settings, promoting patient engagement and adherence to treatment regimens. Moreover, semantic analysis streamlines data interpretation, reducing the burden on healthcare providers and enhancing overall efficiency in healthcare delivery. By harnessing wearable technology and semantic analysis, healthcare systems can achieve improved patient safety, better health outcomes, and advancements in research and population health management.

4. Real-world case studies

4.1 Continuous glucose monitoring (CGM) in diabetes management

Continuous Glucose Monitoring (CGM) devices have emerged as a pivotal tool in the management of diabetes, revolutionizing how patients monitor and control their blood glucose levels. Research, including studies by, consistently demonstrates the effectiveness of CGM in providing real-time glucose data, enabling timely adjustments in insulin dosing and dietary habits. This proactive approach not only enhances glycemic control but also mitigates the risks associated with hypoglycemia and hyperglycemia-related complications.

CGM devices operate by continuously measuring glucose levels in the interstitial fluid surrounding cells, providing patients and healthcare providers with a comprehensive view of glucose fluctuations throughout the day and night. The data collected is transmitted wirelessly to smartphones or dedicated receivers, where it can be analyzed in real-time. This capability allows for immediate interventions based on actionable insights, empowering patients to make informed decisions about their diabetes management strategies.

Furthermore, the integration of CGM data into clinical practice facilitates personalized medicine by tailoring treatment plans to individual glucose patterns and responses. For instance, patients can adjust their insulin doses based on trends revealed by CGM, thereby optimizing glycemic control and minimizing the risk of diabetic complications over time. The continuous nature of CGM also promotes patient engagement in self-care, as they gain deeper insights into how their daily activities and dietary choices impact their blood glucose levels.

In conclusion, CGM represents a significant advancement in diabetes care, offering continuous monitoring capabilities that go beyond traditional intermittent blood glucose testing. By leveraging real-time data analytics from CGM devices, healthcare providers can enhance patient outcomes through personalized diabetes management strategies that promote better glycemic control and reduce the burden of diabetes-related complications.

4.2 Wearable ECG monitors for cardiac monitoring

Wearable ECG monitors have transformed cardiac monitoring by providing patients with continuous insights into their heart health. Studies, such as those conducted by Lam et al. (2019), highlight the integration of real-time data analytics with wearable ECG devices to detect and alert both patients and healthcare professionals to abnormal cardiac rhythms promptly. This capability enables early intervention, potentially preventing serious cardiac events and improving patient outcomes in individuals with chronic heart conditions.

The functionality of wearable ECG monitors involves sensors that continuously record and transmit electrocardiographic signals, capturing cardiac activity over extended periods. This continuous monitoring is particularly beneficial for patients at risk of arrhythmias or those recovering from cardiac procedures, as it allows for the detection of subtle changes in heart rhythm that may indicate underlying cardiovascular issues.

Moreover, real-time data analytics from wearable ECG monitors empower patients to actively participate in their cardiac care by monitoring their heart rhythms and recognizing symptoms of concern. By providing actionable insights, these devices enable timely communication between patients and healthcare providers, facilitating informed decisions regarding treatment adjustments or interventions.

In summary, wearable ECG monitors offer a transformative approach to cardiac monitoring, enhancing patient safety and quality of life through continuous, remote monitoring capabilities. The integration of real-time data analytics with wearable technology underscores its potential to revolutionize the management of chronic heart conditions, fostering proactive healthcare strategies that prioritize early detection and intervention.

4.3 Remote patient monitoring and chronic disease management

Remote patient monitoring (RPM) facilitated by wearable devices has emerged as a cornerstone of modern healthcare, particularly in managing chronic diseases and elderly care. Research, exemplified by studies from Howcroft et al. (2019), illustrates how wearables equipped with sensors for fall detection and activity monitoring enable caregivers to remotely track patients' health metrics and intervene promptly in case of emergencies. The functionality of RPM wearables extends beyond basic monitoring, encompassing real-time data analytics that analyze trends and anomalies in patients' health metrics. This capability enables healthcare providers to detect early signs of deterioration or changes in health status, facilitating proactive interventions that can prevent hospitalizations and improve overall patient outcomes.

Furthermore, RPM wearables promote patient independence and autonomy by enabling individuals to maintain their daily routines while being remotely monitored. Caregivers and healthcare teams can access real-time data insights from wearables, allowing for personalized care plans tailored to individual patient needs and preferences. This personalized approach not only enhances patient satisfaction but also optimizes healthcare resource allocation by focusing on proactive management strategies.

In conclusion, RPM through wearable devices represents a transformative shift in chronic disease management and elderly care, leveraging real-time data analytics to support proactive healthcare interventions. By enabling continuous monitoring and early detection of health issues, RPM wearables empower patients to live independently while ensuring they receive timely and appropriate medical attention when needed.

4.4 Athlete performance monitoring and injury prevention

Wearable technology has revolutionized athlete performance monitoring by providing real-time insights into physiological metrics and movement patterns. Studies, such as those reviewed by Halson (2019), highlight how wearable devices equipped with sensors for monitoring heart rate variability, muscle fatigue, and biomechanics enable sports teams to optimize training regimes, prevent injuries, and enhance overall athletic performance.

The integration of real-time data analytics with wearable devices allows coaches and medical staff to monitor athletes' physiological responses to training sessions and competitions. This continuous monitoring facilitates evidence-based decision-making regarding workload management, recovery strategies, and individualized training adjustments tailored to athletes' specific needs and performance goals.

Moreover, wearable technology supports injury prevention by identifying biomechanical imbalances or fatigue patterns that may predispose athletes to injuries. Early detection through real-time data analytics enables timely interventions, such as modifying training loads or implementing rehabilitation protocols, thereby reducing the risk of injuries and ensuring athletes can maintain peak performance levels throughout their seasons.

In summary, wearable technology plays a crucial role in optimizing athlete performance and injury prevention strategies through continuous monitoring and real-time data analytics. By providing coaches and medical professionals with actionable insights into athletes' physiological responses, wearables contribute to evidence-based sports medicine practices that prioritize athlete health, safety, and performance excellence.

4.5 Telehealth and remote consultations

The COVID-19 pandemic accelerated the adoption of telehealth solutions, leveraging wearable technology to facilitate remote patient consultations and monitoring. Research, as highlighted by studies from Steinhubl (2020), underscores how wearables enabled healthcare providers to remotely monitor patients' vital signs, manage chronic conditions, and adjust treatment plans without the need for in-person visits.

Wearable devices equipped with sensors for measuring vital signs, such as heart rate, oxygen saturation, and respiratory rate, played a crucial role in telehealth consultations during the pandemic. Real-time data analytics from wearables allowed healthcare providers to assess patients' health status remotely, enabling timely interventions and reducing the risk of virus transmission in healthcare settings.

Furthermore, telehealth consultations supported by wearable technology facilitated continuity of care for patients with chronic conditions, ensuring they received ongoing monitoring and medical support despite physical distancing measures. By integrating real-time data analytics, telehealth platforms enabled healthcare providers to make informed clinical decisions, optimize treatment outcomes, and enhance patient satisfaction through personalized care delivery.

In conclusion, wearable technology has proven instrumental in supporting telehealth initiatives during public health emergencies like the COVID-19 pandemic. By enabling remote patient monitoring and consultations, wearables equipped with real-time data analytics have demonstrated their versatility and resilience in healthcare delivery, underscoring their potential to transform patient care and enhance healthcare accessibility in a post-pandemic era.

5. Algorithm and flow chart

Wearable devices have transformed healthcare by enabling continuous monitoring of vital signs and activity levels, providing valuable data for personalized health management. The process of capturing and deriving insights from this data involves several crucial steps outlined in the following flowchart.

5.1 Step 1: Data acquisition from wearable devices/continued: Pre-processing and data cleaning

The process begins with collecting raw data from sensors embedded in wearable devices worn by patients. These sensors capture a wide range of metrics such as heart rate, blood pressure, sleep patterns, physical activity, and more.

Once the data is collected, it undergoes preprocessing and cleaning to ensure its quality and reliability for subsequent analysis. This step involves removing noise, handling missing values, and standardizing formats to facilitate seamless integration into the analysis pipeline.

5.2 Step 2: Feature extraction

Next, relevant features are extracted from the cleaned data. Feature extraction involves identifying key parameters or patterns within the data that are critical for generating meaningful insights. This step aims to reduce the dimensionality of the data while retaining essential information.

5.3 Step 3: Semantic analysis

Semantic analysis is performed to derive meaningful interpretations and insights from the extracted features. This step involves applying advanced analytics techniques such as machine learning algorithms to uncover patterns, trends, and correlations within the data.

5.4 Step 4: Integration with electronic health records (EHRs)

The insights derived from wearable device data are integrated with electronic health records (EHRs) of patients. This integration enriches the existing medical history with real-time data, providing a comprehensive view of the patient's health status over time.

5.5 Step 5: Data fusion and analysis

Data fusion involves integrating wearable device data with other relevant datasets such as genomic data, environmental factors, and historical health records. This integrated data is then analyzed holistically to generate comprehensive insights into the patient's health condition.

5.6 Step 6: Clinical decision support and actionable insights

Based on the analyzed data, clinical decision support systems generate actionable insights and recommendations for healthcare providers. These insights assist in making informed decisions regarding patient care, treatment plans, and interventions.

5.7 Step 7: Visualization and reporting

Findings from the analysis are presented through visualizations such as graphs, charts, and dashboards. These visual representations help healthcare professionals and patients alike to understand complex data patterns and trends easily.

5.8 Step 8: Security and privacy measures

Ensuring data security and privacy is paramount throughout the entire process. Robust security measures are implemented to protect patient information and comply with regulatory requirements such as HIPAA (Health Insurance Portability and Accountability Act) in the United States.

5.9 Step 9: Continuous monitoring and feedback loop

The process concludes with continuous monitoring and a feedback loop mechanism. Feedback from healthcare providers, patients, and outcomes of interventions are incorporated back into the system to improve algorithms, enhance data accuracy, and optimize personalized health management strategies.

This structured approach illustrates a comprehensive framework for processing wearable device monitoring data to generate actionable insights in healthcare. By leveraging advanced data analytics, integrating with electronic health records, and ensuring stringent security measures, healthcare providers can harness the power of wearable technology to deliver personalized, efficient, and effective patient care.

6. Impact of cloud computing and AI/ML on healthcare industry

Cloud computing and generative AI are well positioned to revolutionize healthcare by enabling intelligent semantic analysis through wearable devices. Cloud computing provides scalable and secure storage and processing capabilities, allowing extensive amounts of data from wearables to be collected, stored, processed and analyzed in real-time.

Cloud computing can also help remove the concerns around costs involved in data processing by the pay as you go pricing that is offered by cloud providers. This cloud infrastructure supports continuous monitoring of patients' health metrics such as heart rate, blood pressure, and activity levels, facilitating early detection of anomalies or trends indicative of health risks.

Generative AI algorithms enhance semantic analysis by extracting meaningful insights from this data, identifying patterns, correlations, and predictive models that aid in personalized medicine and clinical decision-making. These technologies enable healthcare providers to remotely monitor patients, deliver timely interventions, and optimize treatment plans based on individual health profiles. By leveraging cloud computing and generative AI, wearable devices become powerful tools in enhancing patient care, improving health outcomes, and driving advancements in preventive healthcare strategies.

Recent advancements in cloud computing and generative AI are poised to catalyze a paradigm shift in healthcare, particularly through their integration with wearable technology.

Cloud computing offers unparalleled scalability and secure storage capabilities, allowing vast quantities of data from wearable devices to be collected, stored, processed, and analyzed in real-time. This infrastructure not only supports continuous monitoring of critical health metrics such as heart rate, blood pressure, and activity levels but also addresses concerns regarding the costs associated with data processing through flexible, pay-as-you-go pricing models offered by cloud providers. This scalability and cost-efficiency are pivotal in enabling healthcare systems to manage and derive actionable insights from the immense volume of data generated by wearable devices.

Moreover, cloud computing enhances the ability of healthcare providers to detect anomalies and trends indicative of health risks promptly. By facilitating real-time analysis of patient data, cloud-based solutions empower clinicians to identify deviations from normal health patterns early, enabling timely interventions that can mitigate the progression of diseases or complications. This proactive approach not only improves patient outcomes but also reduces healthcare costs associated with emergency interventions and hospitalizations.

Generative AI algorithms complement cloud computing capabilities by leveraging advanced semantic analysis techniques. These algorithms sift through complex datasets to extract meaningful insights, uncovering hidden patterns, correlations, and predictive models that inform personalized medicine and clinical decision-making.

By discerning subtle relationships within health data, generative AI enhances the precision and efficacy of treatment plans, tailoring interventions to individual patient profiles and optimizing healthcare delivery.

Furthermore, the integration of cloud computing and generative AI transforms wearable devices into powerful tools for healthcare providers. These devices enable remote monitoring of patients, allowing for continuous assessment of health metrics and proactive management of chronic conditions. Real-time data insights derived from wearable devices empower clinicians to make informed decisions on adjustments to treatment protocols, ensuring personalized care that aligns closely with patient needs and preferences.

In conclusion, the convergence of cloud computing and generative AI represents a pivotal advancement in healthcare technology. By harnessing these technologies, healthcare providers can enhance patient care, improve health outcomes, and drive innovations in preventive healthcare strategies. Moving forward, continued research and innovation in cloud-based solutions and AI algorithms are essential to further unlocking the transformative potential of wearable technology in revolutionizing healthcare delivery on a global scale.

7. Challenges and considerations

Wearable technology has emerged as a promising frontier in healthcare, offering continuous monitoring capabilities that can revolutionize patient care and health management. However, despite its potential benefits, several challenges must be addressed to optimize its effectiveness and ensure widespread adoption in clinical settings.

Key challenges include data privacy, security concerns, accuracy of readings, and cost management, all of which play critical roles in shaping the future of wearable technology in healthcare.

Data privacy and security stand out as paramount concerns when integrating wearable technology into healthcare ecosystems. The collection and transmission of sensitive health data via wearables pose significant risks of data breaches and unauthorized access. Patient confidentiality must be safeguarded through robust encryption protocols, secure data storage solutions, and adherence to stringent regulatory frameworks such as HIPAA (Health Insurance Portability and Accountability Act) in the United States. Ethical guidelines are equally crucial to establish trust among users and ensure responsible data handling practices by healthcare providers and device manufacturers (Steinhubl, 2020).

Moreover, the accuracy and reliability of wearable sensors remain a critical challenge. While wearable devices can monitor basic health metrics like heart rate and activity levels effectively, their ability to sense biological biomarkers continuously and with high sensitivity is still in the early stages of development. Inaccurate readings could potentially lead to false alarms or missed opportunities for early detection of health issues, compromising patient care.

Addressing this challenge requires rigorous validation of sensor technologies against standardized benchmarks and nomenclatures to ensure consistent performance and reliability across diverse patient populations and clinical settings.

Cost management is another significant hurdle in the widespread adoption of wearable technology in healthcare. The initial investment in wearable devices, coupled with ongoing costs associated with data storage, processing, and maintenance, can be prohibitive for healthcare providers and institutions, particularly in resource-constrained environments. Implementing cost-effective strategies, such as leveraging cloud computing for scalable data storage and processing, and adopting innovative business models like pay-as-you-go pricing offered by cloud providers, can help mitigate financial barriers and enhance the sustainability of wearable technology deployments.

Furthermore, ongoing research and development are essential to optimize the effectiveness of wearable technology and unlock its full potential in delivering personalized healthcare solutions. Advances in sensor technology, machine learning algorithms, and data analytics hold promise for enhancing the capabilities of wearables to monitor and predict health outcomes more accurately.

Collaborative efforts between researchers, healthcare professionals, and technology developers are critical to driving innovation, validating new technologies, and translating research findings into practical applications that benefit patients and healthcare providers alike.

While wearable technology presents exciting opportunities to transform healthcare delivery, addressing challenges such as data privacy, security, accuracy, and cost management is imperative to realize its full potential.

By establishing robust regulatory frameworks, adhering to ethical guidelines, validating sensor technologies, and fostering continued research and development, the healthcare industry can overcome these challenges and harness the transformative power of wearable technology to improve patient outcomes and enhance the quality of care across global healthcare systems.

8. Conclusion

In conclusion, the integration of intelligent data analysis techniques, particularly through innovative semantic analytics, marks a transformative advancement in healthcare. By leveraging real-time data from wearable technology and applying sophisticated analytical methods, healthcare providers can deliver personalized care that is tailored to individual patient needs. This approach not only enhances the monitoring of chronic conditions but also facilitates early intervention and prediction of patient outcomes with unprecedented accuracy. These capabilities signify a significant leap forward in healthcare delivery, empowering clinicians to make data-driven decisions that optimize treatment efficacy and patient outcomes.

However, the realization of these benefits hinges on overcoming substantial challenges, foremost among them being data privacy and regulatory compliance. Safeguarding patient information and ensuring adherence to stringent privacy regulations like HIPAA are critical to maintaining trust and ethical standards in healthcare practices. Addressing these challenges requires ongoing investment in robust security measures and collaborative efforts across the healthcare ecosystem to establish unified standards and best practices for data management and governance.

This chapter of the overall book underscores the imperative for continued research and collaboration to further innovate wearable technology and semantic analytics applications in healthcare. By fostering interdisciplinary partnerships and leveraging technological advancements, the healthcare industry can unlock new possibilities for improving patient care and operational efficiency.

Embracing these innovations not only enhances the capabilities of healthcare providers but also empowers patients to actively engage in their own health management, driving toward a future where personalized medicine is the cornerstone of healthcare delivery.

In essence, the convergence of intelligent data analysis and semantic analytics represents a pivotal evolution in healthcare, promising to redefine how healthcare is delivered and experienced. As these technologies continue to evolve, their potential to revolutionize diagnostics, treatment strategies, and patient outcomes remains vast, signaling a promising trajectory toward more effective and sustainable healthcare solutions globally.

AI disclosure

During the preparation of this work the author(s) used [NAME TOOL/SERVICE] in order to [REASON]. After using this tool/service, the author(s) reviewed and edited the content as needed and take(s) full responsibility for the content of the publication.

References

Lam, C.K., et al., 2019. Wearable Technology and Electronic Health Records: A Review. Healthcare Informatics Research.

Steinhubl, S.R., 2020. Emerging technologies for remote monitoring: security, privacy, and ethics. In: IEEE Engineering in Medicine and Biology Magazine.

CHAPTER 5

Current trends in IoT-enabled healthcare for wellness management

Shaurya Gupta[1], Sonali Vyas[2] and Tarandeep Kaur Bhatia[3]
[1]School of Computer Science, University of Petroleum and Energy Studies, Dehradun, Uttarakhand, India; [2]IET, JK Lakshmipat University, Jaipur, Rajasthan, India; [3]Amity School of Engineering and Technology, Amity University, Mohali, Punjab, India

1. Introduction

During recent times the population all over the world has increased dramatically and it will continue more or less in the same trend. Apart from the life expectancy of people around the world has also increased considerably which is resulting in leading a health life for a greater part of the population. Though it requires significant developments in the overall healthcare systems including the overall lifestyle management of people, generally the elderly ones. As there are many demographic changes going on the healthcare expense is also growing in the overall society especially for the elder ones who are more and more susceptible to any kind of chronic disease. Apart from that this sector is contributing directly or indirectly to varied job sectors in the healthcare field. With time the technological developments in the field of IoT (Shah et al., 2024), it has become an indispensable part of the healthcare field. Many devices like smart watches have eased diagnosis and monitoring of illness in an efficient way (Fischer et al., 2024). Many clinical research are in the process of conduction nowadays with the help of certain IoT enabled healthcare devices. They are quite helpful in calculation of blood pressure, heart rate, blood oxygen levels and many more. All these clinical data needs to be communicated via contemporary communication channels in far flung areas in the healthcare sectors (Atadoga et al., 2024). Considering these sorts of healthcare data in association with upgraded communication services with ever expanding technologies which includes all sorts of sophisticated algorithms, IoT enabled remote sensors devices.

2. Literature review

As far as the usage of wearable technologies is considered, it is still undergoing a lot of research and innovation and a maximum number of device or their prototypes are being industrialized nowadays so it reaches the maximum market outreach. The current development scenario focusses on that sector of the population wherein people are finding it quite difficult to lead a sustainable life or are quite elderly or suffering from some kind of chronic diseases (Zahedian Nezhad et al., 2024; Vatin et al., 2024). For a very long period of time

Intelligent Semantic Analysis for Healthcare
ISBN 978-0-443-33512-9
https://doi.org/10.1016/B978-0-443-33512-9.00006-8

the idea of patient's remote monitoring has been deliberated with the help of research and development and nowadays it is quite feasible that the health associated data like heartbeat, ECG are being collected with the help of varied devices apart from high deliverable sensor technologies. Considering a remote centered health monitoring system which might use a smartphone acting as a network gateway responsible for collecting sensor data and then finally transmitting it to a central medicinal center or information processing center. All the sensor devices which are being attached to the body are a part of Body area network. For facilitating communication amongst these sensor devices varied wireless communication protocols are being like ZigBee, Wi-Fi and Bluetooth (Hoque and Khan, 2024). Sensor nodes have extremely limited memory apart from computational capability, therefore the data being collected by sensor nodes must be directly sent to some kind of programmable logic unit for either storage or further processing. Any location enabled technology can help in tracking of patient or that particular health device (Khan and Alkhathami, 2024; Roy and Sadiwala, 2024a). Data being collected at regular intervals from varied devices are being analyzed constantly by medicinal practioners to check recovery of patients from time to time. Even tracking of ambulances, healthcare personnel, and personal caretakers apart from patients is feasible with the help of IoT in accordance with GPS, in certain times the local positioning system may be very efficient in tracking and pin pointing the exact location as it quite efficient in short distance communication (Parihar et al., 2024; Puri et al., 2024; Singh et al., 2024). Even Ultra wide band radio is also favored as it deliver an enhanced time resolution in terms of arrival time of persons in different locations as it helps in tracking the position of people in real time (Islam et al., 2024; Pulimamidi, 2024; Arul et al., 2024). In today's scenario the data from infirmary, medicinal records of patients, and data from sensors form an essential part of BigData and apart from that varied tools and techniques are being used considerably in analyzing these voluminous amounts of data that would have been quite difficult otherwise and it helps in rapid identification, predicting and cure of any kind of illness (Talpur et al., 2024; Quasim et al., 2019).

Fig. 5.1 deliberates the architecture discusses the working of IoT enabled healthcare or wellness monitoring system consisting of varied components like publisher, connection, broker and subscriber which are in turn an amalgamation of sub components (Kante and Ndayizigamiye, 2021; Adil et al., 2024). With the enablement of IoT architecture in smart health care apps the concept of IoT and cloud computing is integrated into the field of healthcare or medicine. The process of transferring data of patient from varied sensor devices to a central medicinal data center and sending that data to medical practioners or doctors is quite defined and managed with the help above architecture (Singh and Kaunert, 2024; Clemente-Lopez et al., 2024; Akinsanya et al., 2024). All these components are interconnected wherein the publisher, broker and subscriber form the basic building components of IoT healthcare management system. The data of patient may be collected with or without the help of sensors which are associated to a publisher network. The varied measures may include the following:

- Heart Rate
- Temperature

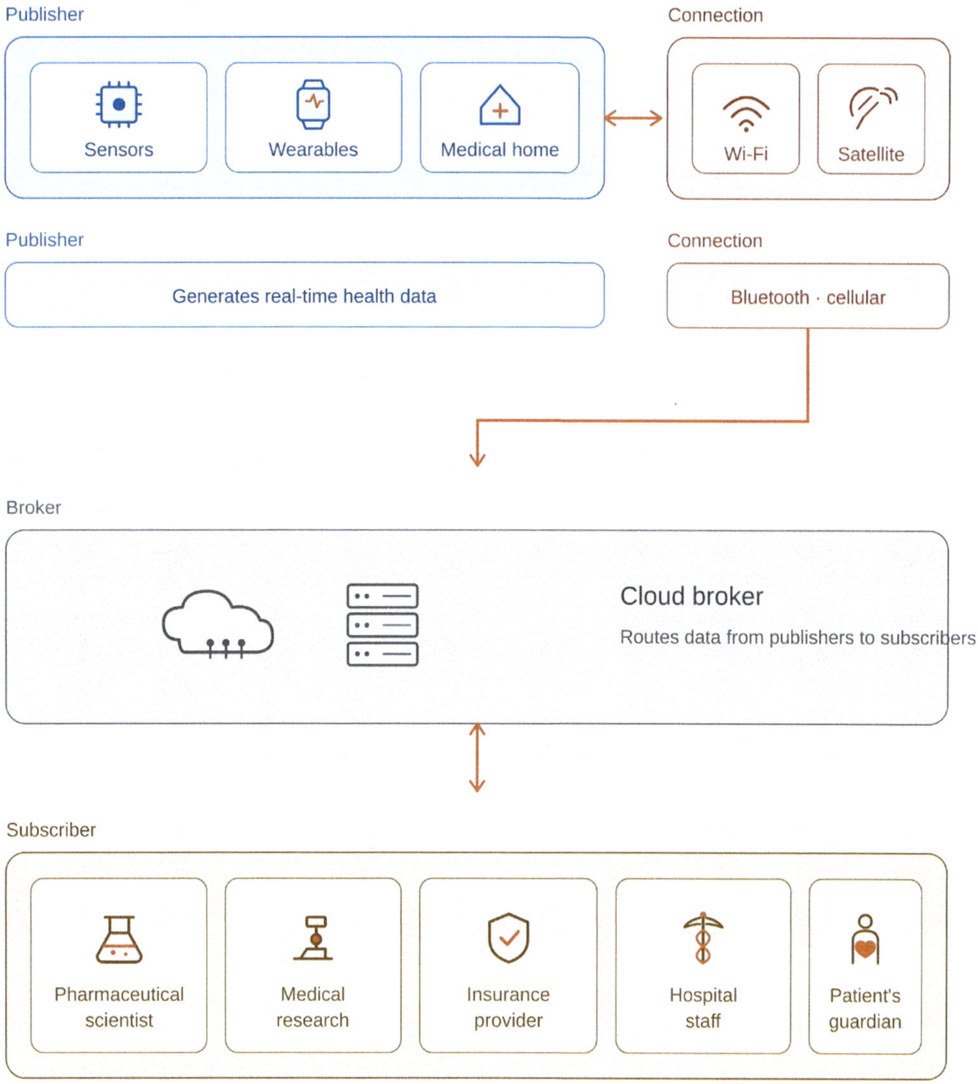

Figure 5.1 IoT enabled wellness monitoring architecture.

- Blood Oxygen
- Blood Pressure

These measures are of very much importance in terms of ECG, EEG, and EMG. With the help of some mediator the information is disseminated to brokers on a periodical basis. It is the complete responsibility of broker in terms of gathering, storing and analysis online data. Patients have the liberty of observing their personal data via some handheld devices (Lakhan et al., 2024; Hussain, 2024). In case of any deviation in terms of patient's health the publisher will take overall responsibility in terms of evaluating and conveying the relevant information.

3. IoT advancement in medicinal field

It is quite important in identifying all sorts of activities which are linked directly in development of novel IoT system, for healthcare monitoring of patients and above all it should comply to all sorts of medicinal standards including all kinds of procedures adopted in varied diagnosis process as all kinds of illness corresponds to a specific medication or treatment which may involve surgery also. discusses the varied factors associated with IoT enablement of health services, like identification technology, communication and location technology as identification of certain ailment along with location of patient and the mode of data communication from patients end to medicinal central server facilitates the effective and efficient IoT advancement in healthcare management (Upadhyaya et al., 2024; Malathi et al., 2024) (Fig. 5.2).

Data or information transformation forms a very preliminary stage of data preparation procedure which is being used in conversion of raw data or information in the form of statistical analysis. Nowadays the processes involves the concept of ETL that is, Extract, Transform and Load which in turn allows analysis and display of data across varied platforms and in different formats. Apart from that numerous healthcare technologies plus tools have been developed and some of the sophisticated tools and technologies will be developed in due course of time and will be catering to the necessities in the area of healthcare, information analysis and other administrative operations (Kaur, 2024).

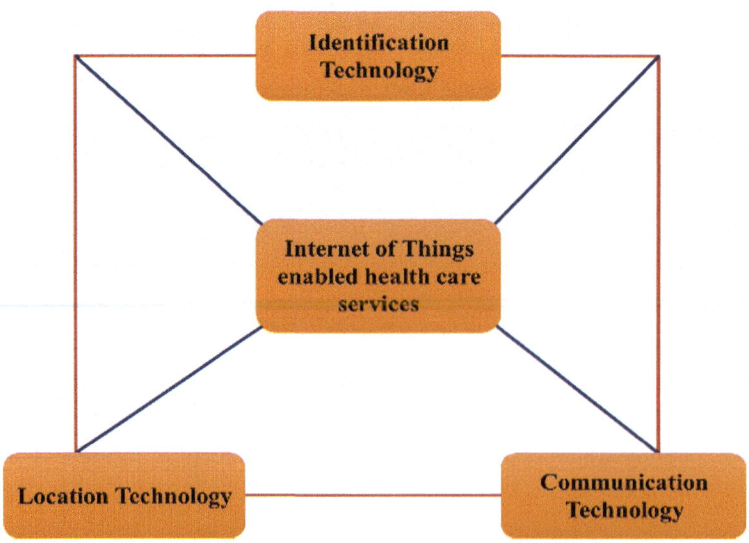

Figure 5.2 IoT advancement in healthcare management.

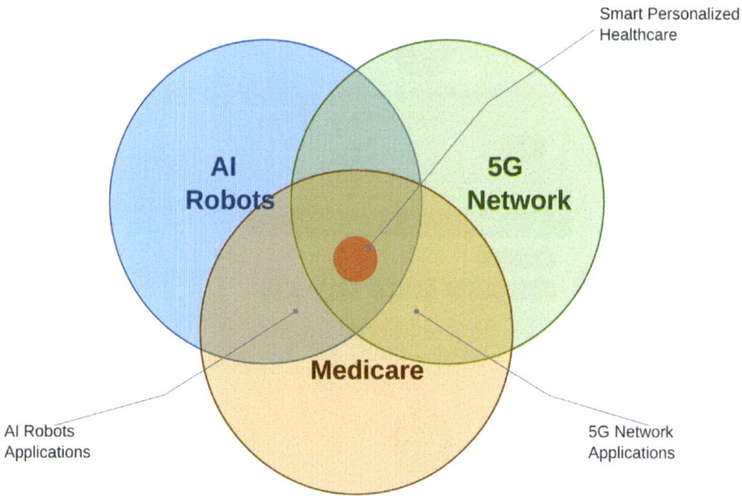

Figure 5.3 5G with AI in healthcare operations.

Fig. 5.3 discusses the role of 5G in healthcare application areas. Any health ailment today remains as one of the major reasons for morbidity burden (ElSayed et al., 2024; Amune et al., 2024). As a result of growth in population and early aging of citizens due to sedentary lifestyles, the burden of any severe health ailment is increasing such as heart failure, hypertension, and diabetes (Khang, 2024; Shafik, 2024; Aziz et al., 2024).

With time the increasing obtainability of some of the inclusive clinical data sets like electronic health records having large information of individuals which is quite useful in terms of development and design of more robust risk prediction models in terms of healthcare (Shah and Khang, 2023; Pandey et al., 2024). In certain cases prominent deep learning architectures are showcasing moderately fair performance in terms of complex EHR datasets, generally for risk prediction of varied health associated ailments (Pavitharani et al., 2024). Contemporary research has displayed a good progress in terms of explanation of varied deep learning models apart from including some of the features of natural language processing plus computer vision including features like saliency map and feature prediction and Fig. 5.4 deliberates regarding the overall 5G application ecosystem prevalent in today's scenario. After the post pandemic era, a lot of countries around the globe are facing deficiency in terms of healthcare services which has given rise to self-detection and management of illness (Roy and Sadiwala, 2024b; Nissar et al., 2024). While stress seems as an inevitable part of our daily lifestyle, that is, sedentary lifestyle. Though it cannot be eliminated but it can be managed and controlled with the help of smart wearable devices, apart from this its physiological effects can also be controlled (Ghoul and Naifar, 2024; Khang et al., 2024).

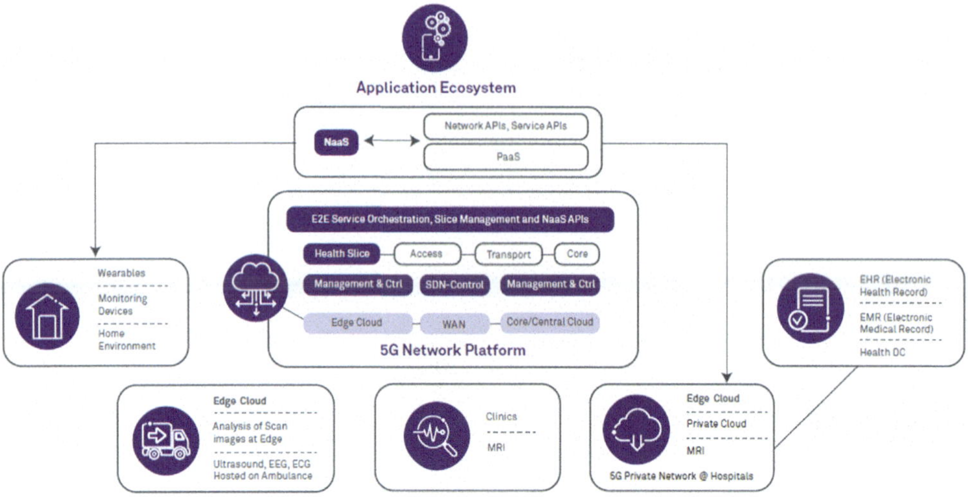

Figure 5.4 5G application ecosystem.

Many research articles discusses the electro dermal activity of wrist and arm bands in terms of prediction of different stress levels of people according to their age group (Kazi, 2024). Varied smart wearable devices are gaining popularity as they are quite useful in detection as well as management of many diseases like high blood pressure, diabetes and many more (Olorunsogo, 2024; Sharmila et al., 2024). Fig 5.5 reflects the overall healthcare management domains which are quite predominant in today's scenario.

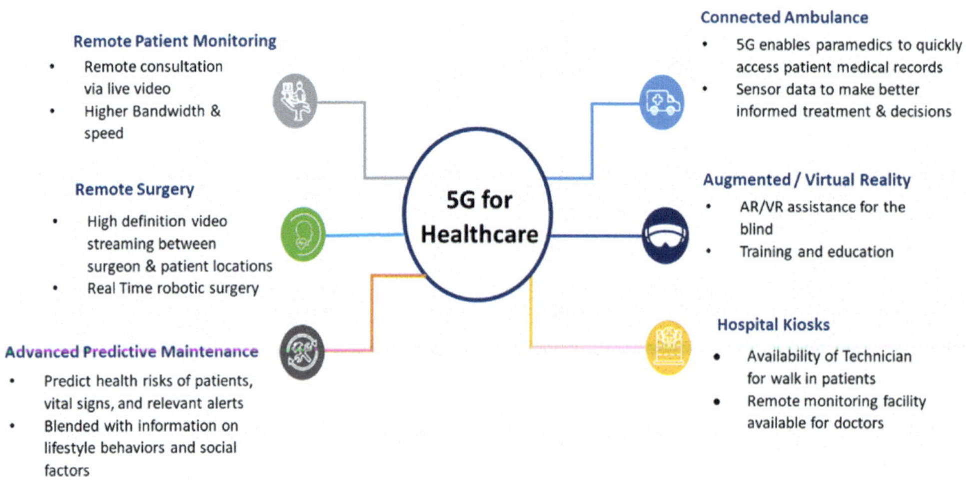

Figure 5.5 Healthcare management using 5G technology.

Researchers are still working on different kindly of deep learning models and enhancing their accuracy with time to time so as to augment the efficacy of results (Mathkor et al., 2024).

4. Challenges and future directions in 5G healthcare

5G mid-bandwidth ranges from 50 to 100 MHz, while 5G high-bandwidth can reach up to 40 GHz. The increased performance of 5G with respect to 4G—delivered through higher frequency bands and lower latency—is offset by an increase in propagation loss. It could be overcome by installing new small cells every approximately 100 m, whereas traditional 4G cells can have several kilometers radius. Inaccurate estimation of coverage negatively impacts the performance of 5G (Rehman and Manickam, 2024). Underestimated coverage leads to overlap with neighboring cells and interference generated by it, while overestimation leads to a received signal that is weak and below the threshold level. Thus, selection of cell sites needs to be exact and well planned, considering many factors. These include the cost involved in acquisition of a site and the availability of neighbor cell sites to have proper uniform coverage. Similarly, cell-site location needs to acquire permission from private property holders or municipality authorities (Rajesh et al., 2024).

It is remarkable that this review did not discuss the role of 5G networking in the development of nanoelectromechanical systems for medical applications, in particular concerning drug delivery for nano-oncology, and how perhaps very large data sets could be monitored in real-time by AI to optimize dosages. These NEMS could be interfaced with external devices (wearables, smartphones, other NEMS, etc.) using 5G and enable new IoNT applications (Saritha et al., 2024; Ahad et al., 2024). As discussed in the introduction, the security and privacy of 5G and its upcoming evolutions remain crucial, as 5G pervades effectively the physical domain. 5G medical devices will involve a large number of connected devices of different types. Wearables, smartphones, edge devices, cloud servers, and hospital/medical center IT infrastructures will thus be connected. Since cybersecurity threats may leverage vulnerabilities among these elements, with the possibility of harm to patients, cybersecurity becomes of the utmost importance and needs to be responsibly addressed with continuous monitoring for breaches and hacking events (Tariq, 2024) in view of the increasing deployment of 5G devices in healthcare.

References

Adil, M., Khan, M.K., Kumar, N., Attique, M., Farouk, A., Guizani, M., Jin, Z., 2024. Healthcare internet of things: security threats, challenges, and future research directions. IEEE Internet Things J. 11 (11), 19046–19069. https://doi.org/10.1109/JIOT.2024.3360289.

Ahad, A., Jiangbina, Z., Tahir, M., Shayea, I., Sheikh, M.A., Rasheed, F., 2024. 6G and intelligent healthcare: taxonomy, technologies, open issues and future research directions. Internet Things 25, 101068. https://doi.org/10.1016/j.iot.2024.101068.

Akinsanya, M.O., Chizoba Ekechi, C., David Okeke, C., 2024. Security paradigmsfor IOT intelecomnetworks: conceptualchallenges and solutionpathways. Eng. Sci. Technol. Int. J. 5 (4), 1431–1451. https://doi.org/10.51594/estj.v5i4.1075.

Amune, A., Chavanke, S., Joshi, S., Kuttarmare, S., Kathane, S., 2024. Integrated healthcare system for saline monitoring, patient communication, and heart disease prediction using IoT and machine learning algorithms. In: International Conference on Emerging Smart Computing and Informatics, ESCI 2024. Institute of Electrical and Electronics Engineers Inc., India https://doi.org/10.1109/ESCI59607.2024.10497196. http://ieeexplore.ieee.org/xpl/mostRecentIssue.jsp?punumber=10497186.

Arul, R., Alroobaea, R., Tariq, U., Almulihi, A.H., Alharithi, F.S., Shoaib, U., 2024. IoT-enabled healthcare systems using block chain-dependent adaptable services. Personal Ubiquitous Comput. 28 (1), 43–57. https://doi.org/10.1007/s00779-021-01584-7.

Atadoga, A., Omaghomi, T.T., Elufioye, O.A., Odilibe, I.P., Daraojimba, A.I., Owolabi, O.R., 2024. Internet of Things (IoT) in healthcare: a systematic review of use cases and benefits. Int. J. Sci. Res. Arch. 11 (1), 1511–1517. https://doi.org/10.30574/ijsra.2024.11.1.0243.

Aziz, R., Jawed, F., Khan, S.A., Sundus, H., 2024. Wearable IoT Devices in Rehabilitation: Enabling Personalized Precision Medicine Medical Robotics and AI-Assisted Diagnostics for a High-Tech Healthcare Industry. IGI Global, India, pp. 281–308. https://doi.org/10.4018/979-8-3693-2105-8.ch018.

Clemente-Lopez, D., Rangel-Magdaleno, J.de J., Manuel Muñoz-Pacheco, J., 2024. A lightweight chaos-based encryption scheme for IoT healthcare systems. Internet Things 25, 101032. https://doi.org/10.1016/j.iot.2023.101032.

Z. ElSayed, N. Elsayed, S. Bay, A novel zero-trust machine learning green architecture for healthcare IoT cybersecurity: review, analysis, and Implementation. arXiv, United States arXiv. 23318422. (2024), doi: 10.48550/arXiv.2401.07368.

Fischer, G.S., Ramos, G.d.O., Costa, C.A.d., Alberti, A.M., Griebler, D., Singh, D., Righi, R.d.R., 2024. Multi-hospital management: combining vital signs IoT data and the elasticity technique to support healthcare 4.0. Internet Things 5 (2), 381–408. https://doi.org/10.3390/iot5020019.

Ghoul, Y., Naifer, O., 2024. A healthcare application based on IoT devices. Wirel. Netw. 30 (4), 2541–2556. https://doi.org/10.1007/s11276-024-03683-8.

Hoque, M., Khan, A., 2024. Embracing IoT for advancing healthcare monitoring: a Comprehensive review and future Outlook. Int. J. Innov. Res. Technol. Sci. 12 (2), 261–265.

Hussain, I., 2024. Secure, sustainable smart cities and the internet of things: perspectives, challenges, and future directions. Sustainability 16 (4), 1390. https://doi.org/10.3390/su16041390.

Islam, Md M., Hasan, M.K., Islam, S., Balfaqih, M., Ahmed, I.A., Alalwan, N., Safie, N., Zaheed, A.B., Thakkar, R., Ghazal, T.M., 2024. Enabling pandemic-resilient healthcare: narrowband internet of things and edge intelligence for real-time monitoring. CAAI trans. intell. technol. https://doi.org/10.1049/cit2.12314.

Kante, M., Ndayizigamiye, P., 2021. Internet of medical things, policies and geriatrics: an analysis of the national digital health strategy for South Africa 2019–2024 from the policy triangle framework perspective. Sci. Afr. 12, e00759. https://doi.org/10.1016/j.sciaf.2021.e00759.

Kaur, J., 2024. Green Healthcare in Smart Cities: Harnessing IoT for Sustainable Transformation of Healthcare Systems Secure and Intelligent IoT-Enabled Smart Cities. IGI Global, India, pp. 334–354. https://doi.org/10.4018/979-8-3693-2373-1.ch016. Available from: https://www.igi-global.com/book/secure-intelligent-iot-enabled-smart/332802.

Kazi, K., 2024. AI-Driven IoT (AIIoT) in healthcare monitoring Using Traditional Design Methods to Enhance AI-Driven Decision Making. IGI Global, India, pp. 77–101. https://doi.org/10.4018/979-8-3693-0639-0.ch003. Available from: https://www.igi-global.com/book/using-traditional-design-methods-enhance/323652.

Khan, M.M., Alkhathami, M., 2024. Anomaly detection in IoT-based healthcare: machine learning for enhanced security. Sci. Rep. 14 (1). https://doi.org/10.1038/s41598-024-56126-x.

Khang, A., 2024. AI and IoT Technology and Applications for Smart Healthcare Systems AI and IoT Technology and Applications for Smart Healthcare Systems. CRC Press, United States, pp. 1–401. https://doi.org/10.1201/9781032686745. Available from: http://www.tandfebooks.com/doi/book/9781040021798.

Khang, A., Abdullayev, V., Hrybiuk, O., Shukla, A.K., 2024. Computer Vision and AI-Integrated IoT Technologies in the Medical Ecosystem Computer Vision and AI-Integrated IoT Technologies in the Medical Ecosystem. CRC Press, United States, pp. 1–439. https://doi.org/10.1201/9781003429609. Available from: https://www.taylorfrancis.com/books/9781003837299.

Lakhan, A., Nedoma, J., Mohammed, M.A., Deveci, M., Fajkus, M., Marhoon, H.A., Memon, S., Martinek, R., 2024. Fiber-optics IoT healthcare system based on deep reinforcement learning combinatorial constraint scheduling for hybrid telemedicine applications. Comput. Biol. Med. 178. https://doi.org/10.1016/j.compbiomed.2024.108694.

Malathi, J., Kusha, K.R., Isaac, S., Ramesh, A., Rajendiran, M., Boopathi, S., 2024. IoT-enabled Remote Patient Monitoring for Chronic Disease Management and Cost Savings: Transforming Healthcare Advances in Explainable AI Applications for Smart Cities. IGI Global, India, pp. 371–388. https://doi.org/10.4018/978-1-6684-6361-1.ch014. Available from: https://www.igi-global.com/book/advances-explainable-applications-smart-cities/301208.

Mathkor, D.M., Mathkor, N., Bassfar, Z., Bantun, F., Slama, P., Ahmad, F., Haque, S., 2024. Multirole of the internet of medical things (IoMT) in biomedical systems for managing smart healthcare systems: an overview of current and future innovative trends. J. Infect. Public Health 17 (4), 559–572. https://doi.org/10.1016/j.jiph.2024.01.013.

Nissar, G., Khan, R.A., Mushtaq, S., Lone, S.A., Hassan Moon, A., 2024. IoT in healthcare: a review of services, applications, key technologies, security concerns, and emerging trends. Multimed. Tool. Appl. https://doi.org/10.1007/s11042-024-18580-7.

Olorunsogo, T.O., 2024. Integrating IOT in pediatric healthcare: a systematic review of current applications and future directions for pancreatic diseases and obesity Tolulope O. Olorunsogo1. Int. J. Res. Med. Sci. 4 (3), 305–318. https://doi.org/10.51594/imsrj.v4i3.923.

Pandey, A.K., Das, A.K., Kumar, R., Rodrigues, J.J.P.C., 2024. Secure cyber engineering for IoT-enabled smart healthcare system. IEEE Internet Things Mag. 7 (2), 70–77. https://doi.org/10.1109/IOTM.001.2300172.

Parihar, A., Prajapati, J.B., Prajapati, B.G., Trambadiya, B., Thakkar, A., Engineer, P., 2024. Role of IOT in healthcare: applications, security & privacy concerns. Intell. Pharm. 2 (5), 707–714. https://doi.org/10.1016/j.ipha.2024.01.003.

Pavitharani, G.P., Joy, R.T., Rajendran, R.K., 2024. AI and IoT for Universal Health and Well-Being across Generations the Climate Change Crisis and its Impact on Mental Health. IGI Global, India, pp. 187–197. https://doi.org/10.4018/979-8-3693-3272-6.ch014. Available from: https://www.igi-global.com/book/climate-change-crisis-its-impact/335121.

Pulimamidi, R., 2024. To enhance customer (or patient) experience based on IoT analytical study through technology (IT) transformation for E-healthcare . measurement. Sensors 33, 101087. https://doi.org/10.1016/j.measen.2024.101087.

Puri, V., Kataria, A., Sharma, V., 2024. Artificial intelligence-powered decentralized framework for internet of things in healthcare 4.0. Trans. Emerging Telecommun. Technol. 35 (4). https://doi.org/10.1002/ett.4245.

Quasim, M.T., Khan, M.A., Abdullah, M., Meraj, M., Singh, S.P., Johri, P., December 2019. Internet of things for smart healthcare: a hardware perspective. In: 2019 First International Conference of Intelligent Computing and Engineering (ICOICE). IEEE, pp. 1–5.

Rajesh, R., Hemalatha, S., Nagarajan, S.M., Devarajan, G.G., Omar, M., Bashir, A.K., 2024. Threat detection and mitigation for tactile internet driven consumer IoT-healthcare system. IEEE Trans. Consum. Electron. 70 (1), 4249–4257. https://doi.org/10.1109/TCE.2024.3370193.

Rehman, S.U., Manickam, S., 2024. Application of smart sensors for internet of things healthcare environment: study and prospects. In: Next-Generation Smart Biosensing. Academic Press, pp. 287–305.

Roy, C.K., Sadiwala, R., 2024a. Smart environment for IoT based disease diagnosis healthcare framework. J. Integr. Sci. Technol. 12 (2).

Roy, C.K., Sadiwala, R., 2024b. Smart internet of things (IoT) based healthcare framework environment for chikungunya disease diagnosis. J. Integr. Sci. Technol. 12 (4). https://doi.org/10.62110/sciencein.jist.2024.v12.779.

Saritha, K., Abhiram, C., Reddy, P.A., Sathwik, S.S., Shaik, N.R., Alzubaidi, L.H., Raj, V.H., Dutt, A., Yadav, D.K., 2024. IoT enabled hospital asset tracking using advanced interdisciplinary approaches. In: E3S Web of Conferences. EDP Sciences, India. https://doi.org/10.1051/e3sconf/202450701007. www.e3s-conferences.org/507.

Shafik, W., 2024. IoT-Enabled Secure and Intelligent Smart Healthcare: Beyond 5G in Enabling Smart Cities Secure and Intelligent IoT-Enabled Smart Cities. IGI Global, Brunei Darussalam, pp. 308–333. https://doi.org/10.4018/979-8-3693-2373-1.ch015. Available from: https://www.igi-global.com/book/secure-intelligent-iot-enabled-smart/332802.

Shah, I.A., Jhanjhi, N.Z., Brohi, S.N., 2024. IoT Smart Healthcare Security Challenges and Solutions Advances in Computational Intelligence for the Healthcare Industry 4.0. IGI Global, Malaysia, pp. 234–247. https://doi.org/10.4018/979-8-3693-2333-5.ch012. Available from: https://www.igi-global.com/book/advances-computational-intelligence-healthcare-industry/332779.

Shah, V., Khang, A., 2023. Internet of Medical Things (IoMT) Driving the Digital Transformation of the Healthcare Sector Data-Centric AI Solutions and Emerging Technologies in the Healthcare Ecosystem. CRC Press, India, pp. 15–26. https://doi.org/10.1201/9781003356189-2. Available from: http://www.tandfebooks.com/doi/book/10.1201/9781003356189.

Sharmila, E.M.N., Krishna, K.R., Prasad, G.N.R., Anand, B., Kwatra, C.V., Kapila, D., 2024. IoMT-Applications, Benefits, and Future Challenges in the Healthcare Domain Advances in Fuzzy-Based Internet of Medical Things (IoMT). Wiley, India, pp. 1–23. https://doi.org/10.1002/9781394242252.ch1. Available from: https://www.wiley.com/en-us/Advances+in+Fuzzy+Based+Internet+of+Medical+Things+%28IoMT%29-p-9781394242238.

Singh, B., Kaunert, C., Vig, K., Gautam, B.K., 2024. Wearable Sensors Assimilated with Internet of Things (IoT) for advancing Medical Imaging and Digital Healthcare: Real-Time Scenario Inclusivity and Accessibility in Digital Health. IGI Global, India, pp. 275–297. https://doi.org/10.4018/979-8-3693-1463-0.ch018. Available from: https://www.igi-global.com/book/inclusivity-accessibility-digital-health/328539.

Singh, B., Kaunert, C., 2024. Integration of Cutting-edge technologies such as internet of things (IoT) and 5G in health monitoring systems: a comprehensive legal analysis and futuristic outcomes. GLS Law J. 6 (1), 13–20.

Talpur, M.S.H., Abro, A.A., Ebrahim, M., Kandhro, I.A., Manickam, S., Ul Arfeen Laghari, S., Dandoush, A., Uddin, M., 2024. Illuminating healthcare management: a comprehensive review of IoT-enabled chronic disease monitoring. IEEE Access 12, 48189–48209. https://doi.org/10.1109/ACCESS.2024.3382011.

Tariq, M.U., 2024. Advanced Wearable Medical Devices and Their Role in Transformative Remote Health Monitoring Transformative Approaches to Patient Literacy and Healthcare Innovation. IGI Global, United Arab Emirates, pp. 308–326. https://doi.org/10.4018/979-8-3693-3661-8.ch015. Available from: https://www.igi-global.com/book/transformative-approaches-patient-literacy-healthcare/336553.

Upadhyaya, A.N., Saqib, A., Devi, J.V., Rallapalli, S., Sudha, S., Boopathi, S., 2024. Implementation of the Internet of Things (IoT) in Remote Healthcare Analyzing Current Digital Healthcare Trends Using Social Networks. IGI Global, India, pp. 104–124. https://doi.org/10.4018/979-8-3693-1934-5.ch006. Available from: https://www.igi-global.com/book/analyzing-current-digital-healthcare-trends/330973.

Vatin, N.I., Pant, R., Mohan, C., Kumar, A., Rajasekhar, N., 2024. IoT-enhanced healthcare: a patient care evaluation using the IoT healthcare test. In: BIO Web of Conferences. EDP Sciences Russian Federation. https://doi.org/10.1051/bioconf/20248601092. www.bio-conferences.org/86.

Zahedian Nezhad, M., Bojnordi, A.J.J., Mehraeen, M., Bagheri, R., Rezazadeh, J., 2024. Securing the future of IoT-healthcare systems: a meta-synthesis of mandatory security requirements. Int. J. Med. Inf. 185. https://doi.org/10.1016/j.ijmedinf.2024.105379.

CHAPTER 6

Intelligent health semantic interoperability for internet of martial art things: From wearables to standards

Amelie Gyrard[1] and Vincent Pham[2]
[1]Machine-to-Machine Measurement (M3), Paris, France; [2]Eqimonie, Paris, France

The key to a healthy life is having a healthy mind.

Richard Davidson

1. Introduction

Since World War II, clinical psychology focused on the diagnosis and treatment of mental disease (Wallace and Shapiro, 2006). Fortunately, understanding and cultivating positive mental health is encouraged by Seligman and Csikszentmihalyi (2000).

Benefits of sports: Sports are increasingly recommended for a healthy lifestyle, reducing stress, improving wellbeing, etc. More and more companies are creating new devices to support drills, such as smartwatches (e.g., Fitbit), and fitness applications (Google Fit, Apple Health), etc. Google Fit is designed to follow the World Health Organization (WHO) and American Heart Association (AHA) instructions such as tracking minutes of activity. Sport brings several benefits such as personal growth (e.g., beliefs), psychological growth (e.g., sporting qualities), social growth (e.g., social support), and physical growth (e.g., strength) (Wadey et al., 2013). Four topics for sports performance are addressed: 1) mental training, 2) emotional states, 3) the effects of personality traits, and 4) cognitive and psychophysiological characteristic (Di Corrado et al., 2021). Sports training involves planning, analyzing and monitoring sport activities and comprises knowledge from multiple domains (e.g., anatomy, physiology, biomechanics, biochemistry, sociology, psychology, etc.) (Deak et al. (2016)). Apple Health[1] focuses on activity, sleep, mindfulness, and nutrition. At Wright State University, Ohio, USA, bicycles are set up in the library[2] to encourage people to exercise even if they do not find the time for

[1] https://www.apple.com/lae/ios/health/
[2] https://www.wdtn.com/news/local-news/wright-state-library-combines-study-and-workout-with-new-bikes/

Intelligent Semantic Analysis for Healthcare
ISBN 978-0-443-33512-9
https://doi.org/10.1016/B978-0-443-33512-9.00021-4

that, bicycles are equipped to be able to work, study and read books at the same time than cycling. Sport and exercise help reduce depression[3] (Blumenthal et al., 2012).

The National Center for Complementary and Integrative Health (NCCIH) in US encourages complementary medicines.[4] Combinations such as psychological and physical approaches (e.g., yoga, tai chi, acupuncture, dance or art therapies) or psychological and nutritional (e.g., mindful eating) are helpful. NCCIH trategic Plan FY 2021–25 Mapping a Pathway to Research on Whole Person Health (Langevin, n.d.) objectives are: 1) Advance fundamental science and methods development. 2) Advance research on the whole person and on the integration of complementary and conventional care. 3) Foster research on health promotion and restoration, resilience, disease prevention, and symptom management. 4) Enhance the complementary and integrative health research workforce. 5) Provide objective evidence-based information on complementary and integrative health interventions. The National Center for Complementary and Integrative Health (NCCIH) was created more than 20 years ago to facilitate the study and evaluation of complementary and alternative health practices.

Traditional Chinese Medicine (TCM) is recognized by the US National Institutes of Health (NIH) as an effective complementary and alternative medicine (Burke et al., 2003).

1.1 Why do we need internet of things for sports?

Multiple body sensors allow a more accurate and more effective deployment of signal acquisitions on human body (Nardi, 2019). IoT devices can: 1) collect data on student' movements and performance, 2) track fitness, help understand the health benefits of practicing sports, etc (Ramanathan et al., 2024).

Internet of Things for Sports (IoTSport) (Ray, 2015) and recreational activities are introduced as an emerging research topic that could benefit health and rehabilitation domains. As an example, an accelerometer can be used to measure the athlete' movement speed. Standards are missing in this research topic.

IoT and AI-based smart fitness (Farrokhi et al. (2021)) solutions are surveyed: Fitness trackers (including wearable and non-wearable sensors), movement analysis, and fitness applications. Smart fitness is classified into 1) IoTbased smart fitness (fitness trackers, movement analysis and fitness applications), 2) Social-IoT and 3) Artificial intelligence (Farrokhi et al., 2021). It is reminded that a healthy lifestyle is based on five important interrelated factors: 1) Physical activity, 2) Food and drink consumption, 3) Sleep quality, 4) Stress, and 5) Social interaction. Sensors can help automatically monitor sleep quality, diet, and physical activity. Recommendations for enhancements are: 1)

[3] https://www.ncbi.nlm.nih.gov/books/NBK279276/

[4] www.nccih.nih.gov/health/complementary-alternative-or-integrative-health-whats-in-a-name.

Advanced Smart Fitness by enhancing AI-based coaching, 2) Accuracy improvement, 3) Adopting an SIoT approach for better decision-making, 4) Integration of sensed data per user in a place, and 5) Improving security and privacy, 6) Reducing cost, 7) Data management and big data collected via smart fitness applications, 8) Establishing other IoT solutions.

1.2 Why do we need martial arts: Kung-Fu, Taichi, meditation, Qi Gong?

Shaolin Kung Fu comes from the monks of the Shaolin Temple in Henan, China (Ishac et al., 2023).

A visionary project will be to measure the impact of meditation, bodily awareness, taichi, and Kung-Fu with sensors. Internet of Things could help to design experiments to measure the effect. We envision experiments to design a physiological database to be measured on trainees, and to consider the context/environment. This work is inspired by scientific research from Peter Wayne for Taichi, Mathieu Ricard, Richard Davidson, Steven Laureys, etc.

Mind-Body Therapies (MBT) include Tai Chi, Qigong, yoga, and meditation to reduce symptoms and improve quality of life (Bower and Irwin, 2016).

The mind and the body represent two distinct natures of the same experience of living, communicating with and influencing each other (Di Corrado et al., 2021). In recent years, the mind-body has a growing interest in psychology and neuroscience applied to sports.

Martial arts students learn numerous soft skills such as self-discipline, perseverance, and temperance (Li et al., 2023).

Given the disadvantages of first-line treatments such as psychotherapy and pharmacotherapy for Major Depressive Disorder (MDD), there is a need to reduce the costs and medication side effects such as sexual dysfunction, weight gain, and sleep disturbance (Kong et al., 2019). Researchers encourage alternative therapies such as Taichi.

Taichi in the martial arts fitness program is suitable for all ages (Li et al., 2023). Taichi is a mind-body exercise combining low to moderate intensity slow movements, deep breathing and mental concentration (Tao et al., 2017). Taichi includes physical, cognitive and social, which are believed to aid in maintaining cognition in older adults. Tai Chi can be an intervention that integrates physical, psychosocial, emotional, spiritual, and behavioral. Tai Chi can help reduce stress and produce anti-inflammatory effects (Kong et al., 2019). Tai Chi employs slow, gentle movements, breathing techniques, and cognitive tools (i.e., attention, imagery) to strengthen, integrate, and relax the body and mind.

Social-psychological outcomes of martial arts practice among youth (Vertonghen and Theeboom, 2010). 27 papers over 350 are studied from various scientific disciplines (sport sociology, psychology, pedagogy, and philosophy).

Problem 1: if you are an athlete, you can be a victim of bad judgment from referees which has a huge impact on your career. Having IoT devices would help in arbitration.

A second use is that devices can help practice the right way (e.g., bone alignment).

Taking inspiration from the sport philosophy, where athletes do not want injuries and in case of injuries, heal faster. Can we use the spirit to avoid diseases and any symptoms in our everyday life?

Robots are taking inspiration from martial arts to practice complex activities, such as practicing nunchucks.

A robot named Taekwondo Personal Trainer (TPT) is designed to be kicked.

Physical activity in all its forms is important and necessary for the promotion and maintenance of mental and psychological health. Martial arts help to enhance our mental qualities such as determination, courage, vigilance, and control of emotions.

Physical fitness refers to the physical ability to perform a sport or activity (Ye et al., 2022). Martial arts are considered a cultural heritage in China (Ye et al., 2022). Chinese martial arts include Shaolin, Taiji, Wing Chun, etc. Martial arts education improve morals such as perseverance, tenacity, and patience. Martial arts education, body organs and tissues can be trained to develop physical qualities such as strength, speed, stamina, and flexibility and improve the body's defenses such as fitness and stamina. Martial arts develops students' independence, autonomy, and self-discipline. To learn to think independently, learn to rely on themselves, learn to control themselves, learn to persist, and learn to motivate themselves.

Wushu is a unique traditional project in China and a treasure in the Chinese nation's cultural heritage (Li et al., 2021a).

1.3 Why do we need ontologies?

The Semantic Web technologies can be used to overcome the heterogeneity of the web by creating a standard method for knowledge representation. Ontologies can help for the integration of heterogeneous information resources (Jokiniemi, 2010). Ontology Development 101 methodology (designed by the creators of the Standford Protégé Ontology Editor tool) encourages to reuse of domain knowledge by reusing ontologies in Step 2 "Consider reusing ontologies".

As an example, BBC, is designing a Sport Ontology[5] to describe competitive sports events. DBpedia, the semantic version of Wikipedia designed MixedMartialArtsEvent class.[6] Additional ontologies for Traditional Chinese Medicine (TCM) will be mentioned in Section 2.5.2.

[5] https://www.bbc.co.uk/ontologies/sport-ontology/

[6] https://dbpedia.org/ontology/MixedMartialArtsEvent

1.4 Research questions, contributions, and structure of the paper

We designed the following Research Questions (RQ):

- RQ1: Which sensors can be used for a healthy lifestyle, to enhance sports performance, by focusing on martial arts-related activities?
- RQ2: How Artificial Intelligence (AI) can be applied to sensor data to enhance healthy martial arts performances?
- RQ3: How to reuse human knowledge expertise shared within scientific publications and ontologies?

Contributions (C):

- C1: Martial arts from a scientific view is reviewed in Section 2: Martial Arts uses AI or sensor technologies as explained in Section 2.1. Taichi benefits are studied in Section 2.2. Meditation is investigated in Section 2.3. Qi Gong and respiration exercices are overviewed in Section 2.4 Traditional Chinese Medicine (TCM) is introduced in Section 2.5: acupuncture, ontologies, and standards.
- C2: A study of sensors relevant for healthy martial arts in Section 3.
- C3: An investigation within health standards, and TCM standards.

Structure of the Paper: Martial arts from a scientific view is reviewed in Section 2. A study of sensors relevant to healthy martial arts in Section 3. Envisioned Experiments are introduced in Section 4. Section 5 concludes the paper and provides future work.

2. Martial arts and related topics background from a scientific view

Martial Arts can use AI or sensor technologies as explained in Section 2.1. Taichi benefits are studied in Section 2.2. Meditation is investigated in Section 2.3. Qi Gong and respiration exercices are overviewed in Section 2.4. Traditional Chinese Medicine (TCM) is introduced in Section 2.5 describing acupuncture, ontologies, and standards.

2.1 AI, IoT, martial arts, and health benefits

Martial Arts can use AI or sensor technologies as explained in this section.

Discussions about nine papers on "Hard" Kung Fu styles are reviewed (Tsang et al., 2008); which provide information about the aerobic intensity of Kung-Fu, body composition (e.g., body mass and percent body fat, bone mineral density), exercise capacity (e.g., muscle strength), other adaptations, and risk of injury.

Conclusion: There is a need to get statistics from associations with registered practitioners and health insurance (with declared injuries). Even with such numbers, it does not reflect minor injuries, but still serious (e.g., several weeks of no practice because of the pain, waiting for the body to rest and cure without official partnerships with doctors).

Advanced mindfulness research and practice in sport psychology (Shaolin martial art) (Kee, 2019) are classified into three categories: Chan-based Dejian mind–body intervention (DMBI), dantian breathing strategies, and internal martial arts practice. Sport psychology highlights the development of mental skills to enhance performance and psychological well-being.

A mobile phone application using IoT and AI can be used to track physical performance and the mental health of Silambam practitioners (Ramanathan et al., 2024). Silambam is a traditional martial art from Tamil Nadu, India. Personalized training programs can be designed to identify and correct technique errors, and track progress over time. IoT sensors such as an accelerometer, and gyroscope, are attached to the Silambam stick to collect data on the user's movements and heart rate and the number of calories burned. Such questions aim to be answered: Who uses Silambam? How often do they use it? What are the benefits they get from using it? What are their pain points?

Deep Learning (LSTM recurrent neural network) algorithm is used on martial arts fitness data (Li et al. (2023)).

IoT-Based Martial Art mat (Ishac et al. (2023)) uses sensors to check the correct Mapu Shaolin Kung Fu stance.

AI-based martial art education governance uses deep learning to track and analyze Wushu practitioners' movements (Ye et al. (2022)). A 4-week experiment evaluates students' learning interests, active participation attitude, independent research ability, and analytical problem-solving ability. A camera is used.

AI-based assistant system for martial arts Wushu competition decision system (Li et al. (2021a)). It can also help with daily training. The system classifies the player's ability such as strikes ground, strikes the body, distance time, strikes kicks attempts, and strikes attempt. The system also measures the player's movement such as kick attempts, punch attempts, side control time, leg kicks, and head kicks. It is not explained where the data comes from, is it measured with a camera or any devices?

Ghazi (2022) address the following research question: "How the effectiveness of AI and strategic planning in building mental modeling to improve sports performance". A total of 15 black belt practitioners of Karate (5 players), Taekwondo (5 players), and Kungfu (5 players) participated in the study by practicing a front kick. The authors study three kinds of look: 1) looking in the eyes of the opponent to read the opponent's intentions and to transfer our determination, strength, and courage to the opponent, 2) looking according to the situation, and 3) the autocratic look. Peripheral vision is mentioned. It is not clearly explained which sensors are used. We could achieve that with a camera or EOG.

Precompetition emotions of martial arts athletes are analyzed (Li et al., 2021b).It is claimed that psychologically can improve the efficiency of athletes by more than 35%.

Deep Strike[7] is an AI-based technology revolutionizing boxing by providing data-driven insights and analysis. Deep Strike measures millions of data points during a fight that it funnels into 50 metrics for each boxer, not only punches thrown, landed, and their quality, but also many other aspects such as footwork, balance, stance, etc.

There is a robot performing nunchaku (but with only one robotic hand) https://www.youtube.com/watch?v=ClieF_kBQGk (Zhao et al., 2018). This research makes it possible to understand the complexity of the movement of a wrist and to reproduce it in robotics, for complex tasks that are still difficult for robots.

KSAS (Kenpo Set Assisting System) (Casas-Ortiz and Santos (2021) is a mobile application for learning American Kenpo Karate, (Blocking Set I). Inertial sensors from an Android device are used. Short-Term Memory (LSTM) neural network algorithm is employed. 29 volunteers were involved to fill in a survey, and 20 volunteers used the app to analyze the movements.

Kick Ass Kung-Fu (Hamalainen et al., 2005) is a physically interactive martial arts training. 46 martial art practitioners participated in the study. Heart rate data was gathered from 20 subjects (rest and after fighting).

2.2 Taichi from a scientific view

Taichi benefits are studied in this section.

Taichi Definition: Taichi in the martial arts fitness program is suitable for all ages (Li et al. (2023)). Taichi and Qigong are practices from traditional Chinese medicine that combine specific movements or postures, coordinated breathing, and mental focus (Bower and Irwin, 2016).

The book "Tai chi, meditation in movement. A medical approach for a millennial art", is written by Peter Wayne, a researcher at Harvard Medical School, passionate about martial arts (more than 35 years of experience practicing and teaching Taichi). Peter Wayne wanted to prove the benefits of Taichi (Wayne, 2018) from a scientific point of view.

6 months Taichi training protocol is analyzed with 27 Taichi experts over 60 healthy older adults (aged 50–79) living within the Greater Boston area, MA, USA from March 2011 to March 2013 and practicing Taichi on average, two classes per week (Wayne et al. - Beth Israel Deaconess Medical Center's Clinical Research Center (BIDMC)). They were also asked to practice a minimum of 30 min, two additional days per week. The study is registered at Clinical Trials clinicaltrials.gov (NCT01340365). Age, gender, BMI, and activity are studied to observe walking.

[7] https://www.linkedin.com/feed/update/urn:li:activity:7120463502019895296?utm_source=share&utm_ medium=member_desktop

Tai Chi can be a mind-body therapy for depression and mood regulation in clinical trials (Kong et al. (2019)). Tai Chi alleviate depressive symptoms, the intervention may modulate the activity and connectivity of key brain regions involved in mood regulation, reduce neuro-inflammatory sensitization, modulate the autonomic nervous system, and regulate hippocampal neurogenesis. Tai Chi experiment with patients with depression is compared with Tai Chi with other mind–body interventions such as yoga and Baduanjin. Tai Chi warm-up exercises that can last 15–30 min. The second section focuses on five core Tai Chi movements following the Cheng Ma Ching Yang-style short form. Participants progressively add these movements over the 12 weeks. The program concludes with 5 min of simple cool-down exercises. The total exercise lasts about 45–60 min.

Tai Chi Chuan and Baduanjin exercise prevent memory deficits (Tao et al. (2017)) in older adults over 12 weeks of both Tai Chi Chuan (Yang style 24-form) and Baduanjin practice (5 days/per week for 60 min per day). 62 (21 in the Tai Chi Chuan group, 16 in the Baduanjin group, 25 in the control group) completed all study procedures over the 102 older adults who participated to the study. Each Tai Chi session included: 1) 10 min of warm-up (for example weight shifting, arm swinging, gentle stretches of the neck and a review of Tai Chi Chuan principles; 2) 30 min of Tai Chi Chuan exercises; 3) 10 min of breathing techniques, and 4) 10 min of relaxation. Tao et al. encourage longer trials with a larger sample size to evaluate better the impact of Taichi. Putamen and insular are brain regions associated with memory and cognition. All MRI scans were applied using a 3.0 T GE scanner (General Electric, Milwaukee, WI, USA).

Perceptions of the Eight Active Ingredients of Tai Chi (Miller and Taylor-Piliae, 2021) are measured: 1) Awareness, 2) Intention, 3) Structural integration, 4) Active relaxation, 5) Strengthening and flexibility, 6) Natural/freer breathing, 7) Social support, and 8) Embodied spirituality. Cognitive and physical measures related to driving performance are analyzed. Fifty-eight Tai Chi practitioners (aged 65 years and older) living in the middle Tennessee area in the US participated to the study between September 2014 and June 2015. Participants performed Tai Chi once or twice a week (at least 30 min each week), and for at least 3 months.

2.3 Meditation from a scientific view

Meditation provides relaxation and enhances well-being (Davidson and Lutz, 2008, Lutz et al., 2008a). Meditation involves training the mind, to focus attention. Mindfulness meditation teaches individuals to bring attention to present-moment experiences with openness, curiosity, and non-judgment (Bower and Irwin, 2016).

Techniques such as Mindfulness-Based Stress Reduction (MBSR) (Kabat-Zinn, 2013, 2016) are scientifically proven with a cohort of more than 18,000 patients, and research on mindfulness meditation benefits on brain and immune system according to Jon Kabat-Zinn and Richard Davidson (Davidson et al., 2003; Kabat-Zinn et al., 2014).

Four types of mental balance: conative, attentional, cognitive, and affective are studied (Wallace and Shapiro, 2006). Questions arise such as whether loving-kindness meditation is a more effective intervention than mindfulness training for depression, or whether it might be optimal to combine those types of practice. Siemens 3T Tim-Trio MRI scanner is used for fMRI (Desbordes et al., 2012).

Meditation's impact on well-being is studied by Sahdra et al. (2011). Data are collected from 60 participants between 21 and 70 years old for 3 months of intensive meditation training in an isolated retreat setting (Retreat 1 in the Shambhala Mountain Center in Red Feather Lakes, Colorado - 30 participants) under the guidance of Dr. Wallace and a wait-list control group that later underwent identical training (Retreat 2–30 participants). Dr. Wallace's meditation training is about: (1) mindfulness of breathing to relax body and mind, and calm thinking and sensory distraction; (2) observing mental events ("settling the mind in its natural state") to enhance attentional stability and vividness; (3) observing the nature of consciousness ("awareness of awareness") to increase the attention; (4) loving-kindness to encourage genuine happiness; (5) compassion; (6) empathetic joy and reduce envy; and (7) equanimity to arouse affection for all beings.

fMRI GE Signa 3.0 T scanner and eye tracking SensoMotoric Instruments (Lutz et al., 2008b University of Wisconsin, USA) are used. EEG is used in Lutz et al. (2004)). Neuroscientific research on Focused Attention (FA) and Open Monitoring (OM) meditation (Lutz et al. (2008a)).

The impact of meditation is demonstrated using fMRI on Tibetan Buddhist monks who already meditating for tens of thousands of hours (Davidson and Lutz (2008), Waisman Laboratory for Brain Imaging and Behavior, University of Wisconsin–Madison, USA). Between 64 and 256 electrodes on the scalp surface measure brain electrical activity.

MBSR Mindfulness meditation (Davidson et al., 2003) demonstrates effects on brain and immune function through an 8-week clinical training program in mindfulness meditation applied in a work environment with 25 healthy employees. Brain electrical activity is measured before and immediately after, and then 4 months after an 8-week training program in mindfulness meditation. Brain electrical activity, or EEG, and EOG (for correcting EEG for eye movements) are employed.

Mind-body therapies (MBT) include Tai Chi, Qigong, yoga, and meditation to reduce symptoms and improve quality of life (Bower and Irwin, 2016).

2.4 Qi Gong: Respiration exercices from a scientific view

Qiqong Definition: Taichi and Qigong are practices from traditional Chinese medicine that combine specific movements or postures, coordinated breathing, and mental focus (Bower and Irwin, 2016).

The benefits of Qigong on mood, anxiety, depression, psychological well-being, general stress management, exercise self-efficacy, and quality of life are reviewed by Wang et al. (2013). Other psychological interventions such as yoga and meditation are excluded; mixed interventions (e.g., acupuncture and Qigong in combination) are excluded. Fifteen studies published between 2001 and 2011 are studied. Duration of the interventions ranged from 70 min to 4 months. Interventions of 3–4 months' duration demonstrate changes while maximizing study enrollment and adherence.

The book "Chinese Medical Qigong Therapy Volume 1: Energetic Anatomy and Physiology" by Jerry.

Alan Johnson (2002), comprises modern and ancient knowledge about Chinese energetic healing techniques. Jerry Alan Johnson is a Master of Martial, Medical, and Qigong training. It explains the body's energetic fields, ancient metaphysical theories of tissue formation, and the evolution of ancient Chinese energetic medicine.

2.5 Traditional Chinese medicine

Traditional Chinese Medicine (TCM) is recognized as an effective complementary and alternative medicine modality by the National Institutes of Health in the United States (Jokiniemi, 2010). The National Center for Complementary and Alternative Medicine in the United States is the Federal Government's lead agency for complementary and alternative healing practices. World Health Organization (WHO) created strategies, guidelines, and standards of botanical medicines to support traditional medicine.

Companies such as Dell or IBM have an interest in TCM (Jokiniemi, 2010). Dell provides consulting services to Chinese hospitals. IBM helps Chinese hospitals standardize their patient records (CHAS: Clinical and Health Records Analytics and Sharing) to identify treatments that combine Western and Traditional Chinese Medicine. China Ministry of Science funded the Unified Traditional Chinese Medical Language System (UTCMLS) project.

2.5.1 Standards Development Organizations such as ITU and ISO working on TCM

Standards Development Organizations (SDO) such as ITU (International Telecommunications Union) and ISO (International Organization for Standardization) are working on Traditional Chinese Medicine. ISO/TC 215 Health informatics has a sub-group named ISO/TC 215/JWG 1 "Joint ISO/TC 215 - ISO/TC 249 WG: Traditional Chinese Medicine (Informatics)".

As an example, ISO/TC 215/JWG 1 "Joint ISO/TC 215 - ISO/TC 249 WG: Traditional Chinese Medicine (Informatics)" defined ISO TS 5346 Health Informatics—Categorial structure for representation of traditional chinese medicine clinical decision support system (TCMCDSS) which lacks a unified semantic information framework, which affects the data exchange and sharing among different institutions and databases. UMLS and HL7 are mentioned.

ISO/CD TS 25083 Health Informatics–Categorial structure of representation of Tuina standard realizes the sharing support of semantic information exchange by defining semantic framework categories and semantic relations. Other physical therapies are out of this scope in this standard.

ITU FGAI4H (Focus Group on Artificial Intelligence for Health) defined "DEL.10.23 TG-AI for Traditional Medicine" which reviews the existing AI solutions for Traditional Medicine. AI in traditional medicine supports healthcare practitioners in recommending integrative medical practices. Traditional Medicine treatments: 1) Ontological and Natural language processing (NLP) are encouraged to extract useful information, 2) Analyzing and integrating data, and 3) Decision support, 4) Predictive analytics, 5) Patient monitoring and feedback, 6) Virtual health assistants, and 7) Research and knowledge discovery.

2.5.2 Ontology for traditional Chinese medicine

Ontology for TCM (Jokiniemi (2010)) address general research questions such as: 1) What is Traditional Chinese Medicine (TCM)? 2) How is TCM used in China and worldwide today? TCM and society research questions: 3) What is the position of TCM in Chinese health care system today? 4) What kind of problems are TCM industry and research facing today? Technological research questions: 5) How can information technology be used in the field of TCM? 6) What are the problems and challenges in developing IT applications for TCM? Service and application development research questions: 7) Why knowledge gathering, knowledge discovery, and knowledge sharing can be challenging in the field of TCM? What kinds of solutions are available for these problems? 8) What is the role of ontology development in TCM? 9) What kind of applications and expert systems exist for TCM and what are their functions? 10) What are the most recent and future trends in TCM IT applications? The importance of five elements (wood, fire, earth, metal and water) is reminded in TCM and that Qi energy and blood, is controlled by the organs. Artificial intelligence such as inference engine could be used to combine single drugs into a new useful drug formula.

An ontology-based system for extracting knowledge of traditional Chinese drugs and formulas from semistructured text (Cao et al. (2004)).

Ontology of Meridian-collateral system (OMCAP) and its (Instances of Meridians, Collaterals, Acupoints and acu-mox Prescriptions) (IMCAP) (Cao and Sui, 2005). The OMCAP consists of 89 categories and 38 slots, and the IMCAP contains 1549 instance frames. It can be used in TCM-related applications such as automated diagnosis and therapy, medical instruction and training, and natural language processing. The ontology describes concepts such as Zang-Fu Organs, Five Elements, and Illnesses. Limitation: Is the ontology share online?

2.5.3 Acupuncture

Acupuncture Definition: Acupuncture is a technique in which several fine needles are inserted into specific points on the patient's body. The aim of acupuncture is to increase circulation and balance of Qi energy. Acupuncture can be used to treat many health problems, such as dysmenorrhea, back pain and headache, and it can also be used as anesthesia (Jokiniemi, 2010).

There are TCM databases such as Acupuncture.com.au (http://www.acupuncture.com.au/education/herbs/herbs.html), Dictionary of Chinese Herbs (http://alternativehealing.org/Chinese_herbs_dictionary.htm) (Jokiniemi, 2010). Acupuncture.com.au provides a list of TCM formulations grouped according to their actions. The Dictionary of Chinese Herbs describes the toxicity and side effects of TCM herbs, treatments and samples of TCM formulas for diseases such as diabetes, cancer, and dengue fever.

3. A study of sensors relevant for healthy martial arts

Most of the time, we cannot find the hardware sensor details used in the paper. We can find detailed information when datasets are available online and with a dedicated paper describing the dataset.

Ideally, we would use those sensors and attach them to the students to collect physiological data: 1) Heart Rate in Section 3.1, 2) Electrocardiogram (ECG) in Section 3.2, 3) Electroencephalography (EEG) in Section 3.3, 4) Blood Pressure in Section 3.4, 5) Body Thermometer in Section 3.5, 6) Electrooculography (EOG) in Section 3.6, 7) Electromyography (EMG) in Section 3.7, 8) Pulse Oximeter, SPO2, Respiration in Section 3.8, 9) Breathing, Respiration in Section 3.9, 10) Skin Conductance - Electrodermal activity (EDA) in Section 3.10, 11) Accelerometer in Section 3.11, 12) Pressure eensor in Section 3.12, 13) Gyroscope in Section 3.13, 14) Blood Glucometer in Section 3.14, 15) Camera, Infrared eensor in Section 3.15, and 16) other sensors to investigate in Section 3.16.

3.1 Heart rate, heart bate—ETSI TR 103 477 V1.3.1 (2023-01) eHEALTH standardization use cases for eHealth

Heart Rate Rules: heart rate thresholds are defined within the standard "ETSI TR 103 477 V1.3.1 (2023-01) eHEALTH Standardization use cases for eHealth": "Patient-X has a heart rate of 150 bpm". A heart rate of 150 bpm for a 50-year-old male riding a bike at 15 km/h up a 10% hill is probably not a health concern, but the same value when the same 50-year-old male is at rest (and has been at rest for 60 min) is very likely a serious health concern.

3.2 Electrocardiogram (ECG)—ISO/IEEE 11073 standards

Electrocardiogram (ECG) Definition: ECG signals represent the heart's electrical activity and can be used to investigate cardiac health (Faust and Ng, 2016). ECG is measured by placing electrodes on the chest, which record the electrical activity of the human heart (Faust et al., 2018). An ECG is a bioelectric signal which records the heart's electrical activity versus time (Lee and Chung, 2009).

Why do we need ECG in martial arts: To measure that there is not too much activity; enough to overcome ourselves, but before it is too dangerous. When we are practicing a Kata/Tao/form, we need to visualize the attack, to become more efficient, we need to use the reptilian brain (instant brain). Fear can generate more efficient movements. Fear can be measured with ECG.

ECG Ontology (Goncalves et al. (Guizzardi, 2009), Brazil), is designed for the cardiological exam ECG and its correlation to the human heart electrophysiology. The ECG ontology supports interactive learning in heart electrophysiology (Gonçalves et al., 2009) for the patient and the physician. The ECG ontology is based on (i) the OBO Relation Ontology, which provides basic relations to be used in biomedical ontologies); (ii) the Foundational Model of Anatomy (FMA), when dealing with the human anatomy for the ECG; (iii) the Ontology of Functions (OF) developed under the umbrella of the General Formal Ontology (GFO) research program to address heart electrophysiological functions. The BioPortal ontology catalog is not mentioned. The Competency Questions (CQs) defined are: CQ1: What essentially composes the ECG record? CQ2: How is the ECG record obtained? CQ3: What in the ECG waveform is object of the physician's analysis for interpreting a correlated heart behavior? CQ4: For all ECG elementary forms, which heart electrophysiological function(s) does (do) it map? CQ5: For all heart electrophysiological functions, which anatomical entity(ies) is (are) able to realize it? CQ6: For all heart electrophysiological functions, which requirements must be satisfied to enable its realization? CQ7: For all heart electrophysiological functions, which goals must be satisfied to accomplish its realization? CQ8: What conditions must be satisfied for the heart to play the role of a blood pump? CQ9: What does the P wave represent in the ECG waveform? CQ10: What does the QRS complex represent in the ECG waveform?

Example: Wearable home health monitoring system (Yao and Warren, 2005), is compliant with ISO/IEEE 11073 standards. Wearable sensors such as ECG, and pulse oximeter) and other sensors such as weight scale, and temperature sensor are used.

Example: Yuan et al. (2020)) experimented on 74275 segments of ECG signal in the training set and 13107 segments in the test set. They only include the signals labeled arrhythmia and normal and split the original signal into segments of length 256. Privacy-sensitive edge computing for healthcare machine learning applications is considered with a federated learning framework to train deep neural networks, partitioned and allocated to IoT devices and a centralized server.

A wearable ECG sensor (Nemati et al. (2012)) is designed which address convenience (direct contact with the skin should be avoided), portability (without disturbance of users' normal lives), and small size.

Deep learning methods for physiological signals focusing on Electromyogram (EMG), Electroencephalogram (EEG), Electrocardiogram (ECG), and Electrooculogram (EOG) to improve the quality of diagnosis are reviewed (Faust et al. (2018), Malaysia–Singapore). 53 research papers, published from January 01, 2008 to December 31, 2017, are analyzed and using VOSviewer http://www.vosviewer.com/. Fast et al. highlight the importance of more freely available datasets. CNN was used 32 times, which makes the most popular DL algorithm in those studies.

Cardiovascular diseases can be consequences of high blood pressure, insufficient physical exercise, poorly balanced diet, smoking, and abnormal glucose levels (Faust and Ng, 2016). Electroencephalogrphy (ECG) based ComputerAided Diagnosis (CAD) can support medical practitioners in monitoring cardiovascular diseases better. Various classification algorithms are references for ECG data. There is no consensus on which algorithms best classify ECG signals to identify cardiovascular diseases.

A wearable smart shirt for patient diagnosis include physiological ECG and three-axis accelerometer to measure physical (Lee and Chung, 2009). The shirt is non-invasive, comfortable, and convenient to wear. The performance test was achieved on a treadmill.

MIT-BIH arrhythmia database is used for heartbeat classification (Luo et al., 2017, China). Stacked denoising auto-encoder (SDA) is used as a Deep Learning method.

Available dataset using ECG: WESAD (Wearable Stress and Affect Detection) dataset.

3.3 Electroencephalography (EEG)

EEG Definition: The EEG is measured by placing electrodes on the person's skull, which observes the electrical activity of the brain. That activity results from firing neurons emitting action potentials (Faust et al., 2018).

Example: Electroencephalography (EEG) signal processing, from acquisition and pretreatment (denoising) to feature extraction, classification, and application is reviewed (Chaddad et al., 2023). Deep learning and machine learning algorithms for classification are explored.

Example: EEG is used to study mental practice (Lutz et al., 2004)

3.4 Blood pressure

This section describes blood pressure hardware devices, datasets, and thresholds to understand if the values are normal or not.

Blood pressure Thresholds: A normal Systolic Blood Pressure (SBP) is below 120 mmHg and a normal Diastolic Blood Pressure (DBP) is below 80 mmHg. Raised Blood Pressure (BP) throughout its range is the most significant cause of death and disability in the world (Lazazzera et al. (2019)). An easy-to-use home BP monitor is needed in many situations, especially for persons with stage 2 hypertension (140–90 mmHg), who need to keep track of their BP at home, which helps to determine if the treatment is working.

Example: CareUp R is a smartwatch developed by Farasha Labs to check blood pressure anywhere the user needs. CareUp has been evaluated with five control subjects at Farasha Labs and 44 patients in the Institut Coeur Paris Center Turin (ICPC) in Paris (Lazazzera et al., 2019). At ICPC, the measures of the CareUp have been compared to two commercial oscillometry-based devices: Thuasne R and Magnien R. The limitation of this device is the need for calibration which can be achieved at the medical office when the physician advises the patient to monitor their own blood pressure, but still not sufficient for hypertension patients (because heart rate change over time).

Available dataset with blood pressure: WESAD (Wearable Stress and Affect Detection) dataset[8] contains data from 15 subjects during a stress-affect lab study while wearing physiological and motion sensors (Schmidt et al., 2018). Raw sensor data was recorded with two devices: a chest-worn device (RespiBAN) and a wrist-worn device (Empatica E4). The RespiBAN device provides the following sensor data: electrocardiogram (ECG), electrodermal activity (EDA), electromyogram (EMG), respiration, body temperature, and three-axis acceleration. The Empatica E4 device provides the following sensor data: blood volume pulse (BVP, 64 Hz), electrodermal activity (EDA, 4 Hz), body temperature (4 Hz), and three-axis acceleration (32 Hz).

3.5 Body Thermometer—IEEE 11073-20601

Available Dataset: WESAD (Wearable Stress and Affect Detection) dataset[9] contains data from 15 subjects during a stress-affect lab study while wearing physiological and motion sensors (Schmidt et al., 2018). Raw sensor data was recorded with two devices: a chest-worn device (RespiBAN) and a wrist-worn device (Empatica E4). The RespiBAN device provides the following sensor data: electrocardiogram (ECG), electrodermal activity (EDA), electromyogram (EMG), respiration, body temperature, and three-axis acceleration. The Empatica E4 device provides the following sensor data: blood volume pulse (BVP, 64 Hz), electrodermal activity (EDA, 4 Hz), body temperature (4 Hz), and three-axis acceleration (32 Hz).

[8] https://ubicomp.eti.uni-siegen.de/home/datasets/icmi18/

[9] https://ubicomp.eti.uni-siegen.de/home/datasets/icmi18/

Example: Hypothermic use case is using body temperature sensor (Frohner et al., 2009) compliant with IEEE 11073-20601. Body temperature data compliant with 11073–20601 is transmitted to the analysis tool. Other data such as weight measurement, acceleration, and inclination measurement are considered. Out-of-hospital emergency care after cardiopulmonary resuscitation is to cool the body of the patient down to a certain temperature–induced hypothermia. BioSig open source software library is used.

3.6 Electrooculography (EOG)

EOG Definition: The EOG measures the corneo-retinal standing potential which occurs between the back and front of the human eye. Two electrodes are placed either left and right of the eye or above and below the eye. The signal enables the detection of eye movements (Faust et al., 2018).

Why do we need EOG in our martial art experiment: Electrooculography (EOG) can help to check that we do not look at the floor which means that we are still focusing on the technique, the memory to remember the movement which is not automatic yet, not memorized by the body yet. In Taichi, we want to check a kind of empty eye tracking. In Taichi and Kung-Fu it is important to have a non-focused eye track (peripheral vision) to be aware of the surrounding environment.

Example: EOG-based activity recognition (Bulling et al., 2009, 2010) used SVM classifier with MATLAB®. An experiment is done with eight participants who study in an office environment using five activity classes: copying a text, reading a printed paper, taking hand-written notes, watching a video, and browsing the web.

Example: SIAMO project provides an electrooculography (EOG)-based system for assisted mobility such as controlling a wheelchair for disabled people (Barea et al., 2002, Spain).

Example: Telepathix system (Kherlopian et al., 2006, USA) understands eye movement commands (looking to the left, looking to the right, and looking straight) to navigate a virtual keyboard (e.g., 6 letters per minute). Spelling trials with 13 different users are performed where the users spell six five-letter words using Left, Right, and Center eye movement commands. The words were zebra, noisy, igloo, input, punch, and pixel.

3.7 Electromyography (EMG)

EMG Definition: The EMG measures the electrical activity of skeletal muscles (Faust et al., 2018); the electrical sensors, known as electrodes, are placed on the skin above the muscles of interest. The EMG signals indicate muscle activation, force produced by the muscle, and muscle state.

Surface electromyography (sEMG) is a non-invasive technique for measuring the electrical signal generated by skeletal muscles and detected on the skin useful in neurophysiological and medical research, rehabilitation, ergonomics, sports and movement science, control of robots and machinery (Deak et al., 2016).

Available data using EMG: WESAD (Wearable Stress and Affect Detection) dataset.

Peter Wayne designed a database (https://physionet.org/content/taichidb/1.0.2/) of 60 beginners practicing Tai Chi, aged 50-79, and another of 27 people with at least 5 years of experience in Taichi. Gait was measured with EMG to understand changes in mobility associated with Taichi training.

Example: The method is based on a fuzzy logic algorithm that has two input variables: the muscle activation area and the mean intensity of muscle activity (Deak et al., 2016).

3.8 Pulse oximeter, SPO2, respiration—X73—ISO/IEEE 11073

Available Dataset with respiration: WESAD (Wearable Stress and Affect Detection) dataset.

Example: Cloud-Based X73 Ubiquitous Mobile Healthcare System (Ji et al., 2014) detects potential medical threats or diseases and sends analysis information to doctors, and community service staff. It is compliant with ISO/IEEE 11073 (X73) to deal with heterogeneous medical sensors such as pulse oximeter, heart rate monitor/pulse, blood pressure monitor, thermometer, weighing scale, and glucose meter. Hardware devices used are not provided. Reasoning such as the DROOLS Java rule engine platform is employed; but rules examples are not provided.

Other technologies are the Hive system for data mining, the distributed framework Apache Hadoop with an efficient Hadoop Distributed File System (HDFS) for data storing, and a Hadoop MapReduce for "big data" processing. Facebook's Hive, acts as a data warehouse and provides data summarization, query, and analysis in the Hadoop distributed environment. Kafka and FlumeNG (to guarantee reliability and fault tolerance of the whole system) are also used. Java Agent Development Framework (JADE), a FIPA open-source software framework; and a Tomcat server are also used.

3.9 Breathing, respiration

Why do we need a breathing/respiration sensor? Breath is really important in martial arts (e.g., Kung-Fu, Qi Gong, Taichi) and other sports (Li et al, 2023).

Example: Wrinkles, blushing, sweating, tearing, and respiration can be detected in emotions (de Melo et al., 2010). The study focuses on 14 simulated affective states having respiratory patterns. Affective states are excitement, relaxation, focus, pain, relief, boredom, anger, fear, panic, disgust, surprise, startle, sadness, and joy. Forty-four participants were recruited. A table is provided about relationships between different kind of emotions and the respiration pattern. The main limitation of this research is to work on simulated affective states.

3.10 Skin conductance—Electrodermal activity EDA

Available dataset using Electrodermal Activity (EDA): WESAD (Wearable Stress and Affect Detection) dataset.

3.11 Accelerometer

Why do we need accelerometer? Wearing an accelerometer helps to measure the speed of movement by athletes when practicing.

Available dataset with Accelerometer: WESAD (Wearable Stress and Affect Detection).

Example: 1 accelerometer and two gyroscope sensors (Nardi, 2019) track movements of the wrist, arm, and shoulder in the user's shoulder using Deep Learning (Deep Convolutional Neural Network) to recognize 5 strength exercises (bench press, bent row, deadlift, lateral rises and overhead press), but also daily activities (sitting or walking). Future work is getting a larger and more dynamic dataset with more users and recognizing more exercises.

Example: Two deep learning architectures are provided to predict the physical activity from inertial data acquired from a smartphone (accelerometer) during the 6-minute walking test (6MWT) (Galán-Mercant et al., 2019): 1) Convolutional Autoencoder (CAE) to extract features from the acceleration data. These features are then classified using a support vector machine; 2) Convolutional neural network (CNN) that firstly extracts features.

3.12 Pressure sensor (e.g., on mat)

Why do we need a Pressure Sensor: A pressure sensor can help to detect martial arts postures (e.g., Mabu stance) (Ishac et al., 2023), provide feedback on form, and detect when the user performs the technique incorrectly.

Example: LifeMat is a IoT pressure-sensitive training mat (Ishac et al., 2023) assists in martial arts training. LifeMat detects martial arts postures (e.g., Mabu stance), provides feedback on form, and detects when the user performs the technique incorrectly. LifeMat detects imbalances and foot position. Serious Game named "Kung Future" assists with remote martial arts training when a teacher is not present. Evaluation from users highlights the need for more feedback from the virtual coach.

3.13 Gyroscope

Example: 1 accelerometer and two gyroscope sensors (Nardi, 2019) track movements of the wrist, arm, and shoulder in the user's shoulder using Deep Learning (a Deep Convolutional Neural Network) to recognize 5 strength exercises (bench press, bent row, deadlift, lateral rises and overhead press), but also daily activities (sitting or walking).

Future work is getting a larger and more dynamic dataset with more users and recognizing more exercises.

3.14 Blood glucometer—Glucose—Glucometer ISO definition, ISO 11073 standard, glucometer rules

A blood glucometer can be used to avoid hypoglycemia when too many sports are practiced.

Glucometer ISO Definition: Continuous Glucose Monitor (CGM) is a device that estimates the concentration of glucose in the blood typically measured from interstitial fluid (ISF) (ISO/IEEE 11073–10425:2024 Personal Health Device Communication—Device Specialization- Continuous Glucose Monitor (CGM) (ISO, 2024)). Glucose is a measurement of glucose concentration in the blood. Typically for CGM, this measurement is made from other body fluids than blood, and thus calibration is required to calculate the blood glucose levels.

ISO/IEEE FDIS 11073-10425 Personal Health Device Communication—Device Specialization- Continuous Glucose Monitor (CGM) enables plug-and-play interoperability. It specifies term codes, formats, and behaviors of glucose level measurement in the body (every 5 min) through a sensor continuously attached to the person.

Glucometer Rules: In a healthy adult male of 75 kg with a blood volume of 5 L, a blood glucose level of 100 mg/dL (5.5 mmol/L) corresponds to a total of approximately 5 g (1/5 oz and equivalent to a commercial sugar packet) of glucose in the blood and approximately 45 g (1.5 oz) in the total body fluid (which includes blood and ISF). Glucose levels rise after meals and are usually lowest in the morning, before the first meal of the day ([)ISO/IEEE FDIS 11073-10425 (ISO, 2024).

Example: uHealth system (Nam et al., 2011), based on ISO/IEEE 11073 PHD standards, enables structuring data exchanged between medical devices (e.g., glucose meter, weight scale, blood pressure monitor, thermometer) and other devices such as smartphones. It is important for aging-related issues such as increased medical expenses, and a lack of medical specialists. IEEE 11073–10417 glucose meter standardization is used (IEEE Std 11073–10417, Health Informatics- Personal Health Device Communication-Device specialization–Glucose meter, May 2009). An example of data exchanged code compliant with the ISO standard is missing.

3.15 Camera, infrared sensor

Example: Leap Motion and Kinect are used for physical rehabilitation (Postolache, 2017). Leap Motion measures muscle activity using thermography. The skin temperature is measured before and after physical rehabilitation task using a thermographic camera FLIR E60.

3.16 Other sensors to be investigated

Other sensors that need to be investigated but not detailed here:

- Weight scale helps to measure weight lost if the goal is to lose weight. Another point to consider is muscle developed which can increase weight.
- Magnetic Resonance Imaging (MRI) is used for instance to measure the impact of meditation (See meditation explanations in Section 2.3).
- Microphone can be used to measure when we shout (e.g., the impact of the Qi in Kung-Fu, the Kiai in Karaté, etc.)
- Positron Emission Tomography (PET) similar to MRI, can be used for instance to measure the impact of meditation (See meditation described in Section 2.3).

4. Envisioned Experiments

Experiments could be on four kinds of class students: 1) Meditation Experiment, 2) Body-Mind Experiment, 3) Taichi Experiment, and 4) Kung-Fu Experiment. This choice is because we have every year intensive training over 1 week with such classes.

To design the cohort: Data about students would be needed: 1) Anonymous ID, 2) Age, 3) Sex: Female/Male, 4) Race/Origin: French, Other, 5) Years of practice in Meditation, 6) Years of practice in Body/Mind, 7) Years of practice in Taichi, 8) Years of practice in Kung-Fu, 9) Belt level, and 10) Diploma (Year and name of Diploma).

Time: 1 week, over a month, over 2 months, over 6 months, over a year.

Spatial: Eqimonie gymnasium, during 1 week of training (in winter indoors in Antibes/France, in summer: outdoors surrounded by nature), We envision the following experiments as examples:

- Taichi experiment and database available online designed by Peter Wayne from Harvard (described in Section 2.2, and Section 4.1).
- One week morning training in summer (July or August): to measure data on students, ideally, the day before the beginning of the 1-week training, and every day. The arrival day (1h body-mind for travel recovery and training introduction). Every day over 5 days: 30 min to 1 h meditation, 1h body-mind, 1h taichi, and 1h kungfu. The departure day (1h body-mind for saying goodbye and quick revision).
- One week morning and afternoon training in winter (first week of January): Schedule over 5 days: 1h Meditation, 1h body-mind, 1h taichi, and 1h kungfu. Almost similar in afternoon but reverse schedule (1h Kung-Fu, 1h taichi, and 1h body-mind). Within the body-mind class, the last 20 min are dedicated to relaxation for a nice recovery and sleep.
- Experiment: Sleeping quality. Sensors such as respiration, etc.
- Long-term experiment from 1 month to years: 1 month, and then 2 months, and then 12-week (3 months) intervention with students during the class over the

year with the more regular students; then 6 months, then one academic year. Then over the years, for remaining students getting more advanced levels.

- Other reusable datasets are mentioned in Section 4.1 described hereafter.

Specific short exercises to focus on, as examples: 1) Exercise 1: breathing Tao/Kata/ form of eight movements to be measured with 1 breathing sensor. 2) Exercise 2: to measure heart rate, 3) Exercise 3: check the peripherical vision with sensors such as EOG in Taichi and/or Kung-Fu, and/or Body-Mind, especially with ball exercises. 4) Exercise 4: balance ability, to check the central gravity point of the body, to get a good balance; relevant in Taichi and Kung-Fu. 5) Exercise 5: muscle strength (e.g., with EMG), and 6) Exercise 6: muscle endurance (e.g., with EMG).

4.1 Useful existing datasets

Key datasets with physiological data such as respiration, blood pressure, body temperature, weight, glucose are summarized in Table 6.1.

Table 6.1 Comparing sensors used within four datasets (Taichi dataset from Harvard, DEAP emotion dataset, heart rate dataset, and WESAD stress dataset).

Sensor type	Taichi dataset Hardvard -Wayne	DEAP dataset emotion - Koelstra et al. (2012)	Heart rate dataset Ahmed (2017)	Stress WESAD dataset Schmidt et al. (2018)
Pulse oximeter SPO_2 = pulse oximeter?	NO	YES respiration pattern	NO	YES RespiBAN -> respiration
Blood pressure monitor	NO	YES blood volume pressure	YES resting blood pressure	YES Empatica E4-> blood volume pressure (BVP, 64 Hz)
Body thermometer	NO	YES skin temperature	NO	YES RespiBAN-> body temperature YES Empatica E4-> body temperature (4 Hz)
Weighting scale	NO	NO	NO	NO
Glucose meter	NO	NO	YES fasting blood sugar	NO

Taichi experiment and database are released by Peter Wayne from Harvard (https://physionet. org/content/taichidb/1.0.2/): 60 beginners practicing Tai Chi, aged 50–79 years old, and another of 27 people with at least 5 years of experience in Taichi (as explained within Section 2.2). Gait was measured with EMG to understand changes in mobility associated with Taichi training.

Stress dataset: The WESAD (Wearable Stress and Affect Detection) dataset[10] contains data from 15 subjects during a stress-affect lab study while wearing physiological and motion sensors (Schmidt et al., 2018). Raw sensor data was recorded with two devices: a chest-worn device (RespiBAN) and a wrist-worn device (Empatica E4). The RespiBAN device provides the following sensor data: electrocardiogram (ECG), electrodermal activity (EDA), electromyogram (EMG), respiration, body temperature, and three-axis acceleration. The Empatica E4 device provides the following sensor data: blood volume pulse (BVP, 64 Hz), electrodermal activity (EDA, 4 Hz), body temperature (4 Hz), and three-axis acceleration (32 Hz). Three different affective states (neutral, stress, amusement) are considered.

The DEAP dataset[11] comprises 32 patients, and is designed for emotion analysis using physiological signals such as GSR, blood volume pressure, respiration pattern, skin temperature, EMG, EOG, and EEG (Koelstra et al., 2017).

The heart rate dataset[12] comprises 270 patients' data (age, sex, chest pain type, resting blood pressure, serum cholesterol, fasting blood sugar, resting electrocardiographic results, maximum heart rate achieved, etc.) (Ahmed, 2017).

There is also the *ECG dataset*[13] (Ganz, 2014).

5. Conclusion and future work

From a scientific view, martial arts using AI or sensor technologies is reviewed; including Taichi, Meditation, Qi Gong, and Traditional Chinese Medicine (TCM). A study of sensors relevant to healthy martial arts is presented. Envisioned experiments are introduced.

Other benefits of this long-term research: the teacher does not need to repeat the same things all the time, an IoT-based virtual teacher could be designed.

The long-term goal is to set up real experiments with the trainees during training (1-week intensive training over 1 week or during class over the year).

[10] https://ubicomp.eti.uni-siegen.de/home/datasets/icmi18/

[11] http://www.eecs.qmul.ac.uk/mmv/datasets/deap/

[12] https://archive.ics.uci.edu/ml/datasets/Statlog+(Heart)

[13] http://www.cs.ucr.edu/~eamonn/time_series_data/

References

Ahmed, F., 2017. An internet of things (iot) application for predicting the quantity of future heart attack patients. Int. J. Comput. Appl. 164 (6), 36–40.

Barea, R., Boquete, L., Mazo, M., López, E., 2002. System for assisted mobility using eye movements based on electrooculography. IEEE Trans. Neural Syst. Rehabil. Eng. 10 (4), 209–218.

Blumenthal, J.A., Sherwood, A., Babyak, M.A., Watkins, L.L., Smith, P.J., Hoffman, B.M., O'Hayer, C.V. F., Mabe, S., Johnson, J., Doraiswamy, P.M., et al., 2012. Exercise and pharmacological treatment of depressive symptoms in patients with coronary heart disease: results from the upbeat (understanding the prognostic benefits of exercise and antidepressant therapy) study. J. Am. Coll. Cardiol. 60 (12), 1053–1063.

Bower, J.E., Irwin, M.R., 2016. Mind–body therapies and control of inflammatory biology: a descriptive review. Brain Behav. Immun. 51, 1–11.

Bulling, A., Ward, J.A., Gellersen, H., Tröster, G., 2009. Eye movement analysis for activity recognition. In: Proceedings of the 11th International Conference on Ubiquitous Computing, pp. 41–50.

Bulling, A., Ward, J.A., Gellersen, H., Tröster, G., 2010. Eye movement analysis for activity recognition using electrooculography. IEEE Trans. Pattern Anal. Mach. Intell. 33 (4), 741–753.

Burke, A., Wong, Y.-Y., Clayson, Z., 2003. Traditional medicine in China today: implications for indigenous health systems in a modern world. Am. J. Publ. Health 93 (7), 1082–1084.

Cao, C., Sui, Y., 2005. Building an ontology and knowledge base of the human meridian-collateral system. In: International Conference on Innovative Techniques and Applications of Artificial Intelligence. Springer.

Cao, C., Wang, H., Sui, Y., 2004. Knowledge modeling and acquisition of traditional Chinese herbal drugs and formulae from text. Artif. Intell. Med. 31 (1), 3–13.

Casas-Ortiz, A., Santos, O.C., 2021. KSAS: An AI Application to Learn Martial Arts Movements in On-Line Settings, Interactive Events at AIED21 (Artificial Intelligence in Education), pp. 1–4.

Chaddad, A., Wu, Y., Kateb, R., Bouridane, A., 2023. Electroencephalography signal processing: a comprehensive review and analysis of methods and techniques. Sensors 23 (14), 6434.

Davidson, R.J., Lutz, A., 2008. Buddha's brain: neuroplasticity and meditation (in the spotlight). IEEE Signal Process. Mag. 25 (1), 176-174.

Davidson, R.J., Kabat-Zinn, J., Schumacher, J., Rosenkranz, M., Muller, D., Santorelli, S.F., Urbanowski, F., Harrington, A., Bonus, K., Sheridan, J.F., 2003. Alterations in brain and immune function produced by mindfulness meditation. Psychosom. Med. 65 (4), 564–570.

de Melo, C.M., Kenny, P., Gratch, J., 2010. Influence of autonomic signals on perception of emotions in embodied agents. Appl. Artif. Intell. 24 (6), 494–509.

Deak, G.-F., Miron, R., Avram, C.C., Astilean, A., 2016. Fuzzy based analysis method of high-density surface electromyography maps for physical training assessment. In: 2016 IEEE International Conference on Automation, Quality and Testing, Robotics (AQTR). IEEE.

Desbordes, G., Negi, L.T., Pace, T.W., Wallace, B.A., Raison, C.L., Schwartz, E.L., 2012. Effects of mindful-attention and compassion meditation training on amygdala response to emotional stimuli in an ordinary, non-meditative state. Front. Hum. Neurosci. 6, 23050.

Di Corrado, D., Quartiroli, A., Coco, M., 2021. Psychological and motor associations in sports performance: a mental approach to sports. Front. Psychol. 12, 629944.

Farrokhi, A., Farahbakhsh, R., Rezazadeh, J., Minerva, R., 2021. Application of Internet of Things and artificial intelligence for smart fitness: a survey. Comput. Netw. 189, 107859.

Faust, O., Ng, E.Y., 2016. Computer aided diagnosis for cardiovascular diseases based on ecg signals: a survey. J. Mech. Med. Biol. 16 (01), 1640001.

Faust, O., Hagiwara, Y., Hong, T.J., Lih, O.S., Acharya, U.R., 2018. Deep learning for healthcare applications based on physiological signals: a review. Comput. Methods Progr. Biomed. 161, 1–13.

Frohner, M., Urbauer, P., Bauer, M., Gerbovics, F., Mense, A., Sauermann, S., 2009. Design and realisation of a framework for device end communication according to the IEEE 11073–20601 standard. In: Proceedings Tagungsband der eHealth, pp. 135–139.

Galán-Mercant, A., Ortiz, A., Herrera-Viedma, E., Tomas, M.T., Fernandes, B., Moral-Munoz, J.A., 2019. Assessing physical activity and functional fitness level using convolutional neural networks. Knowl. Base Syst. 185, 104939.

Ganz, F., 2014. Intelligent Communication and Information Processing for Cyber-Physical Data. University of Surrey. Ph.D. thesis.

Ghazi, M.A., 2022. The Effectiveness of Artificial Intelligence and Strategic Planning in Building Mental Modeling to Improve Sports Performance in Karate.

Gonçalves, B., et al., 2009. An Ontology-Based Application in Heart Electrophysiology: Representation, Reasoning and Visualization on the Web, pp. 816–820.

Guizzardi, G., 2009. An ontological analysis of the electrocardiogram. RECIIS 3 (1), 45–59.

Hämäläinen, P., Ilmonen, T., Höysniemi, J., Lindholm, M., Nykänen, A., 2005. Martial arts in artificial reality. In: Proceedings of the SIGCHI Conference on Human Factors in Computing Systems.

Ishac, K., Bourahmoune, K., Carmichael, M., 2023. An iot sensing platform and serious game for remote martial arts training. Sensors 23 (17), 7565.

ISO, 2024. ISO/IEEE 11073-10425:2024 Personal Health Device Communication — Device Specialization- Continuous Glucose Monitor (CGM), Standard. International Organization for Standardization.

Ji, Z., Ganchev, I., O'Droma, M., Zhang, X., Zhang, X., et al., 2014. A cloud-based x73 ubiquitous mobile healthcare system: design and implementation. Sci. World J. 1, 145803.

Johnson, J.A., 2002. Chinese Medical Qigong Therapy: Vol. 1 Energetic Anatomy and Physiology.

Jokiniemi, J., 2010. Ontologies and Computational Methods for Traditional Chinese Medicine. School of Science & Technology. MS thesis.

Kabat-Zinn, J., 2013. Full Catastrophe Living, Revised Edition: How to Cope with Stress, Pain and Illness Using Mindfulness Meditation. Hachette uK.

Kabat-Zinn, J., 2016. Au cœur de la tourmente, la pleine conscience. In: De Boeck Supérieur.

Kabat-Zinn, J., Davidson, R., Houshmand, Z., 2014. L'esprit est son propre médecin: Le pouvoir de guérison de la méditation. Guy Saint-Jean Editeur.

Kee, Y.H., 2019. Looking east for mindfulness: a glimpse of practices and research on shaolin martial arts and related practices to advance sport psychology. Psych 1 (1), 76–91.

Kherlopian, A.R., Gerrein, J.P., Yue, M., Kim, K.E., Kim, J.W., Sukumaran, M., Sajda, P., 2006. Electrooculogram based system for computer control using a multiple feature classification model. In: 2006 International Conference of the IEEE Engineering in Medicine and Biology Society. IEEE, pp. 1295–1298.

Koelstra, S., Muhl, C., Soleymani, M., Lee, J.-S., Yazdani, A., et al., 2017. DEAP: a database for emotion analysis; using physiological signals. IEEE Trans. Affect. Comput. 4 (Impact Factor), 585.

Koelstra, S., Muhl, C., Soleymani, M., Lee, J.-S., Yazdani, A., Ebrahimi, T., 2012. DEAP: A database for emotion analysis; using physiological signals. IEEE Trans. Affect. Comput. 3 (1), 18–31. https://doi.org/10.1109/T-AFFC.2011.15.

Kong, J., Wilson, G., Park, J., Pereira, K., Walpole, C., Yeung, A., 2019. Treating depression with tai chi: state of the art and future perspectives. Front. Psychiatr. 10, 237.

Langevin, H. Nccih Strategic Plan FY 2021–2025: Mapping a Pathway to Research on Whole Person Health, National Center for Complementary and Integrative Health website.

Lazazzera, R., Belhaj, Y., Carrault, G., 2019. A new wearable device for blood pressure estimation using photoplethysmogram. Sensors 19 (11), 2557.

Lee, Y.-D., Chung, W.-Y., 2009. Wireless sensor network based wearable smart shirt for ubiquitous health and activity monitoring. Sensor. Actuator. B Chem. 140 (2), 390–395.

Li, S., Huo, S., Ke, W., 2021a. Intelligent decision-making system for martial arts competition using deep learning. Mob. Inf. Syst. 2021 (1), 9920751.

Li, A., Wu, H., Liu, Y., 2021b. Prematch emotions and coping styles of martial arts athletes based on artificial intelligence. Mob. Inf. Syst. 2021 (1), 3497581.

Li, C., Zhao, M., Zhao, X., 2023. Status quo and influence of martial arts fitness in pursuit of health using lstm recurrent neural network algorithm. Int. J. Comput. Intell. Syst. 16, 61.

Luo, K., Li, J., Wang, Z., Cuschieri, A., 2017. Patient-specific deep architectural model for ecg classification. Journal of healthcare engineering 2017 (1), 4108720.

Lutz, A., Greischar, L.L., Rawlings, N.B., Ricard, M., Davidson, R.J., 2004. Long-term meditators self-induce high-amplitude gamma synchrony during mental practice. In: Proceedings of the National Academy of Sciences, vol 101, pp. 16369–16373.

Lutz, A., Slagter, H.A., Dunne, J.D., Davidson, R.J., 2008a. Attention regulation and monitoring in meditation. Trends Cognit. Sci. 12 (4), 163–169.

Lutz, A., Brefczynski-Lewis, J., Johnstone, T., Davidson, R.J., 2008b. Regulation of the neural circuitry of emotion by compassion meditation: effects of meditative expertise. PLoS One 3 (3), e1897.

Miller, S., Taylor-Piliae, R., 2021. Perceptions of the eight active ingredients of tai chi exercise among older adults. J. Int. Soc. Chinese Health Pract. 21 (1).

Nam, J.-C., Seo, W.-K., Bae, J.-S., Cho, Y.-Z., 2011. Design and development of a u-health system based on the iso/ieee 11073 phd standards. In: The 17th Asia Pacific Conference on Communications. IEEE, pp. 789–793.

Nardi, P., 2019. Human Activity Recognition: Deep Learning Techniques for an Upper Body Exercise Classification System.

Nemati, E., Deen, M.J., Mondal, T., 2012. A wireless wearable ecg sensor for long-term applications. IEEE Commun. Mag. 50 (1), 36–43.

Postolache, O., 2017. Remote sensing technologies for physiotherapy assessment. In: 2017 10th International Symposium on Advanced Topics in Electrical Engineering (ATEE). IEEE.

Ramanathan, V., Meyyappan, T., Natarajan, G., 2024. IoT and AI for Silambam martial arts: a review. EAI Endorsed Trans. Internet of Things 10.

Ray, P.P., 2015. Internet of things for sports (iotsport): an architectural framework for sports and recreational activity. In: Proceeding of IEEE EESCO, Vizag, pp. 79–83.

Sahdra, B.K., MacLean, K.A., Ferrer, E., Shaver, P.R., Rosenberg, E.L., Jacobs, T.L., Zanesco, A.P., King, B.G., Aichele, S.R., Bridwell, D.A., et al., 2011. Enhanced response inhibition during intensive meditation training predicts improvements in self-reported adaptive socioemotional functioning. Emotion 11 (2), 299.

Schmidt, P., Reiss, A., Duerichen, R., Marberger, C., Van Laerhoven, K., 2018. Introducing wesad, a multimodal dataset for wearable stress and affect detection. In: Proceedings of the 20th ACM International Conference on Multimodal Interaction.

Seligman, M.E., Csikszentmihalyi, M., 2000. Positive Psychology: An Introduction. American Psychological Association.

Tao, J., Liu, J., Liu, W., Huang, J., Xue, X., Chen, X., Wu, J., Zheng, G., Chen, B., Li, M., et al., 2017. Tai Chi Chuan and Baduanjin increase grey matter volume in older adults: a brain imaging study. J. Alzheimers Dis. 60 (2), 389–400.

Tsang, T.W.M., Kohn, M., Chow, C.M., Singh, M.F., 2008. Health benefits of Kung Fu: a systematic review. J. Sports Sci. 26 (12), 1249–1267.

Vertonghen, J., Theeboom, M., 2010. The social-psychological outcomes of martial arts practise among youth: a review. J. Sports Sci. Med. 9 (4), 528.

Wadey, R., Clark, S., Podlog, L., McCullough, D., 2013. Coaches' perceptions of athletes' stress-related growth following sport injury. Psychol. Sport Exerc. 14 (2), 125–135.

Wallace, B.A., Shapiro, S.L., 2006. Mental balance and well-being: building bridges between buddhism and western psychology. Am. Psychol. 61 (7), 690–701.

Wang, F., Man, J.K., Lee, E.-K.O., Wu, T., Benson, H., Fricchione, G.L., Wang, W., Yeung, A., 2013. The effects of qigong on anxiety, depression, and psychological well-being: a systematic review and meta-analysis. Evid. Based Complement. Alternat. Med. 2013 (1), 152738.

Wayne, P., 2018. Tai-chi, la méditation en mouvement. une approche médicale pour un art millénaire.

Yao, J., Warren, S., 2005. Applying the ISO/IEEE 11073 standards to wearable home health monitoring systems. J. Clin. Monit. Comput. 19, 427–436.

Ye, W., Li, S., Liu, S., Zhou, Y., et al., 2022. Application of artificial intelligence technology in martial arts education governance. Discrete Dynam Nat. Soc. 2022 (1), 5606280.

Yuan, B., Ge, S., Xing, W., 2020. A Federated Learning Framework for Healthcare Iot Devices arXiv preprint arXiv:2005.05083.

Zhao, L., Zhao, Y., Patil, S., Davies, D., Wang, C., Lu, L., Ouyang, B., 2018. Robot composite learning and the nunchaku flipping challenge. In: 2018 IEEE International Conference on Robotics and Automation (ICRA). IEEE.

CHAPTER 7

Medical cyber physical systems (MCPS) for patient monitoring and treatment

A. Saranya[1], Rajiv Iyer[1] and Vedprakash Maralapalle[2]
[1]Department of CSE, Amity School of Engineering and Technology, Amity University Maharashtra, Mumbai, Maharashtra, India; [2]L. S. Raheja School of Architecture, School of Architecture, Mumbai, Maharashtra, India

1. Introduction

Cyber-physical systems (CPS) are a relatively new approach that combines computational and communication processes with mechanical actions (Nayak et al., 2024). In the field of healthcare, medical cyber-physical systems (MCPS) are transforming patient management by allowing monitoring and real-time treatment and management strategies.

1.1 MCPS leverages various technologies to provide enhanced healthcare services

- **Networked Medical Devices:** MCPS employs an inter-connected system of medical equipment inclusive of actuators, sensors, and controllers to facilitate data gathering and sharing.
- **Real-Time Monitoring:** Many sensors are used to track the patient's vital signs, physiological data, and conditions of the environment, which supply immediate information for diagnosis and treatment (Möller and Möller, 2016).
- **Closed-Loop Control:** MCPS uses both closed-loop feedback to ensure constant changes to the treatment based on the patient information to prevent adverse event occurrences.
- **Interoperability:** Integrating coordination and knowledge among different devices, systems, and medical personnel, MCPS ensures the patient flow is smooth (Sharma et al., 2022; Ali et al., 2018).
- **Data Analytics:** MCPS utilizes statistical analysis to extract important knowledge from vast amounts of patient data to include individual patient treatment and population health.

1.2 MCPS has a wide range of applications in healthcare, including

- **Chronic Disease Management:** MCPS can assist in managing other chronic diseases such as diabetes, hypertension, and heart diseases through regular check-ups to

Intelligent Semantic Analysis for Healthcare
ISBN 978-0-443-33512-9,
https://doi.org/10.1016/B978-0-443-33512-9.00014-7

determine the patient's health status and the necessary treatment regime (Cabello et al., 2020; Jimenez et al., 2019).

- **Remote Patient Monitoring:** MCPS helps doctors to monitor patients, getting rid of the need for regular visits to the hospital and enhancing the use of healthcare services, including among the poor.
- **Assisted Living:** MCPS can also help elderly patients or people with disabilities to stay as independent as possible by observing their activities, recognizing falls, and offering the necessary help, if necessary (Huertas Celdrán et al., 2018; Puliafito et al., 2021).
- **Surgical Robotics:** MCPS can improve the accuracy, detail, and repeatability of surgical operations through technology integration of robots, image, and navigation systems.
- **Emergency Response:** MCPS can also assist in enhancing emergency response to a situation where the status of patients is relayed to first responders to warrant the right action and attention.

1.3 MCPS challenges

- **Interoperability and Integration:** One of the main challenges the system has to address is easy upgrade and extension of the solution, as well as harmony between various types of devices, systems, and carers (Krishna et al., 2019; Kaur et al., 2020).
- **Data Privacy and Security:** Personal identifiers that MCPS works with are highly sensitive due to the high level of privacy and data security. The following measures should ensure patients' information security and compliance with legal requirements: auditing, data encryption, and access control.
- **Clinical Validation:** MCPS has to be rigorously clinically trialed to be proven safe, effective, and reliable tools. This has a dire need for the development of new testing strategies and the involvement of healthcare practitioners in testing procedures (Törngren et al., 2017; Sajid et al., 2023).
- **Ethical Considerations:** Ask questions about ethical acceptability from the main duties of MCPS to patient self-determination, consent, and data abuse or misconduct. For these reasons, there is a need for policymakers and healthcare organizations to come up with ethical principles that could guide the handling of such matters.
- **Workforce Readiness:** As a consequence of MCPS, transformations occur regarding the roles of healthcare workers and the introduction of skills and knowledge about data processing and patient integration.

Particularly, there should be training solutions and educational courses to ensure the integration of the workforce into the digital era of health care. Through MCPS, what is possible is the delivery of accurate, preventive, and patient autonomy in healthcare. With computation, networking, and physical processes MCPS can enhance the quality of health care for patients, daily costs of health care, and overall living standards for communities and individuals. However, achieving this potential will require addressing the concerns, which include interoperability, data privacy, therapeutic value, standards, and staffing. With the continuing evolution and expansion of MCPS, it is confronted with the challenge of defining the direction of the progression of healthcare.

2. Applications of CPS in patient monitoring

Cyber-physical systems (CPS) can potentially improve patient monitoring in healthcare with the opportunity of real-time, continuous, and individualized care. CPS contains networked instruments which include actuators, sensors, and control systems for data capture and sharing; it also has improved monitoring tools (Maleh et al., 2022).

2.1 Remote patient monitoring

CPS's main use case in patient monitoring is Remote Patient Monitoring (RPM). In Fig. 7.1 CPS makes it easier for medical personnel to monitor patients from a distance and hence, the need for frequent visits to the hospital increases the chances of care especially to the needy in the poor regions. RPM systems are often in the form of wearable or implantable devices that provide a steady stream of data about patients' physiological and environmental status. The information is then relayed to a central monitoring system and it is used to detect any variations or changes in the health status of a patient

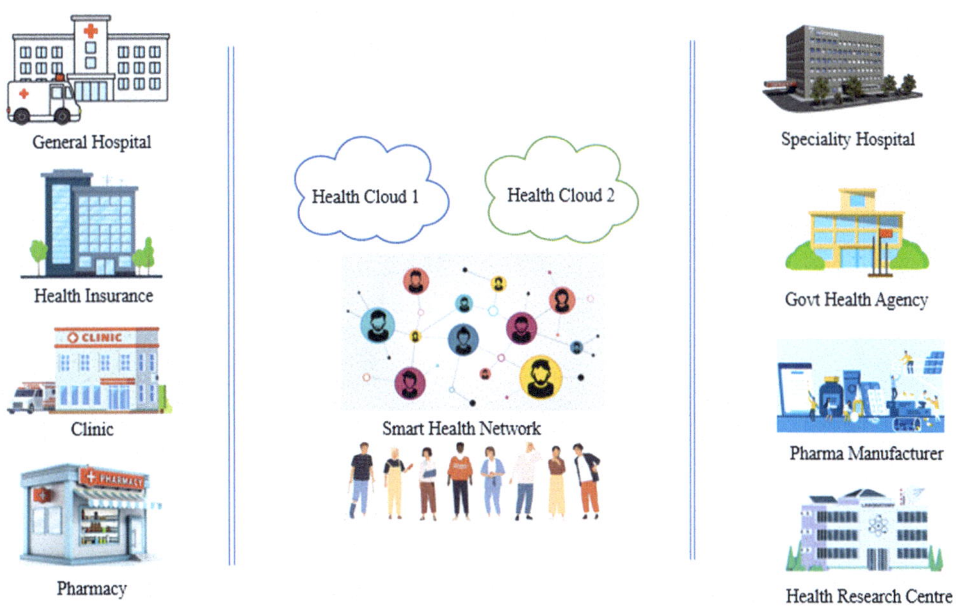

Figure 7.1 Smart healthcare network of cyber-physical systems.

(Tripathy et al., 2024; Oks, 2024). RPM systems can be especially advantageous for those individuals who experience chronic conditions such as COPD, heart failure, or diabetes. By monitoring patients' well-being over time, clinicians can recognize signs of decline and intervene, which could decrease hospitalizations.

2.2 Assisted living

CPS can also promote elderly or disabled patients' self-sufficiency by overseeing their routines, recognizing falls, and offering help when necessary. It commonly consists of a set of sensors in the patient's home including motion sensors, pressure sensors, and door sensors that constantly observe the movements of the patient. This means that if the system records any irregularity like a fall or prolonged inactivity, then the caregivers or the emergency team will be notified immediately. In Fig. 7.2 the assisted living systems can also include smart home technologies like voice-activated assistants or smart appliances that can assist the patient to perform daily activities and hence improving their quality of living. These technologies can also be incorporated into CPS, where medical personnel can obtain a certain amount of information about the patient's health condition and state (Biswas et al., 2023).

2.3 Closed-loop control

Another measure that CPS can use is closed-loop control systems for the automatic regulation of treatment according to the information from patients and to minimize

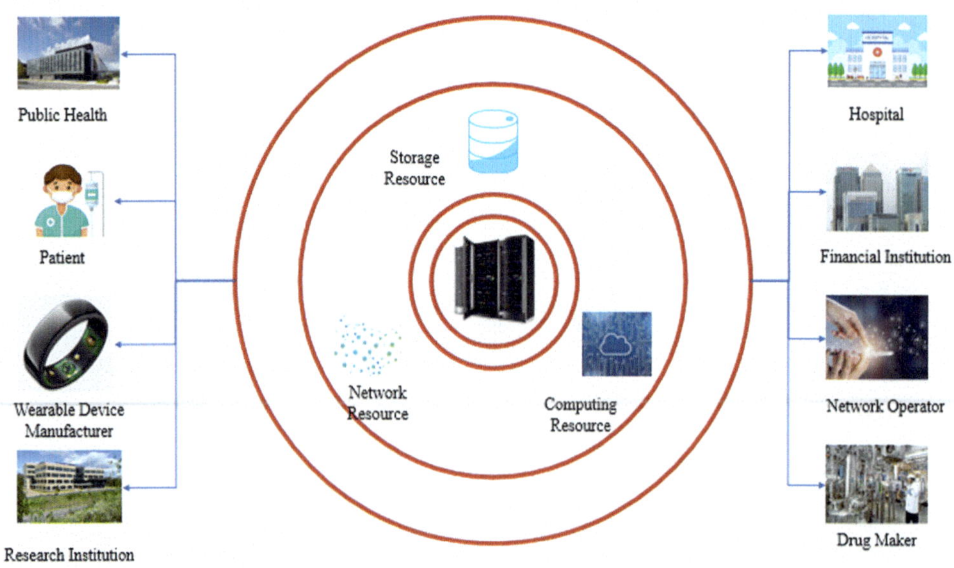

Figure 7.2 HealthCare system assisted by cloud and big data.

adverse events. Such systems include a group of sensors that are always connected to a patient and monitor his/her health state, a control system that analyzes collected data and decides upon the necessary treatment, and a set of actuators that help to implement the required treatment, such as drugs or therapy (Swapnarekha and Manchala, 2024; Jeschke et al., 2017). Due to the necessity of constant changes in the dosage of medication or the duration of the therapy, closed-loop control systems are handy for people suffering from chronic diseases like diabetes or Parkinson's disease. They help to maintain a good health of the patient by continually making adjustments to the treatment plan depending on the real-time data gathered, thus helping to ease the workload of the health care providers. However, achieving this potential can only be done once those obstacles associated with interoperability, data privacy, clinical validation, and ethical concerns have been met (Sharma et al., 2022; Liu, 2024). The trends discussed above point to the fact that, as researchers, healthcare professionals, policymakers, and industry stakeholders continue to address these challenges, the CPS systems in the patient monitoring domain will present new and useful innovations in the future.

2.4 CPS enhances remote patient monitoring

To further improve the current levels of RPM, integrating CPS can help provide continuous, real-time, and individualized care for patients:

- **Continuous Monitoring:** CPS includes a system of connected biomedical instruments like wearables or implantable for capturing patient's biometrics, clinical signs, and environmental data persistently.
- **Personalized Treatment:** Through the constant analysis of the patient's health information, CPS will be able to offer more suitable treatments for each patient depending on their specific wants and requirements. This may help in maintaining a healthy state and reduce the risk of complications accordingly.
- **Improved Access to Care:** Reducing the need for frequent visits to the hospital, RPM systems enabled by CPS can enhance access to care as it is in the rural and other underrepresented populations. They can achieve exceptional outcomes from the treatment by staying in their houses which enhances their life quality and at the same time decrease the cost of the healthcare sector.
- **Early Intervention:** CPS can help healthcare providers to quickly identify any alterations in patient's health data and thus, avoid any adverse outcomes, typically resulting in hospital readmissions.
- **Improved Patient Engagement:** He suggested that the CPS-enabled RPM systems can enhance patient engagement due to the real-time feedback, reminders, and educational materials provided. It can facilitate the effective management of a healthy disease, improve outcomes, and save money on treatment costs.

3. Personalized treatment through cyber-physical systems in healthcare

From multiple resources such as wearable devices, medical equipment, and Electronic Health Records, CPS facilitates the collection, analysis, and storage of vast volumes of patient data to support a patient-centric approach to care. This information is then fed through complex mathematical formulas and artificial intelligence to provide personalized reports and outcomes for each patient.

3.1 Key components of personalized treatment through CPS

- **Data Collection:** CPS makes use of different connected tools to gather information about the state of the human body and mind, temperature, heart rate, level of physical activity, and general health issues.
- **Data Integration:** Clinical data from other clinical sources are coordinated in one system and gives an all-embracing picture of the patient's status.
- **Data Analysis:** The multiple data sources are then processed in advanced analytics and machine learning algorithms to determine the patterns and outcomes of the treatment, as well as the best individual course of care.
- **Personalized Interventions:** CPS can help to intensify or change certain treatment methods, including medication dosage changes, nutritional and behavioral changes, or other treatments that can be employed to increase the usefulness of CPS in patient care.
- **Continuous Monitoring:** One advantage of CPS is that it facilitates constant evaluation of the patient's response to the treatment that can be followed by definitive revisions aimed at increasing the effectiveness of the therapy.

3.2 Benefits of personalized treatment through CPS

The implementation of CPS in healthcare offers several benefits for personalized treatment, including:

- **Improved Patient Outcomes:** When it comes to disease treatment, CPS can mean improved overall patient outcomes, decreased adverse effects, and improved quality of life for patients when treatments are adjusted to match the specific needs of the patient.
- **Reduced Healthcare Costs:** They concluded that through CPS delivered treatment can help prevent problems, reduce admission and ER visits, and efficiently use resources thereby providing a cost-effective plan for the healthcare systems.
- **Enhanced Patient Engagement:** Furthermore, CPS enables patients to have ownership over their medical care by offering them autonomy-supportive information, advice, and resources for self-organization.

- **Improved Clinical Decision-Making:** CPS assists healthcare organizations and personnel in making wise decisions by using real-time data and analytic models along with best practices.

Currently, the implementation of CPS in healthcare is not very advanced, but it can significantly change the approach to targeted therapy:

- **Artificial Intelligence and Machine Learning:** As the methods improve and new form of Artificial Intelligence algorithms are developed, there is great potential that predictions, recommendations and implementing treatment plans will become quite accurate.
- **Edge Computing and 5G:** CPS-based healthcare apps will be able to collect data, analyze this data and make corresponding decisions as well as respond within the shortest time possible due to the implementation of edge computing and 5G networks.
- **Wearable and Implantable Devices:** Due to progress in miniaturization of the sensors and improvement in the sensor technology and sampling rate, there will be a trend toward enhanced wearable and implantable technologies for constant tracking and localized treatments.

It is noted that, through CPS in healthcare, the patient-specific treatment course enables the development of a personalized approach with quality treatment for the patient. On the same note, by using influence of connected gadgets and data analytics coupled with interventional strategies, CPS holds the capacity to enrich the experiences of the patients, cut down the cost of medical expenses, and even increase the efficacy output of the medical sector. On the other hand, several issues such as data privacy, compatibility, acceptance by users, and legal requirements will be key determinants in the next phase of CPS in the healthcare context.

4. Challenges and opportunities in implementing CPS in healthcare

To the extent that MCPS represents a new paradigm of technology-based patient care, it presents a set of opportunities and threats for health care. In Fig. 7.3 CPS in healthcare has incorporated technology including gadgets, surgical tools, and statistics to deliver optimal and individualized care. In this chapter, we will look at the potential issues and benefits of CPS implementation in the healthcare industry.

4.1 Challenges in implementing CPS in healthcare

- **Interoperability and Integration:** CPS in healthcare requires that various devices, systems, and data are integrated and coordinated in optimal ways. However, medical devices are diverse and are available with different data formats and communication protocols which is a very big problem. Standard and framework establishment are

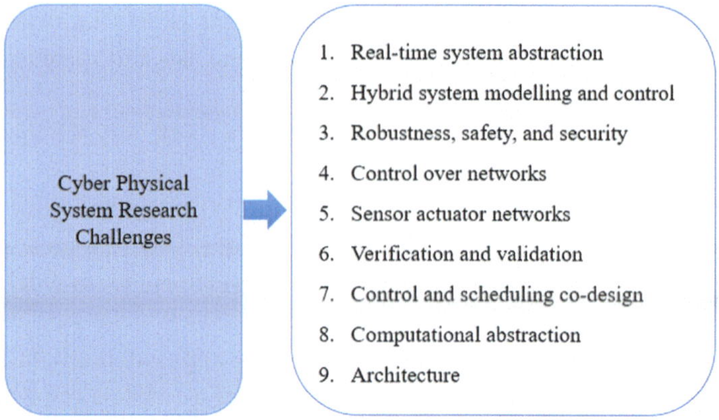

Figure 7.3 Research challenges of cyber-physical system.

seen as necessary to facilitate the different aspects of data sharing and interaction among the CPS system.

- **User Acceptance and Training:** Some of the emerging issues for concerns in the healthcare workforce include the complexity of CPS; Training for users; and Disruption of the status quo in the healthcare provision. There is, however, a need to address these concerns of InfoSec through communication, training, and implementation strategies to support the application of CPS in healthcare facilities.
- **Regulatory Compliance:** Healthcare CPS needs to meet guidelines and standards like Data protecting laws, medical device regulations, and healthcare standards. As mentioned, a major concern when integrating CPS into healthcare organizations is the need to meet diverse and often conflicting regulatory requirements. The ability to cooperate with regulatory authorities and work regarding given standards is critical in this process.
- **Energy Management and Sustainability:** This ensures that there is a free flow of energy within the CPS to support the various activities going on in healthcare. There remains a limitation to resource availability, especially in battery-powered devices and there is a need for efficient utilization of scarce power in CPS-based healthcare systems. To overcome these challenges, efficient components in the design, optimum energy management, and consideration of new sources of energy is a requirement.

4.2 Opportunities in implementing CPS in healthcare

Despite the challenges, implementing CPS in healthcare offers significant opportunities for improving patient outcomes, enhancing healthcare efficiency, and driving innovation:

- **Real-Time Monitoring and Early Intervention:** CPS helps in tracking the patient's status by connecting portable and implantable devices like health sensors.

With data being collected in real-time, medical issues which might otherwise go unnoticed could be addressed early and hence prevent such problems. CPS can also help the organization with remote patient monitoring which will help increase the number of patients attending the facility and also decrease the rate of hospital readmissions.

- **Enhanced Clinical Decision-Making:** CPS makes available to healthcare professionals' timely data, prognostic information, and best-practice suggestions, all of which may facilitate sound clinical decision-making. Using data from multiple sources and using more sophisticated algorithms, CPS can help clinicians make better decisions in relation to diagnosis, treatment choice, and patient progress, which can improve patient outcomes and reduce costs.
- **Improved Care Coordination and Collaboration:** CPS ensures that communication and data sharing between healthcare givers and the patients and other stakeholders can be effectively conducted. As a result, CPS can foster the sharing of timely data and more efficient and effective care delivery, lessening the risk of medical mistakes and enhancing the quality of delivered healthcare. It can also facilitate the connection between different healthcare services like primary, specialty, and rehabilitation services, thus improving client care.
- **Innovation and Transformation:** Adoption of CPS in the healthcare setting creates opportunities for developing new solutions and bringing changes. CPS can also influence the creation of new medical devices, diagnostic tools, and therapeutic methods that would enhance patient care and the clinical effectiveness of cost-effective solutions. CPS can also assist in incorporating new technologies like AI, Robotics, and Virtual Reality into the healthcare industry for innovations in patient treatment and delivery of services. It should be noted that the use of CPS mechanisms in healthcare has its advantages and disadvantages. As for the concerns like data privacy, interoperability, user acceptance, regulatory compliance, and energy management, CPS in healthcare holds great promise. CPS can bring capability for tailor-made and learning healthcare, continuous real-time interventions, better clinical management of patient care, better care coordination, and creativity in the delivery of healthcare services.

5. Security and privacy considerations in cyber-physical systems (CPS) for healthcare

Cyber-Physical Systems (CPS) can fundamentally transform healthcare by introducing intelligence, connectivity, and autonomy in medical devices and systems. CPS also brings new security and privacy threats that need to be considered. The chapter will identify the significant CPS security and privacy concerns in the healthcare setting regarding threat agents, risks, attacks, and countermeasures.

5.1 Security threats and vulnerabilities

5.1.1 Threat models

The threats in healthcare CPS are from a broad spectrum of threat actors with differing intent and skill levels. These include nation-states spying on each other, hackers looking to financially profit, hackers with a political agenda, and insiders with authorized access. Cybercriminals might attack the healthcare CPS to obtain the personal information of patients, compromise essential services, or even harm the patients. Threats are crucial for designing security measures since they define the necessary levels of protection:

- **Vulnerabilities in Connected Medical Devices:** One major weakness is that there exist KEVs in connected medical devices. Claroty's analysis reveals that 63% of the CISA-tracked KEVs occurred in medical systems, and 23% of medical supplies (clinical IoT devices, surgical equipment, imaging systems) contain at least one KEV. These vulnerabilities could be exploited by attackers to compromise the function of such devices and even endanger the lives of patients.
- **Legacy and Unsupported Systems:** A significant number of healthcare institutions currently contain aging healthcare equipment performing unauthorized or unregulated operating systems. These end-of-life systems are now no longer provided security patches which make them easy targets for exploitation. Failure of these legacy devices during critical processes will have adverse impact on patient outcomes in case the legacy devices are compromised.
- **Connectivity and Network Architecture Weaknesses:** The increasing integration of connected medical devices has created new risky entry points and structure flaws. Claroty established that 22% of hospitals use equipment that connects visitor and personal networks with devices that are exposed to the Internet. Some connected surgical devices are also available on the guest networks designed for the patient and visitors. This connectivity increases the potential for exploitable vulnerabilities.

5.1.2 Attacks and impacts

Ransomware and Extortion Attacks: Healthcare institutions are becoming preferred targets for ransomware and extortion, as they cannot afford disruptions. Specifically, the FBI's IC3 reported a 50% rise in ransomware attacks targeting hospitals from the year 2021–2022. Best attacks can paralyze critical patient systems, another study revealed that emergency departments near a ransomware-affected hospital received their patients late, and ambulances were diverted.

5.1.3 Data breaches and theft

Electronic protected health information is particularly at risk of cyber-attacks mainly because of privacy and security breaches referring to identity theft and fraud. Hackers can target EHR and medical images and other HE, as well as steal patient's data directly from the EHR. Security breaches may result in losses, loss of image, or fines against the healthcare firm:

- **Device Tampering and Manipulation:** It can cut off pipes which can supply wrong doses of the prescribed drugs, change treatment programs, or wrong readings. Grievous actors may also use connected devices as a means to access other systems and networks as well. Patients end up receiving incorrect dosages and doses that are wrongly administered harming the patient further and makes treatment delay hence the care providers take legal responsibility for the action of tampering with the devices.
- **Privacy Violations and Unauthorized Access:** Among the vulnerabilities, such options as using weak credentials, default passwords, and no encryption enable hackers to break into the medical device and steal a patient's information. This data could thus be in the hands of the attackers and used for blackmailing, extorting money from the individuals, and impersonating them. Data leaks may lead to a decline in patient trust; fines and sanctions from healthcare regulatory authorities; and potential lawsuits against healthcare organizations.

5.2 Latest advancements in CPS security for healthcare

In the Healthcare CPS, the threat actors can be categorized as follows: nation-state actors, cybercriminals, hacktivists, and malicious insiders. Hackers can attempt to compromise healthcare CPS to gain details of patients, further, make people lose access to necessary treatment, or even harm them. Some of the critical findings include:

- 23% of all connected medical devices have known exploitable vulnerabilities (KEVs).
- Some medical devices use end-of-support operating systems on some devices.
- Some of the devices expose guest and internal networks where hackers can attack.

Enhanced Encryption and Access Control: When applying the enhanced encryption-based security framework for the CPS Cloud, researchers identified the weaknesses inside the cloud-oriented IoMT healthcare systems. To tame costs and deferment, it integrates distributed blockchain data storage with a hybrid computing model. For the sake of security, it also has methods of device authentication, decentralized selective ring-based access restraints, and anonymization of medical data.

Secure Device Design and Development: This element of cost is especially crucial if the manufacturer is to offer safe medical equipment. This includes making sure that the code used was developed then written in a secure way, that the code and its validations were tested properly, and that the security updates were released in time. Devices must also implement the secure boot, Trust Execution Environment (TEE), and other hardware security mechanisms to safeguard devices and software from tampering.

Network Segmentation and Zoning: The following controls should therefore be adopted by healthcare organizations: network segmentation and zoning where medical devices and systems should be isolated from other networks and the Internet. This entails the utilization of firewalls, VLAN, and ACL to enforce the organizational policies on the networks. The devices have been divided into areas depending on the type of device, level of security, and measures put in place to oversee the devices.

Access Control and Authentication: User authentication and access control are two crucial forms of protection for healthcare CPS against unauthorized clients. This is done for security measures like two-factor authentication, Role-Based Access Control, and logging. Devices should also be configured to utilize strong passwords to improve the authentication mechanisms like Remote Authentication Dial-In User Service (RADIUS) and Terminal Access Controller Access-Control System Plus (TACACS+).

Encryption and Data Protection: It is said that patient data is highly sensitive and as such needs high security especially when stored or when in transit. Devices should accept standard encryption such as AES, RSA, and TLS encryption. There is also a need for implementing data loss prevention (DLP) solutions at the healthcare organizations to control the flow of data within those networks.

Vulnerability Management and Patching: A comprehensive risk management program is imperative for discovering and resolving cybersecurity risks in an HCPS. This involves performing vulnerability assessment, the routine search for known threats in devices and the application of security patches and updates as well as threat detection. Healthcare organizations must also perform an inventory of all connected devices for risk assessment and prioritization of remediation activities.

Incident Response and Recovery: As much as possible, organizations and companies must ensure that their security is tight but in the real world, security threats and breaches are always possible. The process of dealing with security incidents must be outlined clearly through a documented response framework containing guidelines on how to identify, control, and remedy an incident in healthcare organizations. This entails issues of backup and recovery to recover from the essential system or data that has been compromised by the attack. It is crucial to remember that, like most tools and models that exist in the organizational safety management system, incident response plans are not fail-proof and require consistent testing and updating. The protection of healthcare CPS is a very complex issue, and its security and privacy cannot be attained using single-measure approaches. Security should be considered in the design and development of medical devices used by manufacturers, and security measures and controls must be implemented, vulnerability management should be established, and adequate responses to security incidents should be put in place by healthcare organizations. Industry, government, and academia must have an agreement for sharing threat intelligence, practice guidelines, and research findings as well. In conclusion, everyone in the healthcare CPS chain has equal responsibility for security and privacy. In the following sections, we describe novel approaches within the CPS domain to address these challenges and, as such, pave the way for realizing the optimal applications of CPS in healthcare settings to provide safer, more effective, and more efficient care for all.

6. Future trends and implications of CPS in healthcare

Cyber-physical systems (CPS) are transforming healthcare as real-time monitoring, analysis, and control of physiological processes are made possible. CPS integrates computation, communication, and physical structures in designing organisms with perception, computing, and action from the bodies of humans. Regarding the healthcare domain, CPS is being used for tasks like remote monitoring, individualized medicine, as well as smart hospitals.

6.1 Trends in healthcare CPS

Increased Integration with IoT and Big Data: CPS is even more intertwined with IoT and big data analytics today. Wearable, implantable, and other IoT devices are enabling the collection of biophysical data on a large scale. CPS can then use the collected data in real-time leveraging on advanced analysis algorithms to detect trends, predict health episodes, and optimize treatment plans.

Personalized and Precision Medicine: CPS is enabling the shift toward patient- and precision-based medicine. Since CPS continuously monitors a person's physiology and integrates this data into the person's genomic and clinical data the healthcare system can identify the risk factors for a specific patient, address the specific disease to the specific patient, and prevent it.

Smart Hospitals and Intelligent Operating Rooms: CPS is being used to establish smart hospitals and smart operating rooms. Real-time location systems, integrated medical devices, and smart workflow can drive hospital operations to be more precise and flawless whilst offering the best to patients.

Robotics and Autonomous Systems: CPS is enabling the development of sophisticated robotic solutions in the healthcare environment. Surgical robots can accomplish intricate operations with increased accuracy and reduced invasiveness. Rehabilitation robots can help patients gain functional independence as a part of their rehabilitation process.

6.2 Implications of healthcare CPS

The adoption of CPS in healthcare has significant implications for patients, providers, and the healthcare industry:

Improved Patient Outcomes: It is argued that CPS can lead to improved patient care through early disease diagnosis, precise and effective treatment provision, and the reduction of medical errors. Telemonitoring can prevent hospitalization and allow early intervention to be made.

Enhanced Patient Experience: Apart from what it provides the patients, CPS can enhance the whole face value of the patient. It is advantageous as patients can order the desired health services online, take care of themselves, and receive recommendations.

Reduced Healthcare Costs: Telemonitoring may have the potential to decrease hospitalization rates and length in a patient population. Anticipation of return on cost through personalized medicine requires immense healthcare promotion since social incurs of ineffective treatments and adverse reactions are in concurring.

New Business Models and Revenue Streams: Through CPS, Digitize is empowering new business models and revenue channels in the global healthcare systems. Device manufacturers can create monthly, yearly, or one-time subscription plans and real-time monitoring services. Employers can deliver wellness-minded incentives and benefits programs to their employees.

Workforce Transformation: Regarding the role and responsibility of the healthcare workforce, CPS will indeed revolutionize CPS through changing the nature of the workforce in the healthcare sector. As a result, providers are going to have to prepare for new expertise in data analyses, algorithm readings, and machine-human interfaces. There are new occupations called CPS engineers and data scientists.

6.3 Challenges and limitations of healthcare CPS

Interoperability and Data Integration: CPS relies on the input obtained from the medical devices, EHRs, and contributions from patients. There are several challenges, however, they are most critical when it comes to interoperability and ensuring data integrity and confidentiality.

Regulatory and Legal Issues: When it comes to legal and regulatory issues, the adoption of CPS in the healthcare sector raises issues regarding data ownership, security, and protection. Providers and manufacturers will also have to face a large number of regulations and will need to adhere to the principle of data protection.

Ethical Considerations: CPS also has ethical issues with informed consent, autonomy, and racism. Lack of awareness is such that the patients may not perceive the implications of data sharing or the probability of mistakes in the algorithms used.

Workforce Readiness: The integration of CPS into the healthcare system means that there will be professionals who must have enough knowledge of CPS technologies and the area. Because of the flexibility in CPS, providers may need specialized training and orientation before implementing CPS into their service delivery systems.

Cybersecurity Risks: CPS is also at risk of data breaches that can threaten patient privacy and interrupt the functioning of the healthcare sector. Security of CPS is an essential factor as patient trust and their data must be protected from cybercriminals.

Cyber-physical systems are revolutionizing the healthcare sector by allowing real-time assessment and adjustment of physiological functions. However, implementing CPS in healthcare to the ultimate extent is not without major challenges concerning the fields of CPS, such as Interoperability, Data Integration, and Regulatory Issues, Ethical Issues, CPS Workforce preparedness, and Cybersecurity. To address such issues and effectively follow the CPS development and deployment roadmap, cooperation between members of the healthcare sphere, technology industry, government, and the patients themselves is going to be critical.

7. Conclusion

CPS is making a transition to personal and precision medicine possible by collecting data on an individual's physiological status in real-time, combining this information with genomic and clinical data, and using advanced analytics to identify biomarkers of risk, select treatments, and even predict and prevent illness. With CPS, smart hospitals, and intelligent operating rooms are being developed. Integrated communication, indoor positioning, and enterprise-wide processes can make a hospital more efficient, avoid mistakes, and enhance patient outcomes. Telemonitoring has the potential of lowering hospitalization rate and length of stay. The practice of personalized medicine will also work to reduce the costs of expensive treatments that do not yield positive results and negative drug interactions as well. CPS will alter the nature of healthcare workforce planning and development by redesigning the roles and duties of healthcare personnel. Providers will be required to seek new knowledge in data analytics, algorithm understanding, and user-interface competence. Some new roles that would appear are CPS engineers, as well as data scientists. Interoperability and quality and security of data is one of the biggest hurdles that CPS faces in the healthcare sector. There will be significant challenges in compliance in terms of regulation for all the providers as well as manufacturers of these products with regards to data protection laws. Regarding ethical considerations, CPS raises concerns about informed consent, patient autonomy, bias, and discrimination. The patient may be unaware of the consequences of data sharing or the potential for errors in algorithm decisions.

The advancement of cyber-physical systems is anticipated to trigger efficiency gains in the healthcare sector with tangible benefits on the lives of patients as well as within the financial input. However, to optimally harness the CPS potential multiple questions remain unanswered regarding the ethical, secure, and fair deployment of these systems. We must unify healthcare providers, IT developers, legislators, and patients to work toward the objective of having CPS as an integral part of healthcare systems that enhance patient satisfaction, reduce costs, and alter their perceptions of health. Their social and ethical loss or gain which is NOT a distant reality as CPSs are on the way to becoming more advanced than they currently are now should not be lost sight of. It is then incumbent upon us to ensure that if at all CPS have to be created, they will have to be created in the right manner and implemented in such a manner to provide equitable outcomes; accord patients dignified treatment; promote patient confidentiality; and make people active, responsible, and stakeholders in management of their health. Therefore, one can sum up that CPS is to remain in the focus and develop further to become an instrument shaping the efficiency and justice of the healthcare system in the future thus contributing to the future of healthcare. Thus, by emphasizing all the benefits these technologies provide and being aware of their dangers and drawbacks, it will become possible to work toward the goal to ensure that everyone receives the care they need for being healthy and happy.

References

Ali, S., et al., 2018. Cyber Security for Cyber Physical Systems, vol 768. Springer, Berlin/Heidelberg, Germany.

Biswas, A., Kumar Mondal, K., Roy, D.G., 2023. A study of smart evolution on ai-based cyber-physical system using blockchain techniques. In: AI Models for Blockchain-Based Intelligent Networks in IoT Systems: Concepts, Methodologies, Tools, and Applications. Springer International Publishing, Cham, pp. 327–346.

Cabello, J.C., Karimipour, H., Jahromi, A.N., Dehghantanha, A., Parizi, R.M., 2020. Big-data and cyber-physical systems in healthcare: challenges and opportunities. In: Handbook of Big Data Privacy. Springer International Publishing, Canada, pp. 255–283. https://doi.org/10.1007/978-3-030-38557-6_12. Available from: https://www.springer.com/gp/book/9783030385569.

Huertas Celdrán, A., Gil Pérez, M., García Clemente, F.J., Martínez Pérez, G., 2018. Sustainable securing of medical cyber-physical systems for the healthcare of the future. Sustainable Comput. Inf. Syst. 19, 138–146. https://doi.org/10.1016/j.suscom.2018.02.010.

Jeschke, S., et al., 2017. Industrial Internet of Things and Cyber Manufacturing Systems. Springer International Publishing.

Jimenez, J.I., Jahankhani, H., Kendzierskyj, S., 2019. Health Care in the Cyberspace: Medical Cyber-Physical System and Digital Twin Challenges. Springer Science and Business Media LLC, pp. 79–92. https://doi.org/10.1007/978-3-030-18732-3_6.

Kaur, M.J., Riaz, S., Mushtaq, A., 2020. Cyber-physical cloud computing systems and internet of everything. In: Principles of Internet of Things (IoT) Ecosystem: Insight Paradigm, pp. 201–227.

Krishna, P.V., et al., 2019. A novel framework for healthcare monitoring system through cyber-physical system. In: Internet of Things and Personalized Healthcare Systems, pp. 21–36.

Liu, H., 2024. Network and communication protocols in cyber-physical systems. In: A Practical Guide on Security and Privacy in Cyber-Physical Systems: Foundations. Applications and Limitations, pp. 25–88.

Maleh, Y., et al., 2022. Blockchain for cyber-physical systems: challenges and applications. In: Advances in Blockchain Technology for Cyber Physical Systems, pp. 11–59.

Möller, D.P.F., Möller, D.P.F., 2016. Introduction to cyber-physical systems. In: Guide to Computing Fundamentals in Cyber-Physical Systems: Concepts, Design Methods, and Applications, pp. 81–139.

Nayak, J., Naik, B., Vimal, S., Favorskaya, M., 2024. Machine learning for cyber physical system: advances and challenges. Intellig. Sys. Refer. Libr. 60. https://doi.org/10.1007/978-3-031-54038-7.

Oks, S.J., 2024. Industrial Cyber-Physical Systems. Springer Fachmedien, Wiesbaden.

Puliafito, A., Tricomi, G., Zafeiropoulos, A., Papavassiliou, S., 2021. Smart cities of the future as cyber physical systems: challenges and enabling technologies. Sensors 21 (10), 3349. https://doi.org/10.3390/s21103349.

Sajid, M., Kumar Sagar, A., Singh, J., Ibrahim Khalaf, O., Prasad, M., 2023. Intelligent Techniques for Cyber-Physical Systems. CRC Press. https://doi.org/10.1201/9781003438588.

Sharma, N.V., Singh Yadav, N., Sharma, S., 2022. Machine Learning and Security in Cyber Physical Systems. Elsevier BV, pp. 171–187. https://doi.org/10.1016/b978-0-12-824557-6.00015-7.

Sharma, U., Nand, P., Chatterjee, J.M., Jain, V., Jhanjhi, N.Z., Sujatha, R., 2022. Cyber-Physical Systems: Foundations and Techniques. wiley, India, pp. 1–315. https://doi.org/10.1002/9781119836636. Available from: https://www.wiley.com/en-ie/Cyber-Physical+Systems%3A+Foundations+and+Techniques-p-9781119836193.

Swapnarekha, H., Manchala, Y., 2024. Application of deep learning in medical cyber-physical systems. In: Machine Learning for Cyber Physical System: Advances and Challenges. Springer Nature Switzerland, Cham, pp. 245–265.

Törngren, M., Asplund, F., Bensalem, S., McDermid, J., Passerone, R., Pfeifer, H., Sangiovanni-Vincentelli, A., Schätz, B., 2017. Characterization, Analysis, and Recommendations for Exploiting the Opportunities of Cyber-Physical Systems. Elsevier BV, pp. 3–14. https://doi.org/10.1016/b978-0-12-803801-7.00001-8.

Tripathy, B.K., Panda, G.K., Sahu, A., 2024. An in-depth analysis of cyber-physical systems: deep machine intelligence based security mitigations. In: Machine Learning for Cyber Physical System: Advances and Challenges. Springer Nature Switzerland, Cham, pp. 27–55.

Further reading

Ramasamy, L.K., et al., 2022. Secure smart wearable computing through artificial intelligence-enabled internet of things and cyber-physical systems for health monitoring. Sensors 22 (3), 1076.

Tyagi, A.K., Sreenath, N., 2021. Cyber physical systems: analyses, challenges and possible solutions. IoT and CPS 1, 22–33.

PART III

The semantic heart: LLMs, knowledge graphs, and interoperability

CHAPTER 8

Enhanced semantic model in healthcare

N. Ambika
Jain College Vasavi Campus, Department of Computer Science, Bangalore, Karnataka, India

1. Introduction

A Semantic Oncology Model is a specialized computational framework that utilizes semantic technologies to process, interpret, and analyze large volumes of cancer-relevant data. This model integrates various types of data, such as genetic, clinical, pathological, and therapeutic information, to enhance our understanding of oncology using ontologies, natural language processing (NLP), and machine learning (ML). The field of cancer research and therapy has entered the era of precision medicine, where the diversity and complexity of data become crucial. To address the challenges posed by the large volume, diversity, and often unstructured nature of oncology-related data, a sophisticated computational framework known as the Semantic Oncology Model was developed. This concept aims to facilitate more accurate and personalized cancer diagnosis, treatment, and research by combining semantic technologies with modern data analytics.

> *"Ontologies offer standardised, structured vocabularies to express concepts and interactions within the oncology domain. They are fundamental components of the Semantic Oncology Model, providing consistent understanding and harmonisation of data from several sources. Natural language processing (NLP) approaches are used to extract meaningful information from unstructured clinical notes, research publications, and other text-based resources. This enables inclusion of knowledge that is not easily accessible in organised formats into the model. Machine learning algorithms can utilise large datasets to find patterns, forecast outcomes, and provide new insights. These models have the ability to learn from past data to forecast results, recommend courses of action, and pinpoint possible biomarkers. The model incorporates information from several sources, such as published research, clinical trial findings, genetic data, and electronic health records (EHRs). Thorough data integration from various sources is essential for a full understanding of cancer biology and therapy."*

The previous ontology-based semantic model (Adel et al., 2022) can combine information from multiple electronic health records (EHRs) by integrating various widely used healthcare data formats and standards. The suggested ontological paradigm consists of two stages. First, an OWL ontology is created for each unique input source, and then these ontologies are combined into a single, updated one. The model includes estimates of the patient's pulse rate and tests for red blood cells, platelets, and white blood cells. The recommendation aims to enhance prior methods to provide patients and/or their

Intelligent Semantic Analysis for Healthcare
ISBN 978-0-443-33512-9
https://doi.org/10.1016/B978-0-443-33512-9.00022-6

guardians with a better understanding of their health, diagnosis, and the role of medication in their recovery. The recommendation aims to enhance prior methods to provide patients and/or their guardians with a better understanding of their health, diagnosis, and the role of medication in their recovery. It improves the understanding by 88.9%.

2. Architecture

The initial layer of the architecture (Zheng et al., 2023), known as the Data Preparation Layer, is responsible for obtaining the system's incoming data. The construction image and external digital data sources are the two distinct data streams that are considered as inputs. The first step in creating a construction picture is to extract its semantics from each image, including the visible construction entities and their associated characteristics and connections. To create a graph representation of the semantically labeled picture, the extracted image semantics and metadata must be translated to an ontology that sufficiently explains the contents of the image. This may be done by creating an RDF data graph. Using the retrieved semantically labeled pictures, the interpretation layer aims to achieve the interpretation of the images. Establishing SII's objective is the first challenge. In addition to the semantically labeled image, additional input such as background information that identifies the various visual aspects for interpretation is needed to achieve further inferences. Formal logic is used in ontology-based inferencing to define rules, which are one type of knowledge representation. To create a semantic Data Transformation Component (DTC), the Integration Layer attempts to merge the external data with picture semantics. Through this process, the database will be enhanced to offer a more thorough description of the real scenario, leading to further applications that are not possible with picture semantics alone. The DTC will be constructed by integrating the layer's input—external ICT data sources represented as RDF graphs, the semantically labeled picture, and inference image semantics during the interpretation phase—into a graph database. The integration process that links the items in the various systems requires a mapping procedure. The linked data concept, which establishes new relationships between related or comparable items in various systems, is adhered to in this mapping process. After integration is completed, helpful applications that support different management requirements may be developed, based on the combined DTC database of picture semantics with external data. Fig. 8.1 represents Oncology-based model.

3. Literature survey

The semantic ontology-based model that was proposed can bring together data from various EHRs. It brings together a variety of well-known healthcare data (Abouelmehdi et al., 2018; Ambika, 2022; Chang et al., 2011) formats and standards. The proposed

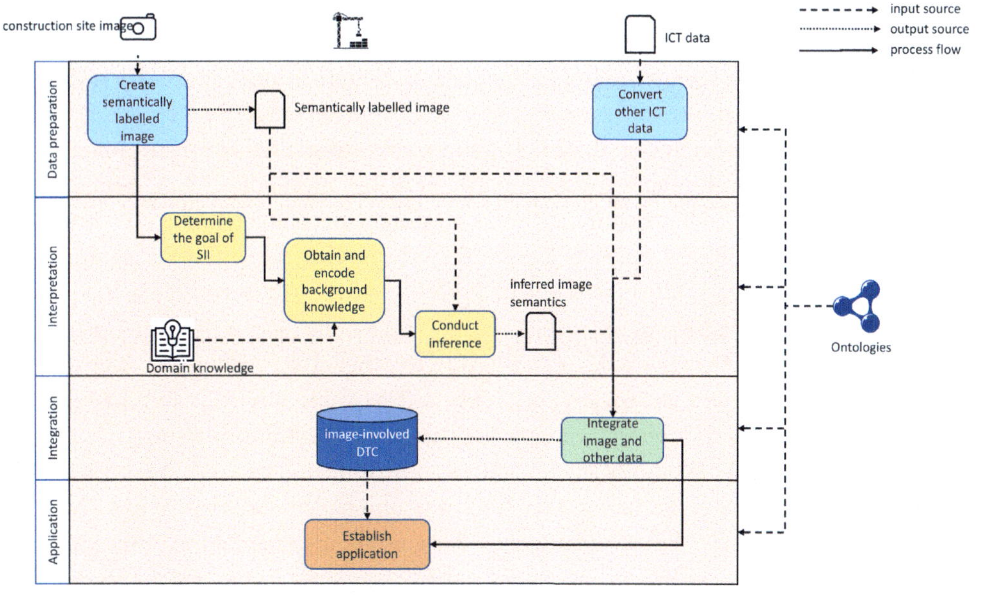

Figure 8.1 Oncology-based model. *(Reproduced with permission. Zheng, Y., Khalid Masood, M., Seppänen, O., Törmä, S., Aikala, A., 2023. Ontology-based semantic construction image interpretation. Buildings 13 (11). MDPI, Basel, Switzerland.)*

ontological model has two phases. The proposed first stage converts each distinct input source into an OWL ontology. In the subsequent stage, it coordinates that large number of ontologies into a consolidated fresh one. 3753 axioms, 2606 logical axioms, 186 classes, 136 individuals, 126 datatype properties, and 257 object properties are included in the integrated ontology (Batbaatar and Ryu, 2019). in a variety of formats, and we want to bring them all together. The subsequent model contains a pulse estimation 17 records in Excel. It incorporates estimations of the patient's pulse rate. The third model is a CSV with 19 records, some of which each patient's primary lab tests. These tests are white blood hemoglobin, platelets, and red blood cells.

As part of the European project ODIN (Luschi et al., 2023) and its potential future expansion, the aim of this scoping assessment is to identify and evaluate ontologies that are currently available and capable of illustrating every use-case describing the hospital environment. On January 13, 2023, the review was conducted using the PRISMA extensions for scoping reviews on the Scopus database. A total of 3225 documents resulting from the database search were evaluated by two reviewers. After additional filtering, 32 publications were left to be examined for the outcomes. The 34 ontologies identified from the recognized papers were also reviewed and discussed. Upon reviewing the titles, 714 of the entries found during the initial search were considered relevant. After an analysis of the abstracts of the publications in this group, 183 items were selected by both reviewers following a second screening. Out of a total of 189 selected records, a third reviewer (EI) resolved the 18 differing viewpoints and selected six publications. After reviewing the entire texts of the remaining papers, the reviewers finally selected 32 documents.

The project (Barisevičius et al., 2018) generated a substantial medical Linked Data Graph (LDG) by combining various publicly available biomedical data sources with several national ones. This required the development of unique RDF converters. It also established the Clinical Knowledge middleware service, which provides the LDG with an abstraction layer through a set of REST services. Additionally, it included a web browser that enables users to search for classes and access information about them, including direct super-classes, relationships to other classes, and more. The project also outlines the Babylonian triaging use case and explains how our hybrid reasoner compares user text to the internal models for triaging and symptom testing.

An ontology model (Matsokis et al., 2010) for semantic maintenance has been developed, emphasizing the gathering and analysis of maintenance data to provide advanced maintenance techniques. These techniques offer various advantages, including the provision of new services, enhanced customer satisfaction, compliance with environmentally friendly laws, and improved performance, reliability, and product quality. Our model integrates the benefits of two previously created models: the PROMISE Semantic Object Model (SOM) and the model created during the PROTEUS project. The SOM is designed to allow closed-loop product lifecycle management, while the PROTEUS model serves as a maintenance-focused platform. Our model facilitates the integration of maintenance activities during the Middle of Life (MOL) of the asset, as well as providing maintenance personnel with remote access to data, the ability to request diagnoses from a distance, and the capacity to interact with intelligent systems for improved maintenance activity management.

The project (Ye et al., 2007) entails an Ontology-Based Approach for Semantic Integration (OBA-SI), which involves defining mappings between concepts using ontologies as conceptual models. The applications are divided into three layers: process, service, and data. OBA-SI aims to combine two software programs that support the Software Configuration Management method, which is a tool for limiting modifications, and the well-known version control system, Subversion. The suggested method addresses integration in three layers: process, service, and data. Three artifacts are crucial during the analysis phase: an ontology for the reference domain to describe the integration domain, the conceptual models of the applications combined with the business's process model that they support, and a remedy as the reference ontology, providing a precise and unambiguous explanation of the domain's entities. It functions as a guide. In the process of integration examination, a model is used to allocate semantics. The conceptual models of the apps that need to be merged can be discovered beforehand or through reverse engineering methods. In response, the business process model should outline the procedures followed by the organization in the integration scenario, including information on workers, activities, inputs and outputs, and supporting tools. Various aspects of integration are covered by these models. Semantics is established at both the data and service levels using the conceptual models of the tools and the reference domain ontology. Although it also aids in service integration, the business process model is primarily used to address process integration. Model elements are given meaning via Vertical Mappings (VMs) in accordance with the reference ontology. Conversely, Horizontal Mappings (HMs) take place between model parts with an emphasis on the integration scenario. Integration objectives and requirements should be established throughout the integration requirements definition process. Integration modeling can begin after the structural models are semantically annotated. The behavioral mappings can be established based on the structural mappings.

The proposed strategy (Dou et al., 2015) is considered to be a viable knowledge management technique that facilitates the application of computer-aided collaborative work (CSCW) in the design applications of the Semantic Web. It is intended for use in a production system designed for automated assembly of electronics connections, which involves a semantic network of intricately designed domain-specific ontologies, conceptual layout, teamwork profiles, and more. The domain ontology that is created can assist users in creating ontology searches to find detailed information about mechanical components for shared reuse, including high-level specifications such as function into specific component details.

CACOnt (Xu et al., 2013) provides semantic logics that can be utilized with rule-based systems to enable context reasoning capabilities. A subtype of the Context Entity is the user model, which describes the user's name, identity, site, and preferences provided by the FOAF ontology in the Personal Profile. A schedule consists of one or more events, each of which has one or more tasks, and each task has one or more activities. On the other hand, the situation describes what the user is doing or their current condition, including their roles, tasks, companions, and surroundings, and connects the Device model with the "owns" attribute and ties to the Environment model for providing Physical Conditions. The device's installed software may be identified by its name, edition, and version, among other necessary attributes, as shown in the Software Profile. Services should be able to sense their surroundings and adjust accordingly. Location data is described by space ontology primarily from three perspectives: the coordinates indicating latitude, longitude, and altitude match the geographic location, while the Spatial Region designates symbolic representations of Space and usually denotes geographical areas under the authority of political entities that are connected to one another. The OWL encoded model CACOnt offers DL-based inference capabilities that can be used with rule-based systems to enhance reasoning skills.

The recommended approach (Ye et al., 2008) for integrating semantics into Supply Chain Management (SCM) is illustrated with a case study of a printer supply chain.

Onto-SCM comprises five modular theories: SC-Structure theory, SC-Activity theory, SC-Resource theory, SC-Item theory, and SC-Management theory. SC-Structure theory defines the representational framework for supply chain structures, strategies, objectives, and procedures. SC-Activity theory provides languages for describing activities, their hierarchy, and dependency relationships within supply chain management systems. The SC-Resource and SC-Item theories specify language primitives that express concepts and relationships related to resources and business items. Finally, the SC-Management theory defines constraint connections in supply chain management and operational processes.

By using inference engines in combination with the suggested formalization of the semantics of individual model parts (Thomas and Fellmann MA, 2009), modeling tools' query functions may be enhanced, opening up new avenues for model validation. A system design and a representative implementation show how the technique is integrated into the IT-based work environments of modelers. The modeling language used was the Unified Modeling Language (UML) and its class diagram. The linking role that the intermediate level, the metadata, has between the model and the ontology is expressed by the vertical stacking of these levels. Here, "metadata" refers to data that has been produced by annotating a model with ontology-related components. The classes Model, Model element, and Model element connector describe concrete models at the model level. Transformation rules, which specify how items at the model level can be changed into elements at the metadata level, are used to describe the models at the metadata level. The Ontology class is the main class at the ontology level. It is a representation of every class in an ontology. The ontology used for the annotation is created and modified using an editor that is directly connected to repository A, which stores ontologies, rules, and models. A layered sBPM framework forms the architecture's core. Basic functions for the components above it, regarding data loading, data interpretation with an inference machine, and data modification and storage, are provided by a semantic data processing component at the bottom layer. The essential features required for the IT support of the semantic process modeling are provided by the components above this tier.

To illustrate how the ontology framework concept enables rapid adaptation and interoperability for semantic computation in Internet of Things (IoT) applications (Ma et al., 2014), it elaborates on its knowledge space. IoT refers to a scenario in which physical items are seamlessly integrated into an information network, and services are offered to communicate with these "smart objects" in order to query and respond to their actions. Entity, dimension, activity, and service are the fundamental ideas of the Internet of Things. Evolution-based evaluation includes two components: conceptualization and application domain evolution. A shifting understanding of concepts or a different usage viewpoint leads to evolution in conceptualization. Similarly, OntoIoT may offer many viewpoints on ontology topology, such as the Entity-centric and Activity-centric viewpoints. The ontology must also be flexible enough to adapt to different application domains. When ontology is re-modeling or expanding, this is brought about by the addition of knowledge in the domains. In the section on preliminary implementation, we have shown how OntoloT may be extended to the realm of smart spaces applications. OntoloT has been effectively expanded into the transportation and healthcare industries, demonstrating its ability to develop in a variety of application areas. This evaluation's outcome demonstrates OntoloT's compatibility and scalability. Logical reasoning, functional design, and logical error checking are components of logical-based assessment. The goal of a logical error check is to identify ontological inconsistencies.

ICT-enabled assisted living (Belani et al., 2022), also known as ambient assisted living, refers to intelligent systems of assistance for a better, safer, and healthier life in the preferred living environment. It encompasses ideas, goods, and services that link and enhance emerging technologies and the social environment with the goal of improving everyone's quality of life at every stage. It is necessary to model the patient's (or the person being helped) satisfaction scale and promptly identify its levels. The methodology involves using the Internet of Things to support ontology-based requirements engineering in the fields of health, aging, and well-being. Instead of creating a new ontology, this approach modifies the existing SAREF4EHAW ontology to better align with the concepts of health, aging, and well-being and to effectively organize the knowledge within this field. The positive scale of the patient (or the person being helped) has to be modeled, and its levels need to be promptly identified. The negative scale of the patient (or the person being helped) needs to be modeled, and its degrees need to be promptly identified. The regular interactions that the patient, or the person being helped, has with his or her companion animals need to be modeled and observed.

The study (Patil et al., 2005) proposes a framework for exchanging product data across different domains during product development. During the design process, a three-dimensional (3-D) CAD model of the product is created using design guidelines. This CAD model serves as input for the manufacturing process. The method generates a comprehensive process plan, known as an NC code, which is used to produce the work item. Capturing important ideas and relationships within the domain of interest is essential during the Ontology phase. This phase also provides textual definitions for the terminology used to describe these connections and ideas. Reasoning methods based on description logic are used to encode the semantics for the PSRL.

An adaptation of the traditional keyword-based information retrieval approach (Fernández et al., 2011) is the ontology-based model. This model involves four primary operations: searching, querying, indexing, and ranking. Semantic annotation is similar to the indexing process. When an SPARQL query is run against a knowledge base (KB), a list of semantic entities that satisfy the query is returned. This stage is entirely Boolean, meaning the returned instances must strictly meet all requirements of the formal inquiry. Once retrieved and rated, the user is presented with papers that have been annotated (indexed) with these instances. Due to the inherent ambiguity in the relationship between a document and the ideas that annotate it, the document retrieval phase is based on an approximate match.

4. Previous model

The semantic ontology-based model, proposed, offers a method for integrating data from various Electronic Health Records (EHRs) by harmonizing different healthcare data formats and standards into a unified ontology. The aim of the model and datasets is to bring together diverse healthcare data into a consolidated format, enabling more efficient data analysis and interoperability across different healthcare systems.

The process involves two main phases:

1. Conversion Phase: Each distinct input source is converted into an OWL (Web Ontology Language) ontology. OWL is a standard for representing rich and complex knowledge about things, groups of things, and relations between things.
2. Integration Phase: The various ontologies created in the first phase are then coordinated and integrated into a single, unified ontology. The integrated ontology includes: 3753 axioms, 2606 logical axioms,186 classes, 136 individuals, 126 datatype properties and 257 object properties These elements together form a comprehensive model that represents the integrated healthcare data. In addition to the ontology model, there are also other datasets mentioned:
   - Pulse Estimation Dataset: This consists of 17 records in an Excel format, which include the patient's pulse rate measurements.
   - Lab Tests Dataset: This consists of 19 records in a CSV format, which include primary lab test results such as white blood cell count, hemoglobin levels, platelets, and red blood cell count.

5. Proposed model

The suggestion improves the previous methodology to provide a better understanding to the patient/guardians. It aims in making the patients aware of their well-being, diagnose of disease, stage and how the medication is going to help them to cure them. Table 8.1 provides the assumptions used in this study.

The system uses a dictionary to that provides mapping to the layman's language. Kaggle data set is used in this work (Breast Cancer Diagnosis Dataset - Wisconsin State). 699 data sets were created, giving an indication of level of disease and the medication prescribed. Table 8.2 provides the notations used in the work.

The procedure is as follows:

- The doctor/executive provides his authentication using his biometrics to login into the system. In the notation (1), the doctor/executive D_i provides biometrics B_i to login into the system. He feeds respective data Data_i into the system. The same is represented in (2).

$$D_i \rightarrow B_i \tag{8.1}$$

Table 8.1 Assumptions used in the study.

• Patients/guardians have to undergo authentication before getting access to the system. In this work, biometric authentication is used.
• Doctors/medical executives who feed the data into the system, will also need to provide biometric authentication.
• The system provides different views according to the user. The doctor is provided with window (enabling medical terms) and the user is provided with layman's language. A brief description is provided to the user.
• Patients/guardians have to undergo authentication before getting access to the system. In this work, biometric authentication is used.
• Doctors/medical executives who feed the data into the system, will also need to provide biometric authentication.
• The system provides different views according to the user. The doctor is provided with window (enabling medical terms) and the user is provided with layman's language. A brief description is provided to the user

Table 8.2 Notations used in the proposal.

Notations	Description
D_i	Doctor/executive
P_i	Patient/guardian
B_i	Biometric
Data_i	Data fed into the system

$$D_i \rightarrow \mathrm{Data}_i \tag{8.2}$$

- The patient logs in by providing his authentication. In the notation (3), the patient P_i logs into the system using his biometrics $B_{i.}$

$$P_i \rightarrow B_i \tag{8.3}$$

- The patient is provided with the description of his diagnosis in layman language. He is given chatbot facility, where he will be given answers to regular questions.

6. Simulation and analysis

The previous ontology-based semantic model (Adel et al., 2022) can combine information from multiple electronic health records (EHRs) by integrating various widely used healthcare data formats and standards. The suggested ontological paradigm consists of two stages. First, an OWL ontology is created for each unique input source, and then these ontologies are combined into a single, updated one. The model includes estimates of the patient's pulse rate and tests for red blood cells, platelets, and white blood cells.

Table 8.3 Simulation details.

Parameters used	Description
Number of doctors used	3
Number of patients used	2
Biometric (length in bits)	32 bits
New data set entry in dictionary	699 records
Number of regular questions activated in the chatbot	20

The recommendation aims to enhance prior methods to provide patients and/or their guardians with a better understanding of their health, diagnosis, and the role of medication in their recovery. The work is simulated in python. Table 8.3 provides the simulation details:

- *Understanding of patient* – The patient will be able to understand his diagnosis better. The system aims in improving the patient's knowledge of his diagnosis. It improves by 88.9% compared to previous system. The same is represented in (Fig. 8.2) Portrays the understanding of both the system.

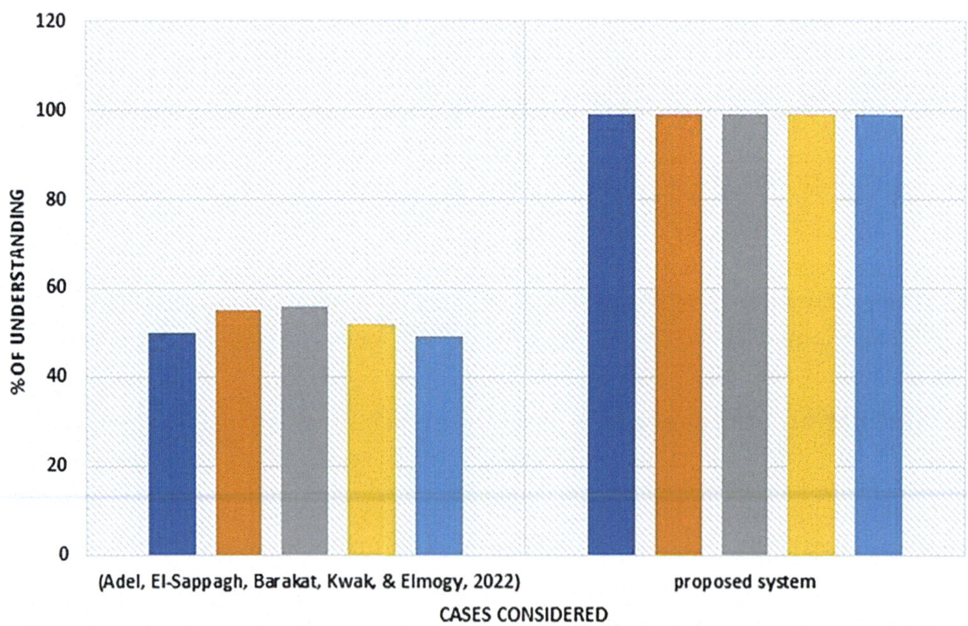

Figure 8.2 Comparison of understanding of both the system.

7. Conclusion

An ontology-based semantic model defines a collection of ideas and their connections, providing a structured framework for representing and integrating data from many sources. This strategy works especially well in fields such as healthcare, where data is dispersed across several standards and formats. An ontology uses concepts (classes), properties (attributes and connections), and instances (individuals) to formally express knowledge within a domain. Ontologies define the semantics and structure of data in a semantic model. Ontologies may be created using the OWL language. It is a powerful tool for integrating and harmonizing complicated, multi-source data, allowing for more efficient data analysis and interoperability by defining and connecting ideas and relationships within a domain, especially in industries like healthcare.

The previous ontology-based semantic model can combine information from multiple electronic health records (EHRs) by integrating various widely used healthcare data formats and standards. The suggested ontological paradigm consists of two stages. First, an OWL ontology is created for each unique input source, and then these ontologies are combined into a single, updated one. The model includes estimates of the patient's pulse rate and tests for red blood cells, platelets, and white blood cells. The recommendation aims to enhance prior methods to provide patients and/or their guardians with a better understanding of their health, diagnosis, and the role of medication in their recovery. It improves the understanding by 88.9%

References

Abouelmehdi, K., Beni-Hessane, A., Khaloufi, H., 2018. Big healthcare data: preserving security and privacy. J. Big Data 5 (1), 1–18.

Adel, E., El-Sappagh, S., Barakat, S., Kwak, K., Elmogy, M., 2022. Semantic architecture for interoperability in distributed healthcare systems. IEEE Access 10.

Ambika, N., 2022. Aiding IoT and Cloud to Control COVID-19: A Systematic Approach. Springer, pp. 349–365.

Barisevičius, G., Coste, M., Geleta, D., Juric, D., Khodadadi, M., Stoilos, G., Zaihrayeu, I., 2018. Monterey, CA, USA 291 306 springer supporting digital healthcare services using semantic web technologies. In: The Semantic Web.

Batbaatar, E., Ryu, K.H., 2019. Ontology-based healthcare named entity recognition from twitter messages using a recurrent neural network approach. Int. J. Environ. Res. Publ. Health 16 (19).

Belani, H., Šolić, P., Perković, T., 2022. Towards Ontology-Based Requirements Engineering for IoT-Supported Well-Being, Aging and Health. IEEE, Melbourne, Australia, pp. 65–74.

Chang, F., Markmiller, J., Yang, J., Kim, Y., 2011. System Health Management: With Aerospace Applications. Structural Health Monitoring, pp. 419–428.

Dou, D., Wang, H., Liu, H., 2015. Semantic Data Mining: A Survey of Ontology-Based Approaches. IEEE, Anaheim, CA, USA, pp. 244–251.

Fernández, M., Cantador, I., López, V., Vallet, D., Castells, P., Motta, E., 2011. Semantically enhanced information retrieval: an ontology-based approach. Web Semant. 9 (4).

Luschi, A., Petraccone, C., Fico, G., Pecchia, L., Iadanza, E., 2023. Semantic ontologies for complex healthcare structures: a scoping review. IEEE Access 11.

Ma, M., Wang, P., Chu, C., 2014. Ontology-Based Semantic Modeling and Evaluation for Internet of Things Applications. IEEE, Taipei, Taiwan, pp. 24–30.

Matsokis, A., Karray, H., Chebel-Morello, B., Kiritsis, D., 2010. An ontology-based model for providing semantic maintenance. IFAC Proc. Vol. 43 (3), 12–17.

Patil, L., Dutta, D., Sriram, R., 2005. Ontology-based exchange of product data semantics. IEEE Trans. Autom. Sci. Eng. 2 (3), 213–225.

Thomas, O., Fellmann Ma, M., 2009. Semantic process modeling–design and implementation of an ontology-based representation of business processes. Bus. Inform. Syst. Eng. 1, 438–451.

Xu, N., Zhang, W., Yang, H., Zhang, X., Xing, X., 2013. CACOnt: a ontology-based model for context modeling and reasoning. Appl. Mech. Mater. 347.

Ye, J., Coyle, L., Dobson, S., Nixon, P., 2007. Ontology-based models in pervasive computing systems. Knowl. Eng. Rev. 22 (4), 315–347.

Ye, Y., Yang, D., Jiang, Z., Tong, L., 2008. Ontology-based semantic models for supply chain management. Int. J. Adv. Manuf. Technol. 37.

Zheng, Y., Khalid Masood, M., Seppänen, O., Törmä, S., Aikala, A., 2023. Ontology-based semantic construction image interpretation. Buildings 13 (11).

CHAPTER 9

Large language models in healthcare semantics

Priyadarsan Patra
Department of Computer Science and Engineering, NIST University, Berhampur, Odisha, India

1. Introduction: GenAI in semantic capture & analysis

In healthcare informatics, semantics refers to the representation of clinical meaning — the mapping between linguistic expressions, structured codes (e.g., SNOMED CT, ICD), temporal states, and clinical intent. Traditional systems relied on symbolic ontologies and rule-based mappings to enforce semantic consistency. Large Language Models (LLMs) introduce a different, often complementing, paradigm: meaning is encoded in high-dimensional probabilistic embedding spaces rather than fixed symbolic schemas. Thus, LLMs shift healthcare semantics from deterministic ontology-bound representations toward context-sensitive, probabilistic-distributional representations of clinical meaning.

A broad category in AI (Artificial Intelligence) that generates content like text, speech, or images based on existing data is called "Generative AI" or "GenAI." A more specific category in GenAI is called "Large Language Model" or "LLM," which is specifically designed for use with human language. GenAI supplies the creative foundation for AI models, and LLMs serve as the "engines" for those models. By utilizing "deep neural networks" that predict words in a sequence, they can summarize complex medical documents or interpret medical jargon. Thus, understanding language models is central to Generative AI and its role in healthcare semantic analysis and analytics. Large Language Models (Brown et al., 2020) and Small Language Models (SLMs) are AI systems trained on large datasets, often comprising billions or trillions of words drawn from books, journals, and other written sources. Models such as GPT-4 or Llama use transformer architectures to predict and generate human-like text by understanding long-range context, patterns, and relationships. Basically, they take an input and pass it through a series of layers of neural networks to perform tasks such as summarizing documents, answering questions, translating languages, and writing code. However, in the medical domain, the challenge extends beyond the grammar of a natural language; it requires aligning these models with the specific structures of biology and medicine to ensure they capture the true meaning of clinical data (Zhou et al., 2025).

In contrast, SLMs are compact versions of LLMs, optimized for efficiency with fewer parameters (e.g., millions instead of many billions). Models like Phi-3 or Gemma run on

Intelligent Semantic Analysis for Healthcare
ISBN 978-0-443-33512-9
https://doi.org/10.1016/B978-0-443-33512-9.00009-3

everyday devices like laptops or smartphones, offering similar capabilities with lower computational demands, reduced energy use, and faster inference—ideal for resource-constrained environments. The table below shows their core differences.

Aspect	LLMs	SLMs
Size	Billions + parameters	Millions of parameters
Strengths	Superior accuracy, complex reasoning	Speed, low cost, edge deployment
Challenges	High compute needs, latency	Limited depth on niche tasks
Examples	GPT-4, Claude, PaLM	Phi-3, DistilBERT, MobileBERT

1.1 From general AI to healthcare applications

LLMs and SLMs connect the general-purpose language processing of transformers with their raw deterministic data representations of a specialized domain. When it comes to healthcare and medicine, the process starts with the natural language processing (NLP) of unstructured data, clinical notes, patient records, research papers, and imaging reports that make up ~80% of healthcare data. Models use it for extracting insights like identifying symptoms from unstructured history text or trial result summaries.

Essential connections are:

- LLMs analyze patient data along with medical literature to propose a diagnosis or a treatment plan, thus speeding up triage (e.g., Med-PaLM).
- Healthcare Analytics: They allow predictive modeling of electronic health record (EHR) data to forecast, for example, which diseases will outbreak, which patients will be readmitted, and which drugs will interact, identifying patterns in data collected from an entire population.
- The SLMs with the requisite efficiency will power on-device apps that facilitate near real-time analytics, like mobile tools for rural clinicians who process local patient data without cloud dependency.

Healthcare analytics uses LLMs to turn raw data into insights. To give an example:

1. Data Integration: Compiling EHR, genomic, and wearable data into stories.
2. Time-series analysis for personalized medicine, or how we can predict sepsis from vital-signs and notes data.
3. Automating the meta-analysis of PubMed literature to detect trends in epidemiology.

This evolution positions LLMs/SLMs as force multipliers for professionals in both high-income and low-income nations—empowering information specialists to curate evidence faster and clinicians to focus on care—while demanding rigorous validation for accuracy, bias mitigation, and regulatory compliance like HIPAA (the U.S. Health Insurance Portability and Accountability Act) and India's DISHA (Digital Information Security in Healthcare Act) and DPDP (Digital Personal Data Protection Act). Fig. 9.1. Shows how typical large language models have been adapted for healthcare applications (Hamid and Brohi, 2024; Busch et al., 2025a; Lin and Kuo, 2025).

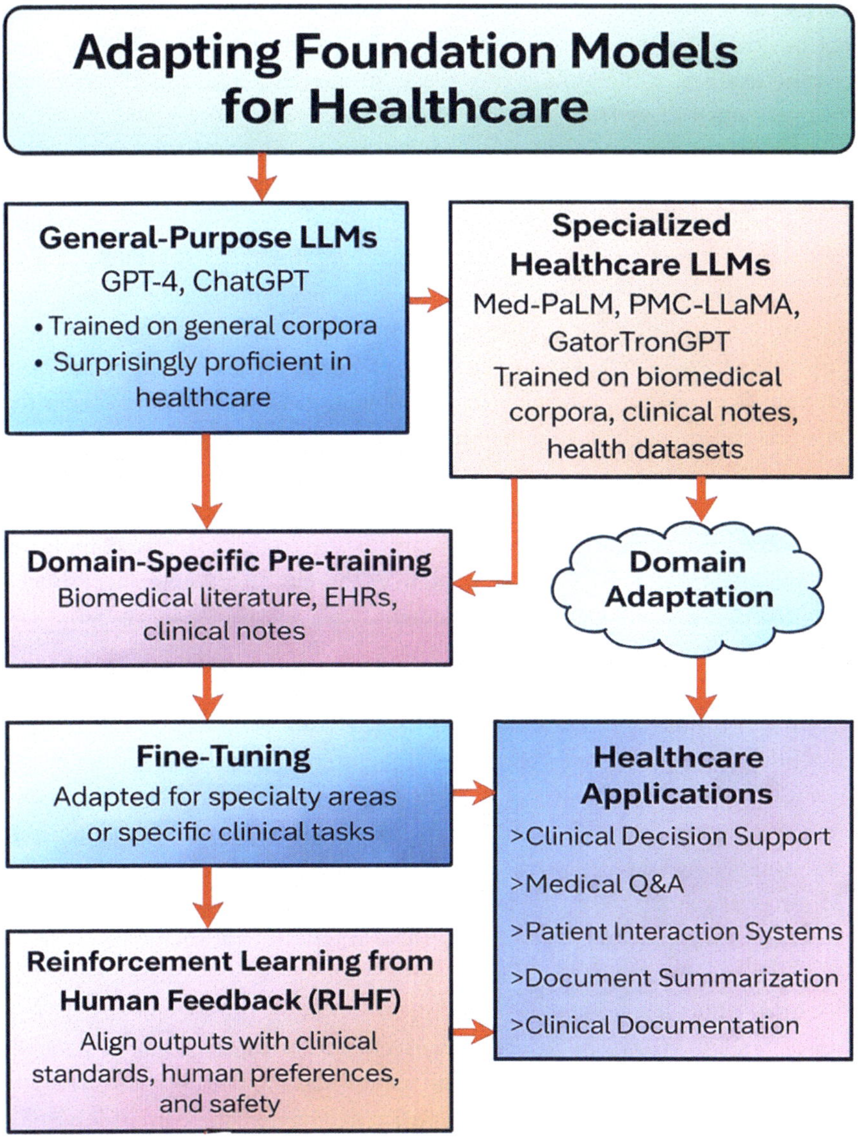

Figure 9.1 Healthcare adaptation of LLMs.

1.2 Role of SLMs in clinical workflows

Small Language Models (SLMs) can serve as an internal reasoning system embedded within healthcare-based, AI-enabled end-user devices, especially when integrated with the retrieval-augmented generation framework. Instead of relying on parametric memory, SLMs can consume clinical artifacts, like practice guidelines, structured patient information, terminologies, or ontologies, and generate structured, machine-readable formats from the information obtained during the retrieval process. At this point, the primary functionality of the SLM is to serve as an internally controlled generation system, with the actual knowledge residing outside in curated sources.

A significant advantage of SLMs, especially in the healthcare setting, is their applicability to privacy requirements with resource constraints. For instance, large-scale LLMs are often heavily dependent on cloud-based infrastructure to facilitate computation, given their high computational demands. However, this raises several concerns, like data residency, transfers, and compliance with existing regulations, especially with the sensitive nature of health-related information. In contrast, SLMs are much more applicable to private settings, given their feasibility of being deployed within the hospital's infrastructure, like servers or edge devices, thereby limiting the exposure of sensitive patient information, especially during the retrieval of information from internal knowledge bases, like the Fast Healthcare Interoperability Resources (FHIR) bundles or practice guidelines hosted within the hospital's infrastructure. Furthermore, several clinical settings are unable to provide the necessary infrastructure to facilitate the computation of 100B+ parameter models with minimal latency, thereby increasing operational costs, especially given the need to ensure alignment with existing clinical protocols and evidence-based practice.

1.3 Generative AI to agentic workflows in healthcare

Clinical NLP (Natural Language Processing) has shifted from extraction-oriented methodologies to generation and synthetic knowledge representation. Generally, traditional NLP has been associated with the extraction of structured information, including entities and relationships (e.g., subject-predicate-object triplets) in clinical texts (Singhal et al., 2023; Moor et al., 2023). This has been very important in standardizing terminology and enhancing interoperability. However, these methodologies have been limited to the use of schemas and ontology-based representations, which can be regarded as a "closed world." This has limited their ability to represent the complexity, ambiguity, and time-dependent nature of patient narratives. For instance, a system may be able to extract a laboratory value such as "the patient has a hemoglobin A1c level of 8.2%," while being unable to represent a statement such as "the patient looks anxious about his blood sugar level," which has important psychosocial and behavioral implications.

In this regard, there has been a notable trend toward generative semantic analysis with the advent of LLMs. The latter shows promise in demonstrating their potential to reason about distributed semantic representations, thereby providing a deeper level of deduction. The emergence of LLMs has been accompanied by the development of a new class of LLMs, namely, SG-LLMs (semantic graph-augmented, safety-guarded).

SG-LLM integrates knowledge from two sources, one being parametric knowledge, which is encapsulated in a large amount of internally acquired information from pre-training, and non-parametric knowledge, or knowledge in its explicit form, which is verifiable and obtained from external sources such as PubMed, UpToDate, and guidelines from relevant institutions (Zakka et al., 2024). The paradigm has changed significantly, moving from a data-gathering tool to a tool for clinical judgment co-reasoning, thus establishing a semantic connection between unstructured clinical text and evidence-based outcomes.

While the initial models of Generative AI were centered on providing 'passive' outcomes, such as summarizing medical charts or answering questions, Agentic Workflows, on the other hand, utilize multi-step reasoning, tool usage, and self-corrective actions to manage patient care proactively. The move from Generative AI to Agentic Workflows, therefore, signifies a significant shift from models that only generate medical information to models that can independently perform clinical tasks. A conceptual illustration of a typical, but simplified, workflow is given in Fig. 9.2.

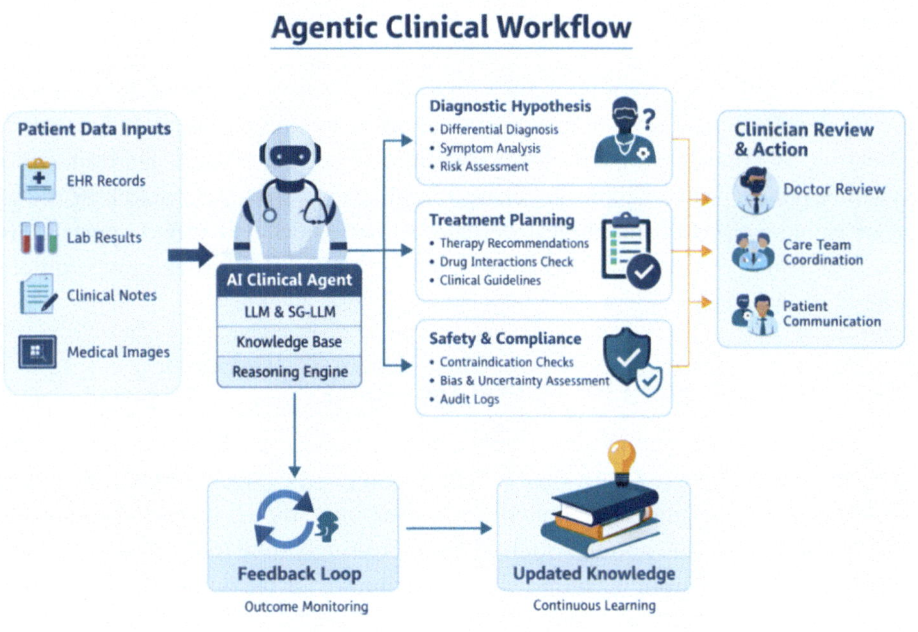

Figure 9.2 A typical agentic clinical workflow Diagram.

1.4 Semantic taxonomy of LLM utility in healthcare

The idea of **"productizing"** Large Language Models (LLMs) is to make them useful across a wide variety of medical disciplines without requiring the user to have extensive technical expertise (Hamid and Brohi, 2024). A wide-ranging taxonomy of LLMs divides their utility into three primary areas of application:

- **Clinical and Professional Support:** LLMs help in automating repetitive tasks such as writing clinical letters, categorizing patients based on the nature of their symptoms, or summarizing electronic health records (EHRs) for discharge notes (Hamid and Brohi, 2024; Lin and Kuo, 2025). Specialized models, such as **GatorTron**, have been pre-trained on over 90 billion words of EHR and clinical text specifically to assist in these complex linguistic tasks (Hamid and Brohi, 2024).
- **Patient Empowerment and Education:** These models function as public-facing tools for translating complex medical terminology for the general public. By providing compassionate and readable responses to health-related queries, they help patients understand complex medical conditions and personalized care plans better (Busch et al., 2025b; Lin and Kuo, 2025).
- **Diagnostic and Specialized Analytics:** In specialties such as radiology, pathology, and ophthalmology, LLMs function as cognitive tools. They assist in formulating diagnoses, providing information on the course of treatment, or retrieving evidence-based information about specific medications (Busch et al., 2025b; Lin and Kuo, 2025).

This chapter discusses various architectural, clinical safety, and regulatory considerations for LLMs in healthcare semantics. It will cover various topics, including prompt engineering to convert probabilistic models into structured clinical reasoning tools, RAG for evidence-based output generation, and layered strategies to address the key issue of hallucinations in the clinical domain. Other topics will include performance metrics, strategic considerations for fine-tuning versus RAG, and the new frontier of agentic clinical workflows.

2. Prompt engineering for clinical reasoning

In a clinical context, "prompt" has a very specific, well-defined role, a set of medical command sequences intended to elicit a particular type of evidence-based, more structured, constrained, or guideline-aligned reasoning from a model. Its purpose is to limit semantic drift, a phenomenon in which LLMs increasingly tend to diverge from factual accuracy and logical reasoning, and to elicit a machine-readable output through prompt engineering.

New forms of prompting for medical LLMs demonstrate that structured prompt sequences can improve both the quality of reasoning and its auditability. For example,

Chain-of-Thought prompting, as demonstrated in (Wei et al., 2022), shows that, by guiding LLMs to reason step-by-step, it is possible to elicit a response that is much more transparent and auditable, akin to steps taken by a medical caregiver in order to diagnose a condition.

In a practical sense, for medical question answering, structured prompting has been evaluated on medical datasets such as MIMIC-IV, demonstrating its potential to elicit a response from a model that meets medical standards, reducing hallucinations. However, most published work has focused on CoT reasoning or retrieval-augmented prompting, and little has been done to rigorously define a unified framework for medical prompting. The new prompts are often built upon a small number of foundational primitives, enabling robust reasoning:

- Role specification: for example, Board-certified senior oncologist, Board-certified oncologist.
- Knowledge domains or guidelines to be followed (e.g., "Reference only the 2025 NCCN guidelines").
- Data related to a patient's context, often structured and represented in a format such as FHIR JSON.
- Output format: directives for the display of the generated output (e.g., "provide the answer in FHIR-compatible JSON").

FHIR is an international standard created by HL7 International that facilitates the sharing of electronic health records (EHRs). As a benchmark for interoperability of healthcare data worldwide, FHIR seeks to break down data silos created by various clinical information systems with incompatible formats. With a framework to share clinical data electronically across multiple platforms, FHIR ensures that information pertaining to a patient can be accessed across systems in a computable manner.

This chapter proposes that a medical prompt can be conceptualized and decomposed into four architectural primitives (reducing tokens reduces LLM inference cost):

Primitive	Token budget	Clinical function
Role	$\leq$10	Calibrates prior probability
Constraint	$\leq$8	Bounds hypothesis space (e.g., "2025 NCCN v5.1")
Context	$\leq$4k	Patient-specific FHIR bundle
Output schema	$\leq$5	Enforces downstream computability (CDS Hooks, FHIR R5)

By combining CoT instructions with architectural primitives, the model's hypothesis space may be narrowed, output variance potentially reduced, and adherence to contemporary medical guidelines improved. While CoT prompting has been validated in prior work, the four-primitives framework represents a conceptual synthesis. Its applicability to clinical datasets such as MIMIC-IV ($n = 2143$) is proposed here as a theoretical extrapolation rather than an empirically validated evaluation.

2.1 Case study: Differential diagnosis reasoning

The impact of semantic engineering on the reasoning of large language models (LLMs) can be readily appreciated by comparing naive questions with questions that have been carefully designed for clinical heuristics. Let us take, for example, a 60-year-old male patient who presents with an acute onset of pain in the right hip, fever (40°C), and moderately high C-reactive protein levels. The naive question of "What causes hip pain and fever?" will most likely elicit a non-specific, non-priority listing of possible etiologies, which will often include a mixture of common degenerative causes and less common, but important, infectious etiologies. The engineered question will, however, prompt a form of Bayesian reasoning, with a focus on "must not miss" etiologies like septic arthritis or osteomyelitis, and will also require the model to support its answer with evidence based on appropriate clinical practice guidelines.

A refined prompt could be articulated as: "Analyze the following: A 60-year-old male patient with acute onset of right hip pain, fever at 40°C, and a moderately elevated level of C-reactive protein. Prioritize infectious causes over degenerative causes. Provide the 'must not miss' diagnoses and format the answer in JSON."

This methodology ensures the clustering of symptoms and the differential diagnosis process. A study conducted on large data sets, for example, MIMIC-IV, with the use of similar refined prompts for the model, has demonstrated an improvement in the accuracy of the results (Kim et al., 2025). Structured prompting and retrieval augmentation have been shown to improve task-specific diagnostic or question-answering accuracy by 15–30 percentage points under controlled benchmark conditions, depending on the task formulation and evaluation protocol. The restriction on the format and the linking with evidence make the output of the model an actionable tool, rather than an information repository.

3. Retrieval-augmented generation (RAG) in medicine

A major architectural solution to address some of the most significant problems (such as knowledge cutoffs and hallucinations) that afflict standalone large language models (Lewis et al., 2020) is Retrieval-Augmented Generation (RAG) (Lewis et al., 2020). Therefore, RAG is used to ground LLM outputs.

3.1 Semantic grounding via retrieval-augmented generation

The problem with pure LLMs is that they have a "knowledge cutoff" after which their "memory" becomes a static snapshot of the past. This is resolved by the Retrieval-augmented Generation (RAG) approach, which allows a generative model to have a "dynamic reference library." By integrating the LLM with live external knowledge sources, RAG ensures that the clinical outputs are not only linguistically fluent but also factually up-to-date and grounded in verifiable data.

3.2 The two-step process of RAG

RAG separates the act of finding information from the act of writing the response, keeping the AI anchored to "ground truths."

- **Retrieval Phase:** Upon receiving a query, the system scans a pre-defined knowledge base—such as clinical guidelines or institutional protocols—to find the most relevant data using similarity matching.
- **Generation Phase:** The LLM receives this retrieved data as context. It then synthesizes its general reasoning capabilities with this specific, up-to-date information to produce a tailored response.

3.3 Why RAG is essential for healthcare

Medical knowledge evolves too rapidly for traditional AI retraining, which is both slow and expensive. RAG addresses these critical healthcare requirements by:

- **Mitigating Hallucinations:** Responses are tied directly to authorized medical literature rather than the model's "best guess."
- **Maintaining Currency:** Systems can access the latest research findings or updated drug formularies in real-time.
- **Contextual Accuracy:** RAG can pull from Medical Ontologies (like SNOMED CT or ICD-10) to recover precise semantic information for healthcare analytics.
- **Personalized Education:** It allows for the rapid synthesis of vast literature into patient-friendly materials adjusted for health literacy levels.

3.4 Key benefits of RAG for healthcare implementation

The benefits of RAG make it a suitable tool for healthcare settings. Unlike black box AI systems, RAG provides clear attribution to source material, allowing healthcare professionals to assess the basis of the AI response. Continuous updates with new research and guidelines can easily be done without an expensive retraining of the model.

Healthcare organizations can create specialized knowledge bases with their chosen clinical materials, ensuring AI answers conform to organizational expectations. The retrieval feature also allows the systematic curation and quality control of information sources.

3.5 RAG's semantic and analytic implications for healthcare

RAG systems can utilize clinical ontologies that semantically enrich interpretations of health data by retrieving background knowledge that enables deeper analytic insight with reference to the overall clinical picture. The respective code and semantic frameworks can be used through RAG for proper navigation of healthcare data standards. A basic understanding of RAG is central to exploring how we can leverage these systems for improving healthcare semantics and analytics, so that AI applications in healthcare are powerful and trustworthy.

In recent years, RAG-based clinical systems have witnessed significant advancements, leading to the development of Multi-Evidence Clinical Reasoning (MECR) frameworks. These are designed to reproduce, with considerable accuracy, how a medical practitioner would review various authoritative sources before making a definitive diagnosis or treatment decision. The MECR process typically starts with *Ingestion,* followed by *Semantic Chunking.*

Instead of splitting text arbitrarily (e.g., every 500 characters), a long text is divided at *meaningful boundaries into chunks*—like sentences, paragraphs, or topic shifts—so each chunk preserves context and coherence. In the present context, authoritative medical sources, ranging from clinical guidelines to drug information and scientific literature, are segmented into semantically meaningful "semantic chunks" of "clinical reasoning units." An important goal of this process is to preserve contextual relationships. For example, a drug's indications and contraindications must be kept together in a single chunk.

The Hybrid Search mechanisms enable the retrieval of relevant information through the integration of various approaches. More specifically, the following approaches are integrated into the proposed method:

(1) Dense vector similarity: Domain-specialized encoders like BioBERT or Clinical-T5 are employed to encode the patient's query and document chunks into dense vector space representations. The similarity in the vector space encodes the concepts and semantic relatedness of the information.

(2) Sparse keyword matching (e.g., BM25): This traditional information retrieval method is employed to achieve precision in the search results through exact term matching. This is particularly important in the retrieval of specific information like drugs, ICD-11 codes, or even genetic information.

The retrieved information, typically the top three to five most relevant evidence snippets, is then integrated with the original patient query. This augmented prompt is given to the model during the *Grounded Generation* step, which explicitly instructs the model to generate a response only based on the provided context (Zakka et al., 2024; Lewis et al., 2020). This constrains the response to be contextually relevant and grounded in actual medical knowledge.

3.6 Mathematical basis of retrieval

The fidelity and clinical utility of medical RAG systems are rigorously quantified through metrics, such as Cosine *Similarity*, which is a mathematical tool for measuring the similarity between two pieces of information by measuring their orientation rather than their size. In real-world applications, any given piece of information, be it a document, a patient's file, or a medical note, can be represented as a vector, a mathematical construct representing a given piece of information in terms of a set of numbers. When two vectors point in the same direction, their cosine similarity is closer to 1, representing a higher level of similarity or relatedness between two given pieces of text. When two vectors point in different directions, or in other words, when they are orthogonal, their cosine similarity is closer to 0, representing a lack of relatedness or similarity.

From a semantic perspective, embedding vectors can be interpreted as geometric encodings of clinical concepts. Two clinical entities — for example, "acute coronary syndrome" and "elevated troponin" — occupy nearby regions in latent space when learned from large corpora. Thus, cosine similarity operationalizes semantic proximity rather than simple lexical similarity. This geometric representation allows LLM-based systems to capture synonymy, abstraction, and contextual nuance beyond what traditional keyword-based systems could represent.

This measure of similarity is highly important in medical applications, allowing users to find similar patient records or align symptoms with medical literature. Thus, the relevance of a retrieved document chunk d to a clinical query q is fundamentally measured by the ***cosine similarity*** τ of their respective vector *embeddings* ($\mathbf{v}q$ and $\mathbf{v}d$) as given below:

$$\text{similarity}(q, d) = \frac{\mathbf{v}_q \cdot \mathbf{v}_d}{\| \mathbf{v}_q \| \| \mathbf{v}_d \|}$$

In some implementations, relatively high thresholds (e.g., cosine similarity $\tau > 0.8$–0.9) are used to prioritize precision. The value is usually derived based on empirical studies and applied as needed. The similarity threshold is a key limiting factor that prevents the incorporation of clinically irrelevant or ambiguous literature into the reasoning process and thereby reduces the likelihood of arriving at false conclusions.

The performance of the system is also validated based on various information retrieval measures. Precision@K is another key performance metric that has received considerable attention. The metric is based on the proportion of the top K documents that are actually relevant to the information query. The state-of-the-art MECR systems reported in recent literature state a high Precision@5 value > 0.90. This high value indicates the high precision with which the systems can retrieve the most relevant information to the query (Zakka et al., 2024; Kim et al., 2025) in the constrained environment. In addition to the similarity scores, the state-of-the-art systems also employ uncertainty quantification techniques such as probability calibration or ensemble-based confidence estimation. These systems can automatically recommend the review of results if the confidence is below the threshold value.

4. Hallucination mitigation in medical LLMs

In the realm of healthcare, large language models can experience hallucinations – gaffes that are plausible-sounding but not based on facts, data, or references. These are mistakes that can lead healthcare efforts astray and put patients at risk. Rather than dismissing these as isolated incidents, we require a systematic and multilayered defense. The approach is defense-in-depth – the use of several layers of defense to reduce the likelihood and impact of these errors. These layers are:

- L1. Self-Reflection (Internal Logic Layer): The model is prompted to reflect upon its own logic and reasoning, employing a Chain-of-Thought or Reflexion pattern. An additional reflection pass is invoked to evaluate intermediate reasoning before producing the final answer, thus preventing errors such as gender-specific dosage contraindications or internal logical inconsistencies (Wei et al., 2022).
- L2. N-Cross Checking (Ensemble Layer): In this approach, multiple parallel answers are generated by multiple independent foundation models, such as GPT-4o, Gemini 2.5, or DeepSeek-R1. A separate Judge evaluates the semantic consistency of these answers. If the models disagree beyond a certain safety margin, hallucinations are detected and alerted, as proposed in (Lyu et al., 2024).
- L3. Knowledge Graph Guarding (Semantic Grounding Layer): This is the most secure approach, wherein the entities generated by the model, such as medications, diagnoses, or ICD-11 codes, are grounded in fixed knowledge graphs, such as DrugBank and SNOMED CT. With the GraphRAG framework, individual claims are validated by traversing multiple hops to detect contraindications that contradict existing knowledge, as proposed in (Mabrouk et al., 2026).

This layer integration reconciles two semantic paradigms: symbolic ontological semantics and probabilistic (distributional) neural semantics. Ontologies such as SNOMED CT encode hierarchical and relational meaning deterministically, whereas LLM embeddings encode meaning probabilistically through contextual co-occurrence. GraphRAG and ontology-grounded systems function as semantic alignment layers, constraining probabilistically-generated outputs within formally defined clinical concept systems. In this sense, hallucination mitigation is not merely a safety technique but a semantic normalization process as well.

4.1 Benchmarking the defense-in-depth Pipeline

Experiments from 2025 to 26 suggest that the transition from large language models to ontology-aware models reduces the hallucination rate significantly. The work of Mabrouk et al. (2026) found that structured clinical semantics could be used to improve the performance of large language models by being incorporated into their reasoning. The accuracy of question answering was greatly improved by using their ontology-based

GraphRAG approach, which was able to attain as much as 98% accuracy during benchmark evaluations. The relative hallucination rate was also reduced by more than 61% when compared to the best-performing LLM (Mabrouk et al., 2026). Here are some data from the literature:

Layer	Technique	Impact (error Reduction)	Latency (typical)	Source
L1	Self-reflection (SR)	~30%	350–500 ms	Wei et al. (2022); Kim et al. (2025)
L2	N-cross-check (N = 3)	~50% consistency gain	1.8–2.2 s	Goel et al. (2025)
L3	KG guard (GraphRAG)	**>61% relative reduction**	2.1–2.5 s	Mabrouk et al. (2026)

Attacking hallucinations represents an important technical challenge, and it also represents an important philosophical change in the way that we think about clinical decisions. Instead of relying on the LLM as the final answer, we are using it as a hypothesis that needs deterministic verification. This perspective on the problem aligns with the Bayesian perspective on diagnosis, in that the evidence provided by the AI system updates the prior probability that the physician had for the diagnosis, without replacing it.

5. Strategic adaptation: Fine-tuning versus RAG

An important technical decision in deploying LLMs in healthcare is whether to use Fine-Tuning, that is, modifying the model's internal parameters, or Retrieval-Augmented Generation (RAG), that is, accessing external knowledge in real-time.

Fine-Tuning is advantageous in embedding domain-specific knowledge, adapting the model's style to the specific communication culture in the healthcare setting, or learning specific medical knowledge for which there exists an abundance of data. Parameter-efficient techniques, such as LoRA (Low-Rank Adaptation) and its quantized version QLoRA (Dettmers et al., 2023), have greatly reduced the computational cost and the problem of "catastrophic forgetting" compared to traditional fine-tuning.

On the other hand, RAG is advantageous for accessing constantly updated knowledge bases, especially in rapidly changing fields, such as oncology, in which treatment recommendations are constantly updated, or in the latest medical research. The advantage of using RAG lies in the ease with which the knowledge bases are updated, without the need for the model's weights and biases to be updated, which involves high computational and time costs.

The choice between these two options, or the combination of both, depends on the evaluation of the Knowledge-Utility Frontier (KUF). The KUF assesses the value that the LLM adds to healthcare against the cost, in terms of computational, regulatory, and maintenance requirements, of the two options. The current best LLM models in healthcare are the combination of both options: the LLM model is first fine-tuned, perhaps using the parameter-efficient techniques, and then the model is used within the RAG system, in order for the model to leverage its medical fluency while at the same time relying on the most current and accurate knowledge available (Zakka et al., 2024; Kim et al., 2025).

5.1 The evolution toward multimodal and agentic architectures

In the real world, it is very rare to find a clinical case where a conclusion is reached solely based on one data source. Instead, a physician naturally synthesizes the patient's history with imaging, laboratory, and genomic data to construct a complete clinical picture. **Multimodal AI** systems aim to mimic this human process by integrating various data streams, such as clinical text, medical imaging, ECG waveforms, and structured EHR data, to produce a single, unified inference (Lin and Kuo, 2025).

The greatest asset of a multimodal AI system is its ability to learn **cross-modal representations** (Lin and Kuo, 2025). Unlike a traditional Large Language Model that operates primarily on text, a multimodal system maps all data types into a coordinated latent space. This allows the system to correlate findings across domains, such as linking a specific radiographic abnormality with a patient's narrative history or laboratory trends. For example, instead of evaluating a chest CT in isolation, a multimodal system can analyze the imaging alongside oxygen saturation curves and radiology notes to more effectively assess a patient's risk of pulmonary embolism.

5.2 Distinguishing multimodal perception from agentic reasoning

It is important to distinguish between **Multimodal AI** and **Agentic AI**, as they represent different dimensions of clinical intelligence:

- **Multimodal AI** is about *perception*: what types of data can the model understand together?
- **Agentic AI** is about *orchestration*: how does the system structure its reasoning, use tools, and execute a workflow?

While a system can be multimodal without being agentic, the most sophisticated clinical assistants are often both. An **Agentic AI** acts as a goal-directed partner that plans, retrieves guidelines, invokes diagnostic tools, and iterates toward a safe outcome (Lin and Kuo, 2025). For instance, a clinical agent might extract structured data from a record, retrieve the latest NCCN guidelines, generate a treatment plan, cross-check for contraindications, and finally format everything into an FHIR-compatible care plan for physician review.

5.3 Architectural frameworks and real-world constraints

Technically, these systems are built using three dominant paradigms or frameworks:

1. **Early Fusion:** Modalities are merged into a shared space before any reasoning occurs.
2. **Late Fusion:** Independent encoders analyze each data type separately, combining their "conclusions" only at the final decision layer.
3. **Cross-Attention:** One modality (e.g., text) dynamically "attends" to specific features of another (e.g., a specific region on a pathology slide).

Despite the promise of unified foundation models, real-world deployment is still constrained by the inherent "messiness" of clinical data. Challenges such as dataset heterogeneity, missing modalities, and domain shifts between institutions mean that these systems require rigorous clinical validation and regulatory oversight before they can be fully integrated into the bedside workflow (Hamid and Brohi, 2024; Lin and Kuo, 2025).

6. Safety, regulatory, and ethical considerations

6.1 Security framework: Reducing adversarial and data risks

As LLMs transition from the lab to the real world, the ability to identify security vulnerabilities becomes crucial to avoid devastating consequences (Hamid and Brohi, 2024). A robust security framework must address a variety of threat areas:

- **Adversarial Manipulations:** Adversarial actors could use **"jailbreaking"** or **prompt injection** to evade safety rules and could potentially manipulate a system to provide dangerous medical advice (Hamid and Brohi, 2024).
- **Privacy and Data Exfiltration:** General-purpose LLMs are vulnerable to **membership inference** and **model inversion** attacks, allowing attackers to exfiltrate sensitive medical information or de-anonymize patient data stored within the model's training weights (Hamid and Brohi, 2024).
- **Clinical Integrity Attacks:** Adversarial actors could use system weaknesses to manipulate clinical information covertly, including altering dosage recommendations and deactivating critical contraindication alerts within an agentic workflow (Hamid and Brohi, 2024).
- **Organizational Defense:** Healthcare is a prime target for cybercrime because it is a historically "soft target" and valuable, sensitive information is at stake (Hamid and Brohi, 2024).

6.2 Output limitations and ethical risks

There are significant **"output limitations"** in the performance of LLMs, as outlined in systematic reviews, which must be addressed by automated audit layers (Busch et al., 2025b). These limitations are classified as follows:

- **Scientific and Factual Inaccuracy:** Output limitations include a lack of reproducibility and completeness (Busch et al., 2025b). There is also a tendency for LLMs to confuse fiction with facts and to generate confident but incorrect **"hallucinations"** (Hamid and Brohi, 2024).

- **Social and Ethical Harms:** There is a persistent tendency to reflect systemic biases related to gender, ethnicity, religion, and age, which can compromise clinical equity (Hamid and Brohi, 2024).
- **Safety and Transparency:** Insufficient data transparency and a lack of optimization for domain-specific models can result in ungrounded clinical responses, creating significant safety issues (Busch et al., 2025b).

As the rules and guidelines continue to evolve and shift—such as the EU AI Act and the latest US-FDA guidance on AI/ML-based medical devices and India's —some types of LLM-based medical tools may be classified as high-risk devices when they influence diagnosis, treatment selection, or patient management. In such cases, they must comply with stringent requirements around safety, performance validation, risk management, transparency, human oversight, and post-market surveillance. Here are the core compliance requirements:

- Transparency and Provenance: Each recommendation and/or summary provided by the LLM must include specific, citable links to the exact source documents upon which the LLM's recommendations are based.
- Bias Auditing: LLMs are subject to rigorous testing for demographic bias. Demographic parity must be checked so the LLM's recommendations are the same across different patient demographics (e.g., by race, gender, age) when the underlying facts are the same.
- Human-in-the-Loop Mandate: LLMs are used as Decision Support Systems (DSSs) in healthcare and are not allowed to make independent decisions regarding patient care. Every action taken by the LLM must be subject to review, with audit trails used for accountability and insurability against malpractice. Some proposals suggest distributed ledger technologies (e.g., blockchains) for tamper-evident audit logs.

7. Performance benchmarking

Recent large language models have demonstrated strong performance on selected medical benchmarks. They perform well on USMLE-type examinations and on certain diagnostic reasoning tasks that require step-by-step reasoning. This represents meaningful progress in retrieving relevant knowledge, connecting intermediate reasoning steps, and answering medical questions when problems are well-defined and evaluated under controlled conditions.

However, high performance on medical benchmarks does not guarantee clinical readiness. Medical benchmarks are standardized evaluations that primarily assess knowledge recall and structured reasoning within well-scoped scenarios. Clinical practice, in contrast, involves incomplete information, ambiguous symptom narratives, evolving patient states, and the medico-legal implications of decision-making. Thus, strong benchmark performance is an important indicator of capability, but it does not, by itself, establish safety, reliability, or suitability for real-world clinical deployment.

The above-mentioned change in the criteria is also part of the wider understanding that being medically proficient is not simply the ability to recall facts. Rather, it is the ability to reason through contexts, navigate uncertainty, and adjust the level of communication to the needs of the patient. As emphasized by Singhal et al. (2023) (Singhal et al., 2023), the ability of Large Language Models to pass standardized medical examinations represents the "table stakes" of clinical AI development. However, the foundational challenge remains the transition from these structured benchmarks to the subtle navigation of clinical ambiguity. To address this, the authors introduced the MultiMedQA framework, arguing that the true measure of a medical system lies in its capacity for long-form reasoning and safety-aligned performance within the inherent complexity of real-world patient encounters.

8. Clinical workflow example: adjuvant therapy selection

To illustrate the application of the defense-in-depth approach in action, we can consider the process of selecting adjuvant therapy for early-stage breast cancer. Below is an illustrative sequence for a 50-year-old woman with ER-positive, HER2-negative, node-negative invasive ductal breast cancer.

The agentic sequence:

1. **Context Assembly (The Data Agent):** The process begins with the automated creation of the patient's "digital twin." An autonomous agent queries the EHR via RESTful APIs to retrieve relevant longitudinal data. This is returned as a structured **FHIR R5 Bundle**, consisting of modular resources: Patient, Condition, Observation, and Diagnostic Report (including Oncotype DX scores).
2. **Completeness Validation (The Audit Agent):** Prior to reasoning, an independent audit agent verifies the "clinical completeness" of the dataset. It checks for the presence of essential variables: tumor size, receptor status, menopausal status, and genomic markers. If data are missing, the system is programmed not to "guess" or hallucinate the missing values; instead, it triggers a request for human clarification.
3. **Evidence Grounding (The Retrieval Agent):** To prevent reliance on static pre-trained knowledge, a retrieval agent grounds the system in real-time medical evidence. It accesses the latest **NCCN Breast Cancer Guidelines**, mapping the specific patient attributes to the corresponding evidence-based clinical pathway.
4. **Constrained Recommendation (The Logic Agent):** Once grounding is verified, the recommendation agent generates therapeutic options. Crucially, this agent operates under a **constrained reasoning** mandate: it is restricted to generating only

those options explicitly supported by the retrieved clinical pathways, effectively bounding the hypothesis space.

5. **Safety Reflection (The Guarding Agent):** The output undergoes a final reflection pass by a guarding agent (integrating **L1 Self-reflection** and **L3 Knowledge-Graph Guarding** layers). The agent cross-references the proposed care plan against the patient's comorbidities—such as assessing cardiovascular risks or potential drug-drug interactions that may not be covered in the primary oncological guidelines.
6. **Human-in-the-Loop (The Care Plan):** The final output is delivered to the oncologist as a structured **FHIR Care Plan** resource. In this architecture, the AI does not execute clinical actions; it serves as a sophisticated "expert drafter." The oncologist remains the ultimate decision-maker and the sole moral agent responsible for the patient's care.

Key components of the workflow:

- **Structured Patient Assembly:** Utilization of FHIR-compatible resources for high-fidelity data input.
- **Evidence Retrieval:** Real-time grounding via retrieval-augmented generation (RAG) from gold-standard guidelines.
- **Constrained Reasoning:** Logic tied explicitly to retrieved sources to prevent "speculative" medical advice.
- **Reflection Pass:** An automated audit to detect clinical contradictions or logical fallacies.
- **Policy Compliance:** Validation against institutional safety standards prior to delivery.
- **Clinical Oversight:** Mandatory human-in-the-loop approval before any clinical action is finalized.

9. Conclusion: Toward agentic clinical workflows

Fundamentally, the evolution described in this chapter is a semantic transformation. Healthcare AI is moving from rigid, ontology-bound meaning representations toward adaptive, context-aware semantic modeling. However, to be used safely in a real-world clinical setting, we need to have a hybrid system that combines probabilistic semantic reasoning with strict, deterministic clinical standards. In the near future, it is expected that LLM-based orchestrators will be able to autonomously perform multi-step tasks, such as retrieving lab results through programmatic interfaces, checking patient information against guidelines such as KDIGO (stands for *Kidney Disease: Improving Global Outcomes*, an international non-profit that develops clinical practice guidelines) for the treatment of chronic kidney disease, and creating initial clinical assessments that require a clinician's co-signature.

As LLMs evolve from chat partners into proactive members of the medical team, the medical priority remains the same: safety. Improvements in prompt engineering, RAG models, and the ability to mitigate hallucinations must be made in the service of this principle. A highly advanced semantic analysis system is of little use in the field if it cannot be trusted. Trust in medical AI systems stems from transparency, reliability, and clear evidence of benefits for patient care.

Traditional clinical NLP systems focused on entity extraction and schema population, effectively mapping text to predefined semantic slots. In contrast, LLM-based systems operate on continuous semantic representations, allowing them to reason across loosely structured narratives, infer latent relationships, and integrate context dynamically. As such systems become more integrated into the field of medicine, they may provide a means of increasing access to medical knowledge, reducing the cognitive burden on medical staff, and improving patient outcomes. However, this potential can only be realized by grounding the development of the system in the reality of the medical field, the regulations of the field, and the ethical implications of the system.

These agent-like systems provide great potential for increasing operational efficiency and reducing potential diagnostic delays. However, the development of the system must be supported by strong safety systems. Such systems ensure autonomous actions are fully reversible, well-documented, and contained within a credentialed policy graph. As LLMs evolve into proactive, integral members of the medical team, the guiding principle remains the same: an unwavering commitment to reliability, auditability, and the medical ethic: "first, do no harm" (primum non nocere) (Zakka et al., 2024).

References

Brown, T., Mann, B., Ryder, N., Subbiah, M., Kaplan, J.D., Dhariwal, P., , … Amodei, D., 2020. Language models are few-shot learners. Adv. Neural Inf. Process. Syst. 33, 1877–1901.

Busch, F., Hoffmann, L., Rueger, C., et al., 2025a. Current applications and challenges in large language models for patient care: a systematic review. Commun. Med. 5, 26.

Busch, F., Hoffmann, L., Rueger, C., et al., 2025b. Current applications and challenges in large language models for patient care: a systematic review. Commun. Med. 5 (26). https://doi.org/10.1038/s43856-024-00717-2.

Dettmers, T., Pagnoni, A., Holtzman, A., Zettlemoyer, L., 2023. QLoRA: efficient finetuning of quantized LLMs. Adv. Neural Inf. Process. Syst. 36.

Goel, A., et al., 2025. Evaluating Large Language Models for guideline-adherent clinical information. J. Med. Internet Res. 27, e78625. https://doi.org/10.2196/78625.

Hamid, R., Brohi, S., 2024. A review of Large Language Models in healthcare: taxonomy, threats, vulnerabilities, and framework. Big Data and Cognitive Computing 8 (11), 161. https://doi.org/10.3390/bdcc8110161.

Kim, J., et al., 2025. Medical hallucinations in foundation models and their impact on healthcare. medRxiv. https://doi.org/10.1101/2025.02.28.25323115.

Lewis, P., Perez, E., Piktus, A., et al., 2020. Retrieval-augmented generation for knowledge-intensive NLP tasks. Adv. Neural Inf. Process. Syst. 33, 9459–9947.

Lin, C., Kuo, C.-F., 2025. Roles and potential of large language models in healthcare: a comprehensive review. Biomed. J. 48, 100868. https://doi.org/10.1016/j.bj.2025.100868.

Lyu, C., Xia, K., Jiang, Y., et al., 2024. LLM-Judge: evaluating LLMs on clinical reasoning. J. Biomed. Inf. 149, 104573. https://doi.org/10.1016/j.jbi.2023.104573.

Mabrouk, M., Zaki, T., Taha, M., 2026. Ontology-grounded knowledge graphs for mitigating hallucinations in clinical question answering. J. Biomed. Inf. 175, 104993. https://doi.org/10.1016/j.jbi.2026.104993.

Moor, M., Naik, N., Masi, M., et al., 2023. Foundation models for generalist medical artificial intelligence. Nature 616 (7956), 259–265. https://doi.org/10.1038/s41586-023-05881-4.

Singhal, K., Azizi, S., Tu, T., et al., 2023. Large language models encode clinical knowledge. Nature 620 (7972), 172–180. https://doi.org/10.1038/s41586-023-06178-3.

Wei, J., Wang, X., Schuurmans, D., et al., 2022. Chain-of-Thought prompting elicits reasoning in Large Language Models. Adv. Neural Inf. Process. Syst. 35, 24824–24837.

Zakka, C., Celi, L.A., Miotto, R., 2024. Almanac: retrieval-augmented Language Models for clinical medicine. NEJM 1 (2). https://doi.org/10.1056/AIOA2300068.

Zhou, J., Li, H., Chen, S., Han, Z., Gao, X., 2025. Large language models in biomedicine and healthcare. Npj Artificial Intelligence 1, 44. https://doi.org/10.1038/s44387-025-00047-1.

Further reading

Hu, E.J., Shen, Y., Wallis, P., et al., 2022. LoRA: low-rank adaptation of Large Language Models. In: International Conference on Learning Representations (ICLR).

Nori, H., King, N., McKinney, S.M., et al., 2023. Can generalist foundation models outperform specialized medical AI? Nature 621, 790–795. https://doi.org/10.1038/s41586-023-06291-3.

Thirunavukarasu, A.J., Huang, S., To, M., et al., 2023. Large language models in medicine. Nat. Med. 29 (8), 1930–1940. https://doi.org/10.1038/s41591-023-02448-8.

CHAPTER 10

Hybrid semantic statistical architectures for explainable and trustworthy clinical decision support systems

Amaresh Kumar Mohanty, Bandhan Panda and Priyadarsan Patra
Department of Computer Science and Engineering, NIST University, Berhampur, Odisha, India

1. Introduction: The trust imperative in intelligent semantic healthcare systems

Artificial intelligence (AI) is becoming a part of clinical practice, which helps with diagnosis and risk classification, as well as treatment planning (Yu et al., 2018; Bajwa et al., 2021; Sutton et al., 2020). Nevertheless, the widespread adoption of AI in healthcare is also hindered by the issues of transparency, interpretability, and credibility (Amann et al., 2020; Babic et al., 2021; Ghassemi et al., 2021). Predictive performance alone, even in high-stakes medical environments, is insufficient for clinical deployment, and explanations need to be semantically meaningful, cognitively deployable, and assisted by established medical knowledge (Holzinger et al., 2019; Miller, 2019; Floridi et al., 2018).

The modern machine learning systems, especially deep neural networks, are highly predictive, and they tend to be non-interpretable (Samek et al., 2019; Lipton, 2018). Their inner representations are non-symbolic and distributed in nature, and it is hard to test whether model reasoning is in agreement with domain knowledge rather than being a random coincidence (Ghassemi et al., 2021; Cabitza et al., 2017).

Explainability in healthcare AI is not merely a technical enhancement but a governance requirement (Floridi et al., 2018; Jobin et al., 2019). Regulatory frameworks, including the principles articulated by the European Commission High-Level Expert Group on AI and provisions embedded within the General Data Protection Regulation, prioritize transparency, accountability, and human oversight (European Commission High-Level Expert Group on AI, 2019; Goodman and Flaxman, 2017; Mittelstadt, 2019).

The currently used explainability techniques, like feature attribution and attention visualization, are mainly post hoc explanations of model reasoning (Ribeiro et al., 2016; Guidotti et al., 2018; Tjoa and Guan, 2020). Although these methods enhance interpretability, they do not assess the semantic coherence between explanations and

Intelligent Semantic Analysis for Healthcare
ISBN 978-0-443-33512-9
https://doi.org/10.1016/B978-0-443-33512-9.00017-2

clinical ontologies and structured medical knowledge (Bodenreider, 2004; Cui et al., 2025; Dissanayake et al., 2020).

This chapter proposes a semantically hybrid approach, combining statistical deep learning with symbolic ontological rules (Rotmensch et al., 2017; Hogan et al., 2021). To model the association between model and clinical knowledge, we create the Explanation Consistency Score (ECS) to measure the correspondence between model-inspired explanations and structured clinical knowledge. Moreover, we introduce a Composite Trust Index (formally defined in Section 18.4) that combines predictive reliability, semantic coherence, and faithfulness of explanations, which makes it possible to assess trust along governance lines (Doshi-Velez and Kim, 2017; Floridi et al., 2018).

The case study of a cardiovascular example enables us to show how the problem of semantic validation can be used to increase the trust levels of intelligent healthcare systems. It is not only necessary to enhance the interpretability, but to operationalize the trust in a cost-effective and governance-friendly way (Topol, 2019; Fehr et al., 2024).

2. Technical contributions of this chapter

The unique contributions of this chapter are summarized as follows:

- **Hybrid Semantic–Statistical Architecture**
  A layered clinical decision support framework integrating supervised machine learning with ontology-driven rule-based reasoning.
- **Explanation Consistency Score (ECS)**
  A formally defined quantitative metric that measures alignment between statistically attributed feature importance and activated semantic rules for a given patient instance. ECS operationalizes cross-layer explanation coherence.
- **Semantic Consistency Framework**
  A structured harmonization mechanism that:
  - Maps predictive features to ontology concepts,
  - Compares feature attribution with rule activation,
  - Detects explanatory discrepancies,
  - Implements a low-consistency escalation protocol for clinician review.
- **Composite Trust Index**
  A composite system-level metric integrating predictive reliability, explanation consistency, and interpretive fidelity to enable structured trust calibration. This is a "trustworthiness proxy".
- **Governance-Integrated System Design**
  Explicit mapping of architectural layers to regulatory and auditability requirements, embedding traceability and oversight mechanisms within system infrastructure rather than post-deployment add-ons.

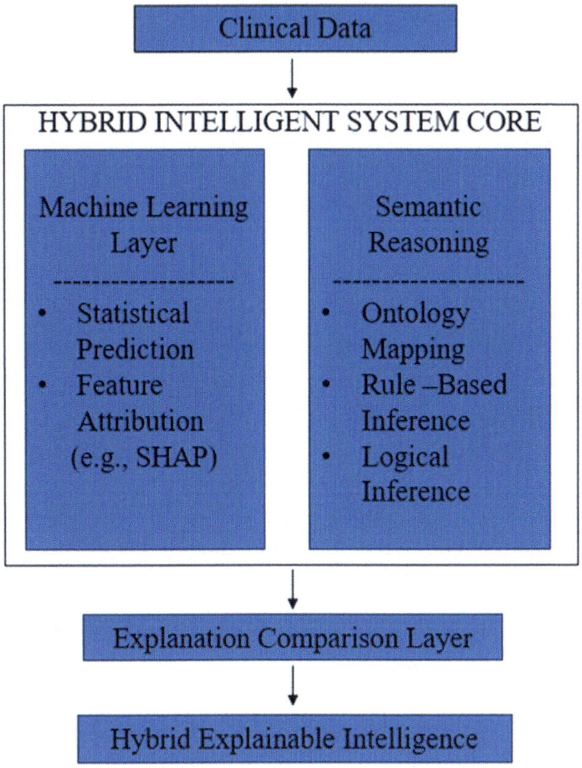

Figure 10.1 Conceptual architecture of hybrid explainable semantic intelligence in healthcare, integrating statistical prediction and ontology-based reasoning. *(Adapted from Arrieta, A.B., et al., 2020. Explainable artificial intelligence (XAI): concepts, taxonomies, opportunities, and challenges. Inf. Fusion, 58, 82–115. https://doi.org/10.1016/j.inffus.2019.12.012; Hogan, A., et al., 2021. Knowledge graphs. ACM Comput. Surv. 54, 1–37. https://doi.org/10.1145/3447772.)*

As the chapter shows through formalization (theory) and validation (empirical), taking a semantic harmonization can increase interpretative coherence without negatively impacting predictive performance. The produced frameworks bring explainable healthcare AI to the level of structurally verifiable trustworthiness instead of a descriptive one (Fig. 10.1).

3. Foundations of explainable artificial intelligence in healthcare

Clinical environments require the incorporation of artificial intelligence, which requires not only accurate systems but also interpretable and legally defensible as well as auditable in terms of their operational processes. Explainability acts as a methodological need and governance tool in high-risk areas like the healthcare sector (Amann et al., 2020; Babic et al., 2021).

In this section, substantive bases of explainability are formalized, they are placed within regulatory frameworks and mechanisms of operational governance that are necessary to ensure trustworthy operation are articulated.

3.1 Conceptual foundations and methodological categorization

Although transparency, interpretability, and explainability are often used interchangeably, they constitute distinct constructs within interpretability research (Lipton, 2018; Guidotti et al., 2018; Sadeghi et al., 2024).

- Transparency refers to the structural comprehensibility of a model (e.g., linear regression, decision trees).
- Interpretability concerns the degree to which a human can understand how input variables influence output.
- Explainability refers to the generation of human-understandable justifications for specific predictions, either intrinsically or through post-hoc analysis.

Two orthogonal dimensions are particularly relevant in clinical AI:

- Intrinsic versus Post-hoc Explainability
- Local versus Global Explanation Scope

To formalize these distinctions, Table 10.1 presents a structured categorization of standard XAI methods used in healthcare.

SHAP (Shapley Additive Explanations) and LIME (Local Interpretable Model-Agnostic Explanations) were widely utilized post-hoc methods that learn an approximation of local decision boundaries using a feature attribution (Ribeiro et al., 2016; Adabi et al., 2016; Berrada, 2018). Nevertheless, as Ghassemi et al. (2021) note, the feature attribution does not imply causal validity and is able to give credible yet false explanations.

Table 10.1 Categorization of explainability methods in healthcare AI.

Dimension	Category	Description	Representative methods	Strengths	Limitations
Model structure	Intrinsic	Model inherently interpretable	Decision trees, logistic regression	Direct transparency	Limited expressive capacity
Model structure	Post-hoc	External explanation applied after prediction	LIME (Ribeiro et al., 2016), SHAP	Model-agnostic flexibility	Potential explanation–model misalignment
Scope	Local	Explains individual patient prediction	LIME, SHAP	Clinically relevant per case	May lack global coherence
Scope	Global	Explains overall model behavior	Feature importance, surrogate models	System-level understanding	May not apply to a specific patient case

Local explanations are especially needed and critical in clinical settings, where there is a possibility of a prediction related to a particular patient instance. Even so, the explanation coherence should not be limited to the results of statistical attribution to be correlated with domain knowledge and structured reasoning.

3.2 Regulatory grounding: GDPR and the EU AI act

Explainability in healthcare AI is not only a methodological concern; it is surrounded by contemporary legal frameworks.

3.2.1 GDPR article 22 Right to explanation

Article 22 of the General Data Protection Regulation (GDPR) offers individuals the right not to be subject to decisions based completely on automated processing that produce legal or similarly significant effects. While discussion exists regarding the precise interpretation of a "right to explanation". GDPR establishes clear requirements for:

- Meaningful information about the automated logic involved,
- The significance and consequences of automated decisions,
- Human intervention and contestability (Goodman and Flaxman, 2017).

In medical contexts, algorithmic decisions affecting diagnosis or treatment recommendations can fall within this regulatory scope. Systems are missing transparent reasoning mechanisms in risk non-compliance.

3.2.2 EU AI Act—High-risk medical AI systems

Under the EU AI Act, clinical decision support systems influencing diagnostic or therapeutic decisions would likely be categorized as high-risk AI systems, thereby necessitating traceability, risk management, and human oversight mechanisms as operationalized in the proposed architecture.

These requirements directly implicate explainability architecture. Observable inference logs, consistency validation, and structured documentation are not voluntary enhancements but compliance-enabling features.

Thus, explainability becomes a regulatory enabler rather than a purely technical add-on. Healthcare AI systems must validate not only predictive validity but also procedural transparency and auditability.

3.3 Governance and traceability workflow

While the chapter references "traceable inference logs," governance compliance needs a formally defined operational workflow. The proposed hybrid architecture implements traceability through the following structured pipeline:

Step-by-Step Explanation Governance Workflow

1. Prediction Generation

   The trained machine learning model computes a probabilistic prediction for a given patient instance.

2. Statistical Explanation Extraction
   Feature attribution methods (e.g., SHAP or equivalent contribution functions) identify top contributing variables for that prediction.
   Output: Ranked feature importance scores.
3. Ontology Mapping and Rule Activation
   Patient features are mapped to ontology concepts.
   Clinical rules are activated over semantic reasoning engines.
   Output: Activated rule set and inference chain.
4. Explanation Consistency Score (ECS) Calculation
   The alignment between top statistical features and ontology rule concepts is computed.
   Output: Quantitative cross-layer consistency metric.
5. Low-Consistency Resolution Protocol (if applicable)
   If ECS falls below a predefined threshold:
   - Automated endorsement is suppressed or flagged.
   - The case is escalated for clinician review.
6. Audit Trail Generation: The system logs:
   - Prediction probability,
   - Feature attribution scores,
   - Activated rules,
   - ECS value,
   - Any override or clinician intervention.

   This structured logging ensures:

- Reproducibility of decisions,
- Post-hoc forensic analysis,
- Regulatory audit readiness,
- Medico-legal defensibility.

Importantly, traceability is implemented as a continuous logging mechanism embedded within the inference execution rather than as a retrospective reporting tool.

3.4 From explainability to governance-embedded accountability

Explainability on your own does not ensure trustworthy AI. Explanations must be:

- Faithful to model behavior,
- Aligned with domain knowledge,
- Documented within traceable workflows,
- Compatible with regulatory expectations.

The hybrid semantic–statistical framework extends conventional XAI by embedding explanation validation, logging, and escalation mechanisms within system architecture. This move transforms explainability from a descriptive output into an assessable governance function.

4. Semantic technologies and hybrid reasoning architecture

4.1 Formal semantic representation in healthcare AI

Semantic technologies provide well established frameworks for encoding domain knowledge in clinical decision support systems. Within the suggested model, the medical knowledge base will be modeled within a formal ontology:

$$O = (C, R, I).$$

where:
C denotes clinical concepts (e.g., Hypertension, Hyperlipidemia, Diabetes Mellitus), R denotes semantic relations (e.g., has-risk-factor, contraindicated-with, indicates), I denotes patient-specific instances.

Web Ontology Language Description Logic (OWL-DL) is used to implement the ontology to deliver certain reasoning that is decided by the Description Logic semantics (Bodenreider, 2004; Hogan et al., 2021; Cao et al., 2024). Rule engines (e.g., SWRL, Jena, or Drools) are used to achieve logical inference, which can be used to check subsumptions, validate consistency, and generate inferences by several hops.

4.1.1 Feature-to-concept mapping

Raw clinical inputs are semantically allied through a mapping function:

$\phi : X \rightarrow C$ where X represents the structured feature space, and C represents ontology concepts.

For example, if systolic blood pressure equals 165 mmHg:

$\phi(165 \text{ mmHg}) = \text{Hypertension\_Stage\_2}$

Given ontology hierarchy:

$\text{Hypertension\_Stage\_2} \sqsubseteq \text{Cardiovascular\_Disorder}$

This transformation ensures that predictive features are grounded in clinically meaningful concepts rather than remaining isolated numerical attributes.

4.2 Hybrid semantic–statistical architecture

The proposed system integrates statistical learning with ontology-driven reasoning through a layered architecture consisting of:

1. Machine Learning Component
   - Risk prediction
   - Probability estimation
   - Feature attribution generation
2. Semantic Reasoning Component
   - Ontology concept mapping
   - Rule-based logical evaluation
   - Multi-hop inference trace generation

3. Explanation Harmonization Core
    - Alignment of ML attribution and semantic inference
    - Conflict detection
    - Consistency scoring (ECS)
4. Trust Calibration Layer
    - Trust Index computation
    - Confidence estimation
    - Consistency verification
5. Ethical & Regulatory Framework Layer
    - Fairness considerations
    - Accountability mechanisms
    - Compliance alignment
    - Audit logging

The conceptual ontology based on harmonization is represented in Fig. 10.2, where the statistical attribution and semantic inference come to a nexus in a central Explanation Harmonization Core and then generate a trust-sensitive clinical decision.

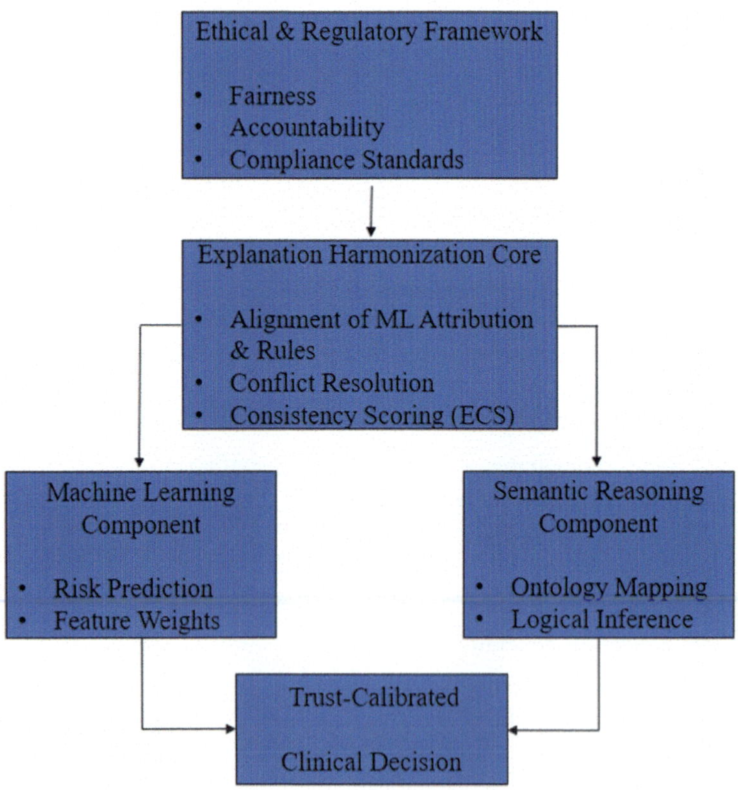

Figure 10.2 Conceptual hybrid semantic–statistical architecture.

As illustrated in Fig. 10.2 that shows the central Explanation Harmonization Core placed among machine learning prediction and ontology-based semantic reasoning, under ethical and regulatory oversight, producing trust-calibrated clinical decisions.

This framework emphasizes ensuring that explainability is not a layer of visualization, but a structural validation process between prediction and decision.

4.3 Semantic contradiction detection

To prevent silent propagation of inconsistent outputs, a formal contradiction rule is introduced. Let:

- $\hat{y}$ denote the ML prediction,
- y_{sem} denote semantic inference output.

A conflict condition is defined as:

$$(\hat{y} = \text{LowRisk}) \wedge \left(y_{\text{sem}} = \text{HighRisk}\right) \rightarrow \text{ConflictFlag}$$

Such rule-based contradiction detection ensures that probabilistic predictions are validated against structured clinical logic before decision release.

4.4 Challenges in hybrid reasoning

Hybrid semantic–statistical systems present specific challenges:

- Inference chain complexity (Miller, 2019),
- Ontology incompleteness and refinement needs (Paulheim, 2017),
- Risk of misleading attribution interpretations (Ghassemi et al., 2021),
- Alignment with clinician mental models (Holzinger, 2016).

These challenges necessitate formal explanation comparison mechanisms, introduced in Section 4.

5. System blueprint and formal data flow specification

The operational outline of the hybrid semantic–statistical clinical decision support system is depicted in Fig. 10.3. The architecture consists of six interacting layers:

1. Clinical Data Layer
2. Predictive Intelligence Layer
3. Semantic Reasoning Layer
4. Explanation Integration Layer
5. Trust Calibration Layer
6. Governance & Compliance Layer

Each layer exchanges formally defined artifacts to ensure traceable and measurable cross-layer validation.

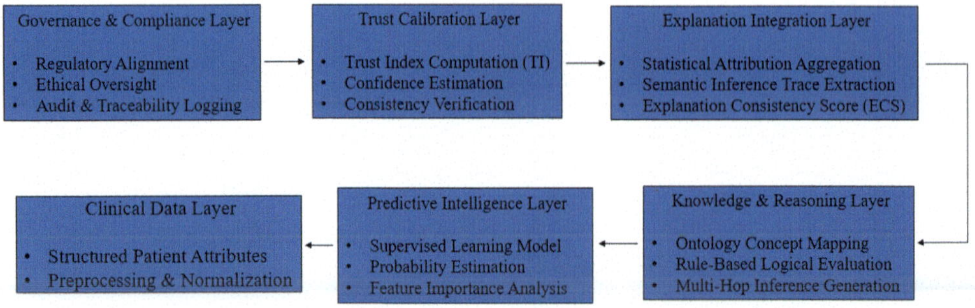

Figure 10.3 End-to-end hybrid semantic–statistical clinical decision support system.

5.1 Formal input–output definitions

Let a patient instance be represented by a structured feature vector:

$$X = (x_1, x_2, \ldots, x_n) \in R^n$$

Prediction Vector

The Predictive Intelligence Layer produces a probabilistic prediction vector:

$$p = f_\theta(X)$$

where:

- f_θ denotes the trained predictive model,
- $p \in$ represents class probabilities (for binary classification, $k = 1$).

This vector encodes the statistical risk estimate for the given patient.

5.1.1 Attribution vector

For the same instance, a feature attribution mechanism generates an attribution vector:

$$a = (a_1, a_2, \ldots, a_n)$$

where:

- $a_i \in R$ denotes the contribution score of the feature x_i,
- a ranks the relative influence of each clinical variable on the prediction.

The top-ranked elements of a from the statistical explanation subset.

5.1.2 Semantic concept set

Instantaneously, the Semantic Reasoning Layer applies ontology mapping and rule-based inference. The mapping function.

$\phi : X \rightarrow C$ transforms raw features into ontology-aligned concepts. Activated rules generate a semantic concept set:

$S_{\text{sem}} \subseteq C$ where:

- C denotes the ontology concept space,
- S_{sem} contains clinically inferred concepts relevant to the patient instance.

This set may include hierarchical expansions obtained through subsumption reasoning.

5.1.2.1 Explanation consistency output

The Explanation Integration Layer receives:

- Statistical explanation subset derived from a,
- Semantic concept set S_{sem}.

It produces a scalar consistency measure:

ECS $\in [0, 1]$ measuring cross-layer alignment as formally defined in Section 4.2.

ECS serves as a validation signal governing subsequent system behavior.

5.1.3 Structured data flow

The system executes the following pipeline:

1. The Clinical Data Layer preprocesses and structures input X.
2. The Predictive Intelligence Layer calculates p and a.
3. The Semantic Reasoning Layer generates S_{sem} through ontology mapping and rule activation.
4. The Explanation Integration Layer computes ECS.
5. The Trust Calibration Layer integrates predictive reliability and consistency into a composite trust indicator.
6. The Governance & Compliance Layer logs all artifacts:
   - Prediction vector p.
   - Attribution vector a.
   - Semantic concept set S_{sem}.
   - ECS value
   - Escalation or override decisions

This validation ensures that explanation validation is not an auxiliary visualization process but a measurable transformation between well-defined mathematical objects across layers.

5.1.4 Architectural implication

By explicitly defining inter-layer inputs and outputs, the architecture guarantees:

- Traceable transformation from raw features to semantic inference,
- Quantifiable alignment between statistical and symbolic reasoning,
- Deterministic logging of decision artifacts for auditability,
- Structured escalation under detected inconsistency.

Thus, the hybrid semantic–statistical framework operates as a formally specified system rather than a loosely coupled combination of predictive and symbolic modules.

5.2 Composite Trust Index (TI): formalization and operationalization

Evaluating intelligent healthcare systems solely on predictive accuracy is insufficient in clinical environments. Trust should be designed as a multidimensional variable representative of statistical reliability, semantic coherence, and explanation faithfulness. In order to operationalize this construct, we use a Composite Trust Index (TI), which is defined as follows, and it consists of three normalized components:

- Predictive Reliability (PR)
- Explanation Consistency Score (ECS)
- Faithfulness (F)

The Trust Index is defined as:

$TI = \alpha PR + \beta ECS + \gamma F,$

where

- coefficients $\alpha, \beta, \gamma \in [0, 1]$.
- $\alpha + \beta + \gamma = 1$ (normalization)

5.3 Predictive Reliability (PR)

Predictive Reliability captures the model's discriminative and calibration performance. It is computed as a weighted combination of normalized AUC and calibration quality:

$$PR = \lambda_1 \cdot AUC_{\text{norm}} + \lambda_2 \cdot (1 - Brier_{\text{norm}})$$

where $\lambda_1 + \lambda_2 = 1$.

All components are min–max normalized across candidate models to ensure comparability.

This formulation ensures that both ranking performance and probabilistic calibration contribute to reliability assessment.

5.4 Explanation Consistency Score (ECS)

The Explanation Consistency Score (ECS) quantifies the degree of semantic alignment between influential model-identified features and domain-validated clinical ontologies. It tests the correspondence of explanations to a conceptual structure that is based on medically grounded statistics, instead of statistical artifacts.

ECS is scaled to the range of [0,1] over the samples of assessment. Notably, semantic plausibility is not a causal validity of ECS. A high ECS indicates structural consistency with domain knowledge but does not establish causal validity.

Let:

- A_K denote the set of top-K statistically significant features identified through attribution analysis,
- S_{sem} denote the set of ontology concepts activated by semantic reasoning.

The Explanation Consistency Score (ECS) is formally defined as:

$$\text{ECS} = \frac{|AK \cap S_{\text{sem}}|}{|AK|}$$

ECS measures structural alignment rather than causal validity; causal consistency would require counterfactual or interventional analysis beyond the scope of the present framework.

where:

- $|A_K \cap S_{\text{sem}}|$ represents the number of overlapping elements between statistically salient features and semantically activated concepts,
- $|A_K|$ represents the total number of selected explanatory features.

Thus:

ECS $\in [0, 1]$ An ECS value closer to 1 indicates high cross-layer alignment, whereas lower values signal structural inconsistency between statistical attribution and semantic reasoning.

A weighted variant of ECS may incorporate attribution magnitude by computing a normalized weighted overlap between statistical feature importance scores and semantically activated concept weights.

5.5 Faithfulness (F)

Faithfulness appraises the extent to which the explanation reflects the model's true decision process. We operationalize faithfulness using perturbation-based consistency:

$$F = 1 - \frac{1}{n}\sum_{i=1}^{n} f(x_i) - f\left(x_i^{\setminus S}\right) \vee$$

where

- $f(x_i)$ is the original prediction,
- $f\left(x_i^{\setminus S}\right)$ is the prediction after removing features identified as explanatory set S,
- n is the sample size.

High faithfulness suggests that removing explanatory features significantly alters model output, demonstrating causal relevance within the model's internal structure.

5.6 Weight selection and calibration

Weights α, β, γ may be determined using:

1. Expert elicitation (clinical governance panels)
2. Multi-objective optimization
3. Sensitivity analysis under deployment constraints

In safety-critical contexts, higher weight may be allotted to ECS and Faithfulness relative to raw predictive performance.

In the cardiovascular case study, empirical ECS values were consistently above 0.65. Accordingly, a practical heuristic threshold was defined as:

$$\tau_{\text{low}} = 0.5$$

Such that:

- If ECS $<0.5 \rightarrow$ mandatory clinician review,
- If $0.5 \leq$ ECS $<0.7 \rightarrow$ warning flag issued,
- If ECS $\geq 0.7 \rightarrow$ automated recommendation permitted.

These thresholds may be recalibrated in larger clinical deployments through sensitivity analysis and governance oversight panels.

5.7 Interpretation

- TI ≈ 1: High reliability, semantic coherence, and faithful explanations
- TI ≈ 0: Low statistical or structural trustworthiness

The TI provides a relative metric across models but does not substitute for regulatory or clinical validation. It is a governance support instrument, not a standalone certification measure.

6. Evaluation and governance-integrated validation

6.1 Case study: Cardiovascular risk prediction

To evaluate the practical feasibility of the hybrid semantic–statistical framework, a cardiovascular risk prediction case study was conducted using the Cleveland Heart Disease dataset (Dua and Graff, 2019).

The dataset contains:

- 303 patient records
- 13 structured clinical attributes
- 1 binary target variable (presence/absence of heart disease)

Clinical attributes include age, sex, chest pain type, resting blood pressure, serum cholesterol, fasting blood sugar, ECG results, and exercise-induced angina.

The class distribution is comparatively balanced (165 non-disease; 138 disease), permitting stable classification evaluation.

All records are publicly available and de-identified.

6.2 Predictive performance

Predictive performance was assessed using 5-fold stratified cross-validation.

The ROC curve of the hybrid model is shown in Fig. 10.4.

The model achieved a mean AUC of 0.932, indicating strong discriminative capacity. Notably, the semantic incorporation did not affect predictive accuracy, and it moderates the thesis of interpretability-performance trade-offs (Adadi and Berrada, 2018).

Mean AUC comparison is presented in Fig. 10.5.

A paired t-test ($P = .859$) illustrates no statistically significant variation in predictive performance, confirming that semantic reasoning preserves classification strength.

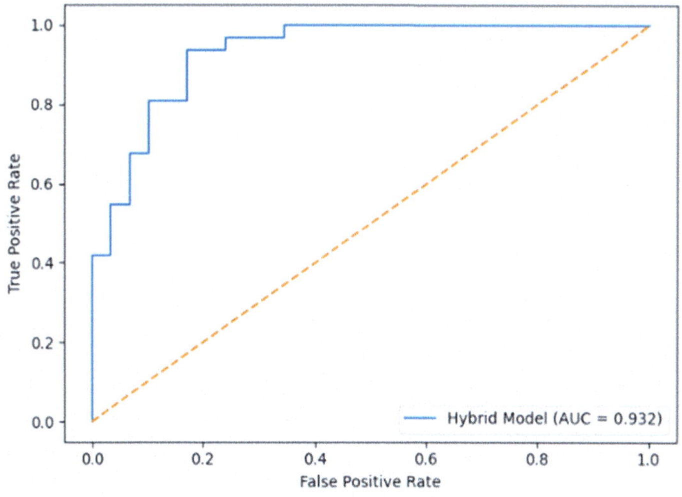

Figure 10.4 Receiver operating characteristic (ROC) curve of the hybrid semantic–statistical model for cardiovascular risk prediction (AUC = 0.932).

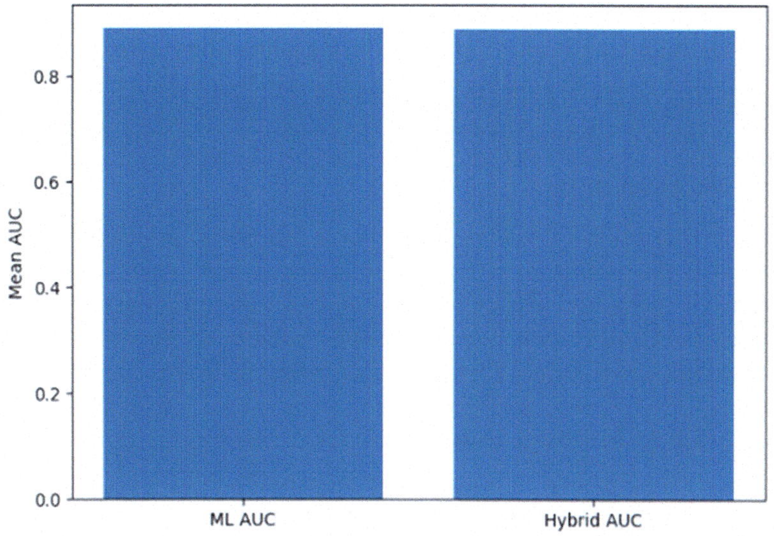

Figure 10.5 Mean cross-validated AUC comparison between baseline ML and hybrid model.

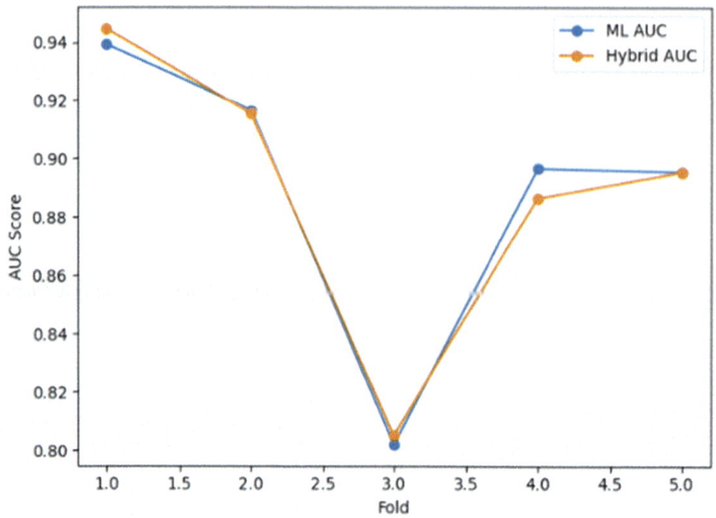

Figure 10.6 Distribution of Explanation Consistency Score (ECS) across validation folds.

6.3 Explanation consistency analysis

Beyond predictive metrics, cross-layer alignment was assessed using the Explanation Consistency Score (ECS) defined in Section 4.

ECS distribution across folds is shown in Fig. 10.6.

The fact that the values of ECS are steadily high means that the statistically salient features and the ontology-derived ones are in great harmony with clinical concepts, which deal with the issues of post-hoc reliability (Doshi-Velez and Kim, 2017; Ghassemi et al., 2021).

6.4 Trust calibration results

The composite Trust Index (defined in Section 4) was computed for both baseline and hybrid models (Fig. 10.7).

The hybrid architecture displays better structural trustworthiness because of cross-layer validation and clarification consistency. This adheres to the view of governance-driven connotations of trustful AI (Floridi et al., 2018; Fehr et al., 2024).

6.5 Governance–architecture alignment

To operationalize regulatory compliance, architectural mechanisms were mapped to governance functions (Table 10.2).

This configuration reflects regulatory expectations for high-risk medical AI systems (European Commission High-Level Expert Group on AI, 2019; Price and Cohen, 2019).

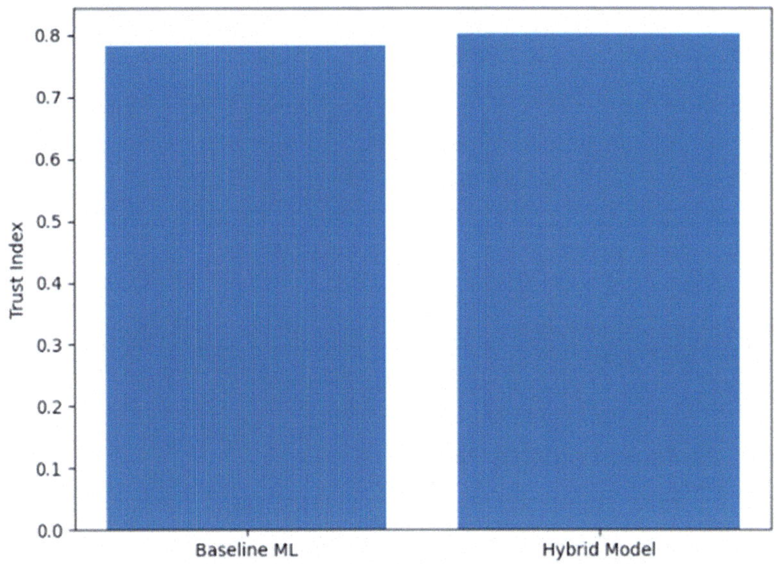

Figure 10.7 Trust index comparison between baseline ML and hybrid semantic–statistical model.

Table 10.2 Alignment between architectural layers and governance mechanisms.

Architectural layer	Technical function	Governance objective	Compliance contribution
Clinical data layer	Structured preprocessing	Data integrity	Supports traceable input handling
Predictive intelligence layer	Probability estimation & attribution	Performance validation	Enables benchmarking & reproducibility
Semantic reasoning layer	Ontology mapping & rule inference	Traceability	Provides domain-grounded reasoning
Explanation integration layer	ECS computation	Consistency auditing	Detects cross-layer inconsistencies
Trust calibration layer	Trust index computation	Risk calibration	Quantifies system-level reliability
Governance & logging layer	Inference storage	Auditability & accountability	Enables post-hoc review and medico-legal defensibility
Human-in-the-loop interface	Override capability	Human oversight	Prevents automation bias

7. Implementation framework for hybrid semantic–statistical systems

The above sections define the harmonization mechanism of the architectural design and formal explanation of the proposed framework. The section translates those principles to a model of implementation, which can be used in the real-world clinical decision support settings.

The implementation proceeds in four sequential phases:

1. Formal ontological representation
2. Predictive model integration
3. Explanation harmonization
4. Trust calibration and governance logging

7.1 Formal ontological representation

To formally integrate statistical inference with structured medical knowledge, the healthcare knowledge base is represented as an ontology:

$O = (C, R, I)$

where:

- C (Concepts): clinical entities such as Hypertension, Tachycardia, Diabetes Mellitus
- R (Relations): semantic relationships such as causes, indicates, contraindicated-with, and has-risk-factor
- I (Instances): patient-specific cases or observations

This representation enables domain-grounded reasoning rather than purely correlational inference (Bodenreider, 2004; Dissanayake et al., 2020).

7.1.1 Mapping function

A mapping function transforms raw feature inputs into ontology-aligned concepts:

$$\phi: X \rightarrow C$$

where $X = \{x_1, x_2, \ldots, x_n\}$ represents structured clinical attributes.

For example:

- Raw input: systolic BP = 165 mmHg
- Mapping rule: if BP $\geq$ 160 $\rightarrow$ Hypertension_Stage_2
- Ontology hierarchy:

$$\text{Hypertension\_Stage\_2} \sqsubseteq \text{Cardiovascular\_Disorder}$$

Thus,

$$\phi(165 \text{ mmHg}) = \text{Hypertension\_Stage\_2}$$

This semantic alignment ensures that numerical inputs are interpreted within clinically meaningful concept hierarchies.

7.1.1.1 Inference rules

Semantic reasoning operates over rule subsets:

$$R_s \subseteq R$$

Each rule $r_j \in R_s$ may take the form:

- If Hypertension $\wedge$, Hyperlipidemia $\rightarrow$ Elevated_Cardiovascular_Risk
- If Pregnancy $\wedge$ ACE_Inhibitor $\rightarrow$ Contraindication

Such rules may be implemented using:

- Description Logic reasoning (OWL-DL)
- Rule engines such as SWRL
- Knowledge graph reasoning frameworks

This combination enables logical implication, subsumption checking, and conflict detection.

7.1.2 Consistency validation

Let:

- y denote the statistical model prediction
- y_{sem} denote the semantic inference outcome

Cross-layer consistency requires:

$$y \approx y_{\text{sem}}$$

Discrepancies between probabilistic prediction and semantic inference are not immediately suppressed but are propagated to the Explanation Comparison Layer (Section 4) for ECS computation.

This mechanism prevents silent contradictions between statistical attribution and clinical logic.

7.2 Predictive model integration

The predictive component is defined as:

$$\hat{y} = f_\theta(X)$$

where f_θ is a trained supervised learning model with parameters θ.

An instance-level attribution approach (e.g., SHAP, gradient-based attribution, attention mechanisms) is produced by feature attribution methods and explanation vectors that will be reconciled with the output of semantic reasoning.

As highlighted in interpretability research, attribution scores alone do not ensure causal faithfulness (Doshi-Velez and Kim, 2017; Ghassemi et al., 2021). Therefore, harmonization remains crucial.

7.3 Hybrid semantic–statistical explanation algorithm

Algorithm 1: HSSE Framework

Input: Patient feature vector X, Ontology O, Trained model f_θ

Output: Prediction $\widehat{y}$, Harmonized explanation, Trust Index (TI).

Procedure:

1. Compute semantic mapping:

$$S_{\text{map}} \leftarrow \phi(X)$$

2. Generate statistical prediction:

$$\widehat{y} \leftarrow f_\theta(X)$$

3. Extract feature attribution scores.
4. Activate semantic rule set:

$$R_s \leftarrow \text{Reason}\left(O, S_{\text{map}}\right)$$

5. Derive the semantic inference outcome, y_{sem}.
6. Compute the Explanation Consistency Score (ECS) as defined in Section 4.
7. Apply threshold logic:
   - If ECS $< \tau_{\text{low}}$, escalate to human review.
   - If $\tau_{\text{low}} \leq$ ECS $< \tau_{\text{high}}$, issue a supervised warning.
8. Compute Trust Index (Section 4).
9. Log prediction, attribution, activated rules, ECS, and trust metrics.
10. Return $\widehat{y}$, explanation report, and TI.

This procedure integrates probabilistic prediction with ontology-based validation within a unified pipeline.

7.4 Computational considerations and real-time constraints

Ontology reasoning complexity depends on description logic expressivity.

- Full OWL-DL reasoning may exhibit exponential worst-case complexity.
- OWL-EL ensures polynomial-time reasoning for subsumption, making it suitable for near real-time decision support in cardiovascular risk assessment.

To reduce computational overhead in real-time systems:

- Frequently triggered rule chains may be precomputed.
- Concept mappings can be stored.
- Rule depth may be constrained to clinically relevant paths.
- Incremental reasoning strategies can update only modified instances.

Given the structured and moderate dimensionality of cardiovascular datasets, reasoning latency remains compatible with near real-time decision support.

7.5 Governance-integrated logging

To satisfy high-risk AI regulatory expectations (European Commission High-Level Expert Group on AI, 2019; Price and Cohen, 2019), the system stores:

- Prediction probabilities
- Feature attribution rankings
- Activated semantic rules
- ECS values
- Consistency status
- Escalation decisions

Structured inference logging improves auditability, traceability, and medico-legal defensibility (Fehr et al., 2024).

8. System-level implications and clinical integration

The empirical validation of the hybrid semantic–statistical framework demonstrates that explanation harmonization can be integrated without compromising predictive robustness, and it can be unified in predictive systems without being a statistical performance-compromising factor. But, in addition to quantitative assessment, the greater meaning of this architecture is that they suggest clinical integration and governance decisions.

8.1 Interpretability as structural validation

Traditional post-hoc explanation methods deliver localized interpretive artifacts. In contrast, the proposed framework embeds cross-layer validation directly into the system architecture. Explanation is therefore transformed from descriptive visualization into a structural consistency check.

There are two implications of this shift:

1. Reduction of silent model-knowledge contradictions.

Statistical models can give great significance to inappropriate features that are not clinically plausible. Cross-layer harmonization explicitly detects such inconsistencies before clinical decisions are issued.

2. Enhancement in domain-grounded reasoning.

The system generates clinically interpretable explanations by balancing statistical saliency to concepts generated based on ontology coherence instead of using correlational inference.

The outcome is a quantifiable interface between information-driven forecasting and an organized group of medical theories.

8.2 Human–AI collaboration

The protocol of escalation presented above allows a human intervention that is not done automatically. When explanation inconsistency goes beyond established threshold values, the system does not command clinical judgment but rather defers to it.

This mechanism mitigates two well-documented risks in clinical AI:

- Automation bias (over-reliance on algorithmic outputs),
- Under-trust caused by opaque decision logic.

By explicitly exposing reasoning conflicts, the framework supports informed clinician engagement rather than passive acceptance.

8.3 Governance-ready clinical deployment

Healthcare AI systems progressively operate within high-risk regulatory categories. The hybrid architecture explicitly embeds governance-aligned mechanisms within its operational structure through:

- Persistent inference logging,
- Explicit cross-layer validation,
- Structured audit trails,
- Escalation traceability.

Unlike purely black-box models, the system preserves explainability artifacts as structured system outputs rather than ephemeral visualizations. This facilitates post-hoc review, regulatory auditing, and institutional accountability.

8.4 Generalizability beyond cardiovascular risk

The architectural logic is not domain-specific, though it has been proven to be correct on structured cardiovascular data. The harmonization framework may be further developed to:

- Oncology risk prediction,
- Adverse drug reaction detection,
- Multi-morbidity assessment,
- Clinical pathway optimization.

The key requirement is the availability of a domain ontology with sufficient concept coverage and rule formalization.

8.5 From trust perception to trust quantification

One of the key contributions of the framework is that trust is operationalized as a system-level property. Instead of trying to equate trust with accuracy, the architecture incorporates predictive reliability and cross-layer semantic coherence.

It is important to distinguish (1) structural trustworthiness, measurable through consistency and fidelity, and (2) perceived clinician trust, shaped by longitudinal interaction and workflow integration. The present framework addresses the former while providing foundations for the empirical study of the latter.

9. Limitations and future research directions

Despite its structural advantages, the proposed hybrid semantic–statistical framework presents several limitations and open research questions.

9.1 Scalability of ontology-based reasoning

As ontologies increase in size and conceptual depth, reasoning complexity grows correspondingly. An increase in the depth of concepts leads to increased reasoning overhead. Although the worst-case complexity of description logic profiles is reduced by using lightweight profiles, the current ortho-scale biomedical ontologies can add latencies to real-time decision processes.

Future work should explore:

- Incremental reasoning architectures,
- Hybrid symbolic–embedding representations,
- Parallelized rule evaluation engines,
- Selective rule activation strategies based on risk thresholds.

9.2 Ontology coverage and knowledge incompleteness

Semantic validation is only as reliable as the ontology it relies upon. Incomplete or outdated knowledge bases may generate misleading inference traces or fail to detect clinically relevant contradictions.

Research directions include:

- Automated ontology refinement,
- Knowledge graph updating mechanisms,
- Integration of evidence-based guideline repositories,
- Cross-institutional ontology harmonization.

9.3 Robustness under distributional shift

Statistical models may degrade under population drift or demographic imbalance. Although the harmonization layer detects cross-layer inconsistency, it does not automatically correct distributional bias.

Future investigations should address:

- Bias-aware consistency thresholds,
- Fairness-aware harmonization metrics,
- Subgroup-level explanation consistency analysis,
- Longitudinal trust monitoring across patient cohorts.

9.4 Evaluation standardization for hybrid systems

While predictive metrics are standardized, explanation consistency evaluation lacks unified benchmarks. The development of:

- Public cross-layer explanation datasets,
- Standardized ECS validation protocols,
- Comparative evaluation across attribution methods, would advance reproducibility and comparability.

9.5 Regulatory evolution and clinical validation

There are pending regulatory requirements for high-risk AI systems. Formal authentication systems have to evolve. Areas of testing that still require testing are large-scale clinical trials and real-world pilot deployments.

- Workflow integration,
- User cognitive load,
- Longitudinal trust calibration,
- Institutional audit processes.

Future research must therefore integrate computational validation with human-centered evaluation.

9.6 Dataset scale and external validity

The empirical validation presented in this chapter is conducted using the Cleveland Heart Disease dataset, which comprises 303 patient records with 13 structured clinical attributes. Although this dataset is widely used for benchmarking cardiovascular risk prediction models, its relatively limited sample size constrains statistical generalizability and may not fully capture the heterogeneity observed in real-world clinical populations.

Small-scale datasets inherently restrict the evaluation of structural consistency behavior under demographic variation, comorbidity diversity, and institutional practice differences. In particular, cross-layer alignment metrics such as the Explanation Consistency Score (ECS) may exhibit different stability properties when exposed to larger and more heterogeneous cohorts. Additionally, ontology rule activation patterns could vary under broader clinical distributions, potentially affecting threshold calibration and trust index sensitivity.

Future research should therefore prioritize validation across large-scale, multi-center clinical datasets encompassing diverse patient populations, healthcare infrastructures, and documentation standards. External validation studies are essential to assess the robustness of explanation harmonization under distributional shift, as well as to examine subgroup-level consistency behavior across age, sex, and comorbidity strata.

Prospective deployment in real-world clinical workflows would further enable longitudinal monitoring of explanation stability, governance traceability performance, and clinician override patterns. Such validation is necessary to transition the proposed framework from controlled experimental evaluation toward regulatory-ready, operational clinical integration.

10. Conclusion

In this chapter, a solution to this limitation of a strictly statistical model was discussed by embedding semantic reasoning into intelligent healthcare systems. This chapter argues that predictive accuracy, while necessary, is not sufficient to have one trust a clinical judgment. It has to be based on trust founded on reliability, semantic coherence, and faithful explanation.

Explanation Consistency Score (ECS) is a metric that we have proposed to compare the compatibility of model-generated explanations and domain ontologies. ECS offers a numeric evaluation of semantic plausibility, which allows one to compare model architectures on a systematic basis.

Based on ECS, the Composite Trust Index, whose three components are predictive reliability, semantic coherence, and explanation faithfulness, is collectively evaluated. This formulation can be used to gauge the AI systems in high-stakes medical settings, which tend to favor governance.

The case study demonstrates that the hybrid semantic–statistical architecture maintains statistically equivalent predictive performance ($P = .859$) relative to baseline models, while substantially improving structural trustworthiness through cross-layer validation and explanation consistency. However, a number of drawbacks still exist. The results were assessed with the help of one clinical dataset, and it needs to be extended to different populations. Moreover, ECS records structural coherence and not causal validity; future research should entail incorporating it with causal inference models.

Intelligent healthcare systems must evolve toward semantically informed accountability rather than performance-based examination. The formal integration of ontological structures into machine learning workflows enables AI systems to move beyond mere predictive performance toward structurally interpretable explanations and governance-sensitive accountability.

Future research should prioritize scalability analysis, cross-institutional validation, and systematic integration with regulatory auditing. To advance intelligent healthcare semantics, there must be a deeper convergence of machine learning and formal knowledge representation. Specifically, validation across multicenter clinical datasets is required for external generalizability and robustness across heterogeneous populations and regulatory readiness.

References

Adadi, A., Berrada, M., 2018. Explainable artificial intelligence for healthcare: a survey. IEEE Access 6, 52138–52160. https://doi.org/10.1109/ACCESS.2018.2870052.

Amann, J., Blasimme, A., Vayena, E., et al., 2020. Explainability for artificial intelligence in healthcare. BMC Med. Inf. Decis. Making 20, 310. https://doi.org/10.1186/s12911-020-01332-6.

Arrieta, A.B., et al., 2020. Explainable artificial intelligence (XAI): Concepts, taxonomies, opportunities, and challenges. Inform. Fusion 58, 82–115.

Babic, B., Gerke, S., Evgeniou, T., Cohen, I.G., 2021. Beware explanations from AI in health care. Science 373 (6552), 284–286. https://doi.org/10.1126/science.abg1834.

Bajwa, J., Munir, U., Nori, A., Williams, B., 2021. Artificial intelligence in healthcare: transforming the practice of medicine. Future Healthc. J. 8 (2), e188–e194. https://doi.org/10.7861/fhj.2021-0095.

Bodenreider, O., 2004. The unified medical language system (UMLS): integrating biomedical terminology. Nucleic Acids Res. 32, D267–D270. https://doi.org/10.1093/nar/gkh061.

Cabitza, F., Rasoini, R., Gensini, G.F., 2017. Unintended consequences of machine learning in medicine. JAMA 318 (6), 517–518. https://doi.org/10.1001/jama.2017.7797.

Cao, J., Fang, J., Meng, Z., Liang, S., 2024. Knowledge graph embedding: a survey from the perspective of representation spaces. ACM Comput. Surv. 56 (6). https://doi.org/10.1145/3643806. Article 159.

Cui, H., Lu, J., Xu, R., Wang, S., Ma, W., Yu, Y., Yu, S., Kan, X., Ling, C., Zhao, L., Qin, Z.S., Ho, J.C., Fu, T., Ma, J., Huai, M., Wang, F., Yang, C., 2025. A review on knowledge graphs for healthcare: resources, applications, and promises. J. Biomed. Inf. 169, 104861. https://doi.org/10.1016/j.jbi.2025.104861.

Dissanayake, P.I., Colicchio, T.K., Cimino, J.J., 2020. Using clinical reasoning ontologies to make smarter clinical decision support systems: a systematic review. J. Am. Med. Inform. Assoc. 27 (1), 159–174. https://doi.org/10.1093/jamia/ocz169.

Doshi-Velez, F., Kim, B., 2017. Toward a Rigorous Science of Interpretable Machine Learning. https://doi.org/10.48550/arXiv.1702.08608.

Dua, D., Graff, C., 2019. UCI Machine Learning Repository. School of Information and Computer Sciences, University of California, Irvine. http://archive.ics.uci.edu/ml.

European Commission High-Level Expert Group on AI, 2019. Ethics Guidelines for Trustworthy AI. https://doi.org/10.2759/346720.

Fehr, J., Citro, B., Malpani, R., Lippert, C., Madai, V.I., 2024. A trustworthy AI reality-check: transparency of AI products in healthcare. Front. Digit. Health 6, 1267290. https://doi.org/10.3389/fdgth.2024.1267290.

Floridi, L., et al., 2018. AI4People: an ethical framework for a good AI society. Minds Mach. 28, 689–707. https://doi.org/10.1007/s11023-018-9482-5.

Ghassemi, M., Oakden-Rayner, L., Beam, A., 2021. The false hope of current approaches to explainable AI in healthcare. Lancet Digit. Health 3 (11), e745–e750. https://doi.org/10.1016/S2589-7500(21)00208-9.

Goodman, B., Flaxman, S., 2017. European Union regulations on algorithmic decision-making. AI Mag. 38 (3), 50–57. https://doi.org/10.1609/aimag.v38i3.2741.

Guidotti, R., et al., 2018. A survey of methods for explaining black box models. ACM Comput. Surv. 51 (5). https://doi.org/10.1145/3236009. Article 93.

Holzinger, A., et al., 2019. Causability and explainability in medicine. Wiley Interdiscip. Rev.: Data Min. Knowl. Discovery 9 (4), e1312. https://doi.org/10.1002/widm.1312.

Hogan, A., et al., 2021. Knowledge graphs. ACM Comput. Surv. 54, 1–37. https://doi.org/10.1145/3447772.

Holzinger, A., 2016. Interactive machine learning for health informatics: when do we need the human-in-the-loop? Brain Informatics 3 (2), 119–131. https://doi.org/10.1007/s40708-016-0042-6.

Jobin, A., Ienca, M., Vayena, E., 2019. The global landscape of AI ethics guidelines. Nat. Mach. Intell. 1, 389–399. https://doi.org/10.1038/s42256-019-0088-2.

Lipton, Z.C., 2018. The mythos of model interpretability. Commun. ACM 61 (10), 36–43. https://doi.org/10.1145/3233231.

Mittelstadt, B., 2019. Principles alone cannot guarantee ethical AI. Nat. Mach. Intell. 1, 501–507. https://doi.org/10.1038/s42256-019-0114-4.
Miller, T., 2019. Explanation in artificial intelligence: insights from the social sciences. Artif. Intell. 267, 1–38. https://doi.org/10.1016/j.artint.2018.07.007.
Paulheim, H., 2017. Knowledge graph refinement: a survey. Semant. Web 8, 489–508. https://doi.org/10.3233/SW-160218.
Price, W.N., Cohen, I.G., 2019. Privacy in the age of medical big data. Nat. Med. 25, 37–43. https://doi.org/10.1038/s41591-018-0272-7.
Ribeiro, M.T., Singh, S., Guestrin, C., 2016. "Why should I trust you?" Explaining the predictions of any classifier. In: Proceedings of KDD. https://doi.org/10.1145/2939672.2939778.
Rotmensch, M., et al., 2017. Learning a health knowledge graph from electronic medical records. Sci. Rep. 7, 5994. https://doi.org/10.1038/s41598-017-05778-z.
Sadeghi, Z., et al., 2024. A review of explainable artificial intelligence in healthcare. Comput. Electr. Eng. 118, 109370. https://doi.org/10.1016/j.compeleceng.2024.109370.
Samek, W., Montavon, G., Vedaldi, A., Hansen, L.K., Müller, K.-R., 2019. Explainable AI: Interpreting, Explaining, and Visualizing Deep Learning. Springer. https://doi.org/10.1007/978-3-030-28954-6.
Sutton, R.T., et al., 2020. An overview of clinical decision support systems. npj Digit. Med. 3, 17. https://doi.org/10.1038/s41746-020-0221-y.
Tjoa, E., Guan, C., 2020. A survey on explainable artificial intelligence (XAI): toward medical XAI. IEEE Transact. Neural Networks Learn. Syst. 32 (11), 4793–4813. https://doi.org/10.1109/TNNLS.2020.3027314.
Topol, E., 2019. High-performance medicine: the convergence of human and artificial intelligence. Nat. Med. 25, 44–56. https://doi.org/10.1038/s41591-018-0300-7.
Yu, K.H., Beam, A.L., Kohane, I.S., 2018. Artificial intelligence in healthcare. Nat. Biomed. Eng. 2, 719–731. https://doi.org/10.1038/s41551-018-0305-z.

CHAPTER 11

Semantic clinical decision support systems: challenges and future scope

Pradnya S. Kulkarni[1] and Selena Firmin[2]
[1]Dr. Vishwanath Karad MIT World Peace University, Pune, Maharashtra, India; [2]Federation University, Ballarat, VIC, Australia

1. Introduction

Clinical decision support systems (CDSS) are systems designed to help physicians in making clinical decisions. These systems analyze data from electronic health records (EHRs) and other clinical sources to provide evidence-based recommendations and guidance for patient care. CDSS aim to advance the healthcare quality by helping to treat, and manage patients. Clinical decision support systems (CDSS) are important for many reasons and help to improve the quality, efficiency, and safety of patient care. Clinicians are provided with accurate, evidence-based recommendations by CDSS, enabling more accurate diagnosis and effective treatment. CDSS can detect health problems early by analyzing patient data, allowing for prevention and timely intervention (Akshay Krishna et al., 2024). CDSS integrates with electronic health records (EHRs) to advance clinical operations and reduce the time for recording and storing patient information. They save time and resources by helping eliminate tests and repetitive processes. CDSS assists doctors by providing different diagnoses as per the symptoms of patient and medical history. They recommend appropriate treatment, dosage and precautions based on the patient's needs. This system helps prevent drug abuse by alerting doctors to drug interactions, allergies, and contraindications. They provide immediate alerts for laboratory values, abnormal vital signs, and other emergencies. These systems promote consistency of patient care across clinics and facilities by ensuring adherence to clinical guidelines and best practices. They help reduce changes in maintenance, providing better and more reliable results. They also provide important resources to doctors by providing them with the latest medical research, advice, and treatment methods. They support continuing education and professional development by keeping physicians informed about new developments in their profession. CDSS supports public health measures and chronic disease management by analyzing big data to identify trends in public health. They support preventive care by identifying high-risk groups and recommending screening and interventions. By improving resource utilization and reducing unnecessary tests and procedures, CDSS helps save costs and improve healthcare performance. They help healthcare organizations meet quality metrics and

Intelligent Semantic Analysis for Healthcare
ISBN 978-0-443-33512-9
https://doi.org/10.1016/B978-0-443-33512-9.00001-9

performance standards that are important for cost-effective care models. CDSS can increase patient compliance with treatment plans and engage patients in care by providing advice and personalized care. They support shared decision-making by providing clear, evidence-based treatment and management options to patients and physicians.

Semantic Clinical Decision Support System (Semantic CDSS) represents an advanced approach to CDSS that uses technology to enhance the understanding, integration, and use of medical information. These systems use data structures and ontologies to enable more accurate, relevant, and context-sensitive medical decision making. Semantic CDSS uses ontologies to persistently represent clinical information and terminology. Ontologies facilitate the assimilation and interpretation of data from different sources. The use of clinical guidelines based on clinical standards such as ICD, LOINC, SNOMED CT improves understanding of clinical language and enhances collaboration among healthcare professionals. Semantic CDSS integrates with electronic health records (EHRs), facilitating the amalgamation of patient records from multiple sources. By applying data models and semantic systems. These systems can exchange and interpret data between different healthcare systems and platforms. Semantic CDSS provides recommendations to patients taking into account their unique medical history, genetic information, and current health conditions. They use context-aware algorithms to interpret data in clinical situations, thereby increasing the accuracy and precision of decision support. These systems make the decision making process more complex by using advanced thinking systems to generate new information from existing data. Integration with machine learning and AI technology enables continuous learning and improved treatment recommendations based on real data. Semantic CDSS provides a user-friendly interface that presents complex information and recommendations in an easily interpretable format. They are designed to fit seamlessly into healthcare professionals' workflows, minimize disruption, and increase usability.

2. Historical background

The concept of using computers to assist in clinical decision-making emerged in the 1960s. The first systems were code-based and focused on specific tasks, such as diagnosing specific diseases or drug use. Table 11.1 shows the history of how the CDSS systems are developed over several years. MYCIN was one of the first successful CDSSs (Van Melle, 1978). It uses the correct method to diagnose the disease and recommends the use of antibiotics. Firstly, Mycin's effectiveness is limited to bacteria and certain areas of the antibiotic. The interface is not user-friendly, making it difficult for doctors to integrate it into their work. Also the initial hardware limitations limited the speed and efficiency of the system. For this reason, MYCIN is not widely used. The INTERNIST-1 and Quick Medical Reference (QMR) (Miller and Masarie, 1990) systems were major advances in the 1980s and 1990s. In The results produced by Internal Medicine-I system

Table 11.1 History of development of CDSS.

ID	Year (s)	DSS trend	Example	Focus/Impact
1	**1960s–1970**	The development of early rule-based systems	MYCIN	Basic logic to provide diagnostic support
2	**1980**	Expert systems	INTERNIST-1	Structured knowledge bases
3	**1990**	Evidence-based medicine	Quick Medical Reference (QMR)	Integration of guidelines and standards into CDSS
4	**Early 2000**	Semantic web technologies	Health evaluation through logical processing (HELP) system	Better data interoperability
5	**Mid 2000**	Creation of medical ontologies	SNOMED CT, LOINC, and ICD	Facilitate semantic interoperability and data sharing across systems
6	**Late 2010**	Use of AI and machine learning to enhance decision support, predictive analytics and personalized recommendations	IBM Watson for oncology	Sophisticated analysis and prediction
7	**2020**	Real time decision support, predictive analysis	Epic systems Corporation's CDSS	Interoperability standards (FHIR) and integration with the internet of medical things (IoMT)

are complex and often produce a long list of diagnoses that can confuse doctors. Further it requires detailed manual entry of patient records which is time-consuming. QMR assists doctors with generating hypotheses in complex patient cases. However, effective use of QMR systems requires extensive training. These systems use a database of symptoms and diseases to help physicians diagnose complex conditions. The ARROW system, developed by Harvard University, provides a clinical judgment for the administration of anticoagulants. The ARROW system is very unique and does not include any other aspect of patient care. This system was integrated with electronic health records (EHRs) in 1990, providing easy access to patient information and increasing the accuracy and precision of support systems. The acceptance of data standards such as Health Level Seven (HL7) and language standards such as LOINC, SNOMED CT improves collaboration and data exchange between CDSS and other healthcare systems. The use of artificial intelligence (AI) and machine learning into CDSS enables more comprehensive data analysis, predictive modeling, and personalized recommendations. The 2000s saw the emergence of commercial CDSS solutions with large EHR systems, increasing adoption in healthcare. Examples include IBM Watson Health (Ferrucci et al., 2013) and Epic decision support tools. Thus early systems were limited by hardware devices that were not automation friendly, difficult integration with other systems, knowledge required to update data, data using manual entry, and more complex features.

The introduction of the technology represents a major advancement in CDSS. Semantic CDSS uses ontologies and data structures to enable context-aware, personalized, and interactive decision-making. Modern CDSS can analyze large amounts of data, such as genomic data, functional tools, and patient-reported outcomes to ensure good and clear recommendations. The adoption of cloud computing and mobile healthcare has expanded the accessibility and capacity of CDSS, facilitating timely decision making and the use of telemedicine. In 2002, Systematic Nomenclature of Medicine - Clinical Terminology (SNOMED CT) (Wang et al., 2002), a comprehensive medical system that allows for accurate and consistent representation of medical information, was created. In 2004, the W3C recognized the ability to create complex systems to represent clinical information, thus enhancing the capabilities of CDSS. In 2010, the US government announced a program to promote the adoption of EHRs and to encourage the integration of CDSS with EHRs to enable immediate decisions about care. Medical change accelerated in 2014 with the introduction of HL7 which aims to provide standards for electronic health information exchange and to promote better integration and interoperability of CDSS. Fast Healthcare Interoperability Resources (FHIR) is a standard for exchange of medical information. In 2019, It was improved with better data structures and languages. In 2020, CDSS was built by combining ML, AI and NLP. It could therefore provide higher level contextual information. Examples include IBM Watson Health and Google's DeepMind Health (Powles and Hodson, 2017). Later, the concept of intelligence explanation emerged, which aims to make intelligence-focused CDSS more transparent and explain and solve the "black box" problem in intelligence.

3. Key components of CDSS

There are five key components when we consider CDSS (Refer Fig. 11.1). They are Knowledge base, Inference engine, Communication Mechanism, Data Management, User Training and Support. The knowledge base is the core of any CDSS, comprising medical knowledge such as clinical guidelines, drug databases, and diagnostic criteria. It

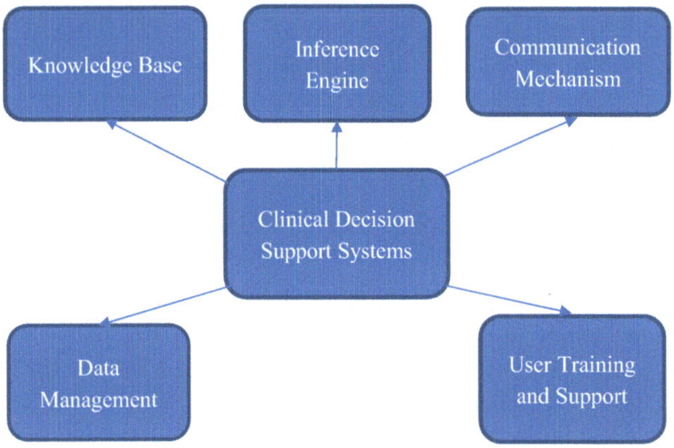

Figure 11.1 Key components of clinical decision support systems.

can be created using various sources including textbooks, peer-reviewed articles, and expert input. Knowledge representation methods include rules, ontologies, and machine learning models. Chen et al. (2020) highlights the significance of updating the database of knowledge in order to ensure the relevance and accuracy of CDSS recommendations. Garg et al. (2005) discusses the different methodologies for knowledge representation and their impact on system performance and adaptability. Knowledge representation is a critical component of CDSS, enabling these systems to process and utilize complex medical knowledge effectively. Various knowledge representation systems have been tried for CDSS. Rule-based systems use "if-then" statements to represent knowledge. These systems are straightforward and interpretable, making them popular in early CDSS. Shortliffe and Cimino (2006) describe rule-based systems as one of the earliest forms of knowledge representation in CDSS, noting their simplicity and ease of implementation.

The Inference engine uses rules appropriate to the knowledge base to generate recommendations. It interprets the data and matches it with the knowledge base to provide better understanding. Shortliffe and Cimino (2006) provide a detailed description of various cognitive processes. Kawamoto et al. (2005a) emphasize the importance of inference systems in providing timely and relevant decision support. Musen et al. (2014) discusses how formal processes are limited by their rigidity and complexity when dealing with unfamiliar or novel situations. Ontologies represent knowledge through structured content and relationships. They provide a simpler and more complete way to model medical knowledge. Rector et al. (1997) highlighted the use of ontologies to represent complex medical terminology and its potential to facilitate collaboration between different systems. Ashburner et al. (2000) discussed the Gene Ontology project as an example of the use of ontologies to model biological knowledge in various domains. Deep learning models, are becoming important in knowledge representation. These models learn from large amount of data, patterns can be identified, and predictions can be made. Research investigated the role of ML in CDSS, highlighted its capacity to improve diagnostic accuracy, and personalized treatment recommendations (Chen et al., 2020). Miotto et al. (2016) investigated how deep learning can be used to predict patient diseases by learning from EHRs. Bayesian networks represent the relationship between variables, allowing CDSS to deal with uncertainty and perform visualizations based on missing data. Heckerman introduced Bayesian networks as an effective, diagnostically relevant method for inference under uncertainty (Heckerman et al., 1992). Lucas et al. (2004) discussed the use of Bayesian networks for CDSS, particularly in the context of disease progression models. Fuzzy logic systems control biases by allowing partial membership in classes, making them useful for modeling inconsistencies in medical data. Bellazzi and Zupan (2001) explore the use of fuzzy logic in CDSS in terms of its ability to make people think.

Communication mechanism component ensures the effective delivery of CDSS recommendations to clinicians. It includes the user interface and the methods for integrating the CDSS with electronic health records (EHR). Bates et al. (2003) examines the challenges of integrating CDSS with existing healthcare IT infrastructure. Sittig discuss the design principles for effective CDSS interfaces that enhance usability and user acceptance (Sittig et al., 2008a). The clinicians interact with CDSS using the user interface. A well-designed UI should present information in an intuitive, user-friendly manner, minimizing cognitive load and supporting efficient decision-making. Alerts and reminders are common features of CDSS, designed to prompt immediate action or draw attention to critical information. These can include drug interaction warnings, reminders for preventive care, or alerts for abnormal test results. Phansalkar et al. (2013) provide guidelines for optimizing alert systems to maximize their effectiveness while minimizing the fatigue caused by unnecessary alerts. van der Sijs explore the balance between effective alerting and alert fatigue, noting that excessive or poorly designed alerts can lead to desensitization and decreased compliance (van der Sijs et al., 2006). Delivering contextually relevant information that is personalized to the specific patient and clinician is essential for the effectiveness of CDSS communication. This includes tailoring recommendations based on patient history, current medications, and clinician preferences. Kawamoto et al. (2005b) review various parameters that are responsible for the success of CDSS, emphasizing the importance of context-aware and personalized communication. Berner examine the impact of personalized alerts on clinician adherence and patient outcomes (Berner et al., 2009). The increasing use of mobiles in medical settings has triggered the development of CDSS components that support mobile and remote access. This enables clinicians to receive decision support anytime and anywhere, enhancing the flexibility and responsiveness of care. Ventola (2014) discusses the use of mobiles in healthcare and its implications for CDSS communication, highlighting the benefits of mobile access to real-time decision support. Kruse analyze the how CDSS is effective in improving clinical decision making (Kruse et al., 2017). Effective communication mechanisms within CDSS lead to several benefits, including improved adherence to clinical guidelines, enhanced efficiency in clinical workflows, and better patient outcomes. Clear, timely, and relevant communication ensures that clinicians are able to take decisions quickly. In this way errors are reduced and the overall quality of care is improved. Black review the effect of CDSS on healthcare workflow efficiency and the quality of decision-making (Black et al., 2011).

Data management tools are critical to CDSS operations. This includes data extraction, transformation, and loading processes to ensure the system has access to efficient, up-to-date patient information. Jasper discussed the challenges of integrating large amounts of data and the importance of good data (Jaspers et al., 2011). Hersh emphasized the role of information process standards and interactions in effective information management (Hersh, 2007). Seamless integration with EHR systems is essential for effective communication in CDSS. This integration provides immediate access to patient information and informs decision support. Middleton et al. (2013) discussed the benefits of EHR integration, such as increased accuracy of recommendations and increased operational efficiency. As more data sources expand, including EHRs, imaging data, genomic data, and wearable devices, the challenge is to use these materials effectively and efficiently. Effective integration of diverse information begins with designing and enabling interactions. Using standards such as HL7, FHIR facilitates the exchange of information between different systems. Research has shown that standardization reduces variability in CDSS and increases the accuracy of data interpretation (Mandel et al., 2016). Data storage solutions combined with big data sources such as Spark and Hadoop can handle large and diverse data. These technologies support the collection, storage, and multiple source data analysis, thereby allowing CDSS to analyze huge amounts of data (Raghupathi and Raghupathi, 2014). ML and AI are essential to manage large amounts of data. ML algorithms can learn from many types of data, including structured and unstructured data, to identify patterns and provide recommendations. Deep learning models, in particular, have successfully interpreted complex data such as medical images and disease genomes, improving the accuracy and efficiency of CDSS (Esteva et al., 2019). Integrating different information creates a patient record, thereby increasing the accuracy of CDSS recommendations. CDSS can provide more comprehensive and personalized information by integrating data from EHRs, test results, imaging studies, and genetic information (Murphy et al., 2014). Using diverse data helps improve the diagnostic and predictive capabilities of CDSS. For example, combining clinical data with genomic data can facilitate the identification of genetic patterns for various diseases, facilitating early and personalized treatment plans (Topol, 2019). Data collected from multiple sources supports clinical research and development by providing a rich database. Researchers can better identify needs, analyze theories, and develop new treatments, ultimately leading to improved healthcare.

Adequate user training and ongoing support is needed in CDSS, ensuring that physicians are comfortable using the system and interpreting recommendations. Berner emphasize the need for regular training appropriate for different user groups. Goldstein discussed the impact of user support on CDSS adoption and continued use (Goldstein et al., 2012). The dashboard serves as the centerpiece of the CDSS user interface and provides an overview of important information. This dashboard includes summary of patient demographics, medical history, current status, and treatment plan. Instant alerts for important issues such as invalid results or urgent care needs. List of important tasks such as monitoring, medication review, and diagnosis also shown on dashboard. The patient profile page includes medical history, current medications, lab results, charts, and medical notes. Decision support tools include patient-centered clinical guidelines, risk calculators, drug interaction analysis, and diagnostic tools that narrow down differential diagnoses based on symptoms and test results. Good data visualization can help doctors quickly interpret complex data. Notable features include visual representations of key trends in key measurements, lab results, and other measurements, and a quick data visualization or effort to help identify areas of concern such as heatmaps. Dashboard also includes a tool that compares current patient data to documented or historical data. Communication tools enhance collaboration among healthcare professionals. These include secure communication for team communication, tools for virtual consultation and remote patient care (telemedicine) where multiple practitioners can contribute and review data. Great care should be taken when designing the CDSS user interface. Simple and clear navigation paths reduce intellectual load. Unified design concepts and terminology must be used throughout the interface. Make sure that the interface can be used by people with different intelligence and physical abilities. New information and notifications should be evaluated immediately. The interface should be adjusted to different devices and sizes. Moreover, Customization of dashboards must be implemented based on user's interests and performance. Provide immediate feedback on tasks, such as confirming data entry or acknowledging alarms. Further, all patient information should be encrypted during transmission and at rest. Medical information is sensitive information, so we must restrict access based on user responsibility. The system should keep detailed records of user activity to ensure authentication and compliance (Liao et al., 2015).

4. Current state of semantic CDSS

Semantic CDSS utilize semantic technologies to enhance clinical decision-making by interpreting and leveraging the meaning and relationships within medical data. Semantic CDSS integrates various types of data such as data from wearable devices, genomic data, patient imaging data, data from electronic health record, lab results by taking help of ontologies. The inference engine makes use of the ML models and data to predict the

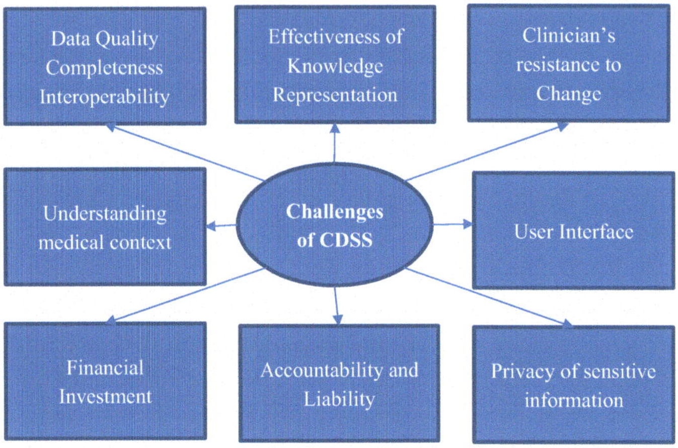

Figure 11.2 Key challenges of semantic clinical decision support systems.

patient diagnosis. Fig. 11.2 show the various features and components of current semantic CDSS. There are already many successful example of CDSS. We will briefly discuss a few of these systems. IBM Watson for Oncology is a well-known CDSS that uses AI which helps oncologists in making treatment decisions. It analyzes medical records, medical procedures, and patient information to provide personalized treatment recommendations. Gordan et al. (2019) provides information about Watson for Oncology improves the quality of treatment plans and enables faster decision making by providing evidence-based options. EPIC's Best Practice Advisory (BPA) system integrates with EPIC's EHR to provide decision supportin real time. BPAs are designed to alert physicians to important medical information and potential issues, such as drug interactions or the need for preventive care. Wright suggested that BPA could increase adherence to medical procedures and reduce medication errors (Wright et al., 2014). Cerner Millennium is an EHR program with advanced CDSS. Provides decision support in operational areas through alerts, alerts, and clinical procedures. Baysari demonstrated that Cerner Millennium's CDSS has the capacity to improve healthcare and better support patients by providing timely and relevant information (Baysari et al., 2018). UpToDate is a CDSS that offers evidence-based medical information to healthcare professionals. It covers a variety of medical topics and provides advice based on the latest research. Users report that UpToDate is especially helpful in making care decisions and continuing medical education (Isaac et al., 2012). According to Isaac, use of UpToDate has been associated with improved patient outcomes and reduced length of stay.

SEBASTIAN (Semantic Data Integration Framework) is a CDSS designed to integrate and analyze various medical data. SEBASTIAN makes medical decisions easier and more convenient by integrating patient information by utilizing semantic web technology. According to Martinez-Romero, SEBASTIAN's ability to integrate different information increases clinical outcomes, improves accuracy and treatment planning (Martínez-Romero et al., 2017). Several semantic clinical decision-making methods for cardiovascular diseases are currently available. Huang proposes a semantic-enabled agent-oriented distributed clinical decision model, showcasing success in integrating semantic technology for clinical decision support systems in a distributed environment (Huang et al., 2021). Persson Lindell developed a system called CDS-FH for cardiovascular risk assessment (Lindell et al., 2022). It was used in 44 primary care hospitals in one city in Sweden. In Paper (Jones and Wilson, 2023) a Natural Language Processing based decision support system built from a repository of knowledge drawn from referral letters written between primary care doctors and specialist medical consultants is developed, which translates pre-processed referral letters into a semantic matrix of document vectors and a set of vocabulary features, based solely on the words used within each referral letter.

5. Challenges in semantic CDSS

Semantic CDSS leverage advanced technologies such as ontologies, NLP and ML to provide more accurate and context-aware clinical recommendations. Despite the promise of these systems, several challenges hinder their widespread adoption and effectiveness. Fig. 11.2 provides main challenges of semantic CDSS.

One of the foremost challenges in semantic CDSS is ensuring the completeness and quality of data. Semantic systems rely on accurate, comprehensive, and well-annotated data to function correctly. Incomplete or erroneous data can result in incorrect recommendations, jeopardizing safety of the patient. According to Meystre the heterogeneity and variability of clinical data sources exacerbate this issue, as inconsistencies and gaps in data are common (Meystre et al., 2010).

One of the fundamental issues is that the data is good and honest. Incomplete or incorrect data can lead to incorrect estimates and recommendations. Processes such as data cleaning, verification and decision making are important for solving these problems (Kahn et al., 2016). Processing sensitive medical information increases security and privacy concerns. Ensuring CDSS comply with regulations such as HIPAA and applying strong encryption and data management systems are important for protecting patient information. Advanced security techniques and anonymization techniques are often used to reduce the risk (Ramanathan et al., 2013). Integration of data from various sources involves complex processes such as information sharing, transformation and reconciliation. Heterogeneity of data input and semantics causes serious problems. Although it requires careful installation and maintenance, middleware solutions and integration platforms are often used to facilitate this process (Mili et al., 2018). Ensuring that data from different systems can be integrated and interpreted consistently requires a strong integration model. However, the lack of international recognition of standards such as

HL7 FHIR causes problems. Benson and Grieve stated that different information and languages used in different systems prevent information exchange and integration (Benson and Grieve, 2016).

Good knowledge representation is essential for the success of semantic CDSS. This includes encoding medical procedures, medical information, and patient information in a machine-readable format. Creating accurate and comprehensive ontologies that capture the complexity of clinical knowledge is challenging. Musen argued that the changing and evolving nature of medical knowledge requires continuous updating of ontologies, which is resource-intensive and time-consuming (Musen, 2015).

NLP plays an important role in semantic CDSS and has the capacity to extract important information from uncomplicated medical documents. However, NLP faces many challenges, such as understanding the meaning and nuances of medical terminology, handling abbreviations, and managing polysemy (more than one meaning of a word). Demner-Fuchman pointed out that accuracy is difficult to achieve in clinical NLP as clinical data is complex and varied (Demner-Fushman et al., 2009).

For a semantic CDSS to be effective, it must be accepted and trusted by practitioners. Resistance to change, concerns about the overreach of automated systems, and perceived threats to self-regulation may affect consumers. Sittig emphasized that collaboration with health professionals in the design and implementation of the system, provision of adequate training, and demonstration of the reliability and validity of the system are important in overcoming these problems (Sittig et al., 2008b). One of the greatest challenges of CDSS is its integration into existing clinical programs. Systems that require significant changes in the way doctors work daily may encounter resistance. Research shows that poor integration of CDSS can lead to increased workload and affect performance, thus reducing the effectiveness of the system (Ash et al., 2007). The design of the UI is important to healthcare professionals' success in adopting CDSS. Any confusion or lack of visibility can hinder usability and lead to ineffectiveness. Jaspers emphasized that the usability of CDSS interfaces should be rigorously tested and improved with input from practitioners to ensure that they meet the needs of end users (Jaspers et al., 2011). CDSS often generates lot of alerts, many of which may be irrelevant or unimportant. This can lead to alert fatigue, where physicians become indifferent to alerts and begin to ignore them. Research found that alert fatigue is a significant problem, with physicians ignoring most alerts and possibly missing important alerts (Poly et al., 2020). Improving the relevance and specificity of alerts is critical to combating fatigue caused by alerts. To be effective, alerts must be appropriate to the context and content of the treatment. Wright have shown that updating alerts according to the patient's condition and prioritizing them according to urgency can help reduce alert failure (Wright et al., 2014).

Ensuring regulatory compliance and resolving ethical issues are key issues in implementing CDSS semantics. These systems must comply with regulations such as HIPAA that require patient information to be protected. Additionally, ethical issues related to patient consent, data privacy, and algorithmic resources must be taken care of. Goodman argue that it is necessary to establish clear guidelines and strong regulatory standards to address legal and ethical issues (Goodman, 2016).

The reliance of CDSS on a large amount of patient information raises concerns about the privacy of medical information. Protecting patient information from unauthorized access is critical. Rindfleisch argued that maintaining the confidentiality of electronic medical records is a significant challenge due to the data leakage risk and access by unauthorized person (Rindfleisch, 1997). Data protection implemented by CDSS includes the use of strong cybersecurity measures to prevent hacking, data theft, and other security threats. Health information is a prime target for cyberattacks due to its high value. The HIPAA has set data standards for maintaining security, but compliance can be difficult and costly (Appari and Johnson, 2010). Informed consent in the context of CDSS includes ensuring that patients understand how their information will be used and the role of the CDSS in their care. Simon argue that patients should be informed about the use of information and advice they receive from the CDSS to ensure they understand its benefits and risks (Simon et al., 2009). Patients have the right to make decisions regarding their care. The CDSS should promote rather than diminish patient autonomy by providing recommendations that enhance rather than replace decision-making. The challenge is to ensure that patients know that CDSS recommendations are advisory and not a substitute for medical judgment (Goddard, 2012). Determining who is responsible for decision-making with the help of the CDSS is a difficult issue. Liability issues arise when CDSS recommendations lead to adverse outcomes. Responsibility may lie with the provider, the CDSS developer, or both. According to Silverman, clear guidelines and legal standards are needed to address accountability and responsibility issues when implementing CDSS (Silverman and Skinner, 2017). Legal implications of using CDSS in healthcare include the need for unlawful claims and legal protections for providers and developers. Clear guidelines for the implementation of CDSS can help reduce legal risks and provide a framework for resolving liability issues (Froomkin et al., 2019). Adherence to these standards is necessary for the effective and safe use of CDSS. Regulations such as HIPAA in the USA and the GDPR in Europe set strict rules regarding data privacy and security. CDSS may also require certification from regulatory agencies such as the (FDA) for clinical use. The approval process can be long and difficult, requiring documentation of the safety and efficacy of the process. According to Droulet, obtaining regulatory approval is a key issue that may delay the deployment of CDSS (De Roulet et al., 2020). The ethical and legal issues surrounding CDSS are important and multifaceted. Solving these issues requires a comprehensive approach that includes strong data protection and security measures, clear disclosure of consent, algorithmic integrity to mitigate risks, accountability and responsibility, and adherence to regulatory standards.

The development and implementation of CDSS requires a significant financial investment. These costs include software development, integration with existing medical records, and necessary purchases. According to Bates, the initial cost of implementing CDSS can be a major concern for smaller organizations due to smaller budgets (Bates et al., 2003). CDSS often requires modification to meet the needs of various clinical settings. This reform may involve significant additional costs to streamline processes for local clinical practices and ensure compatibility with EHRs. Sittig highlighted that integration with EHRs and other healthcare reporting systems is difficult and expensive, requiring significant expertise and resources (Sittig et al., 2008a). The ROI for CDSS may be difficult to demonstrate. While CDSS may improve clinical outcomes and efficiency, these benefits are difficult to quantify in economic terms. Chowdhury noted that improving patient outcomes and reducing adverse events can lead to cost savings, but these benefits often take a long time to be realized and short-term measures of return on investment are not good (Chowdhury et al., 2006). Indirect costs such as physician training and ongoing care should be considered when assessing ROI. In contrast, indirect cost savings such as improving physician productivity and reducing readmissions can be significant but are often difficult to measure. Jones found that CDSS can lead to long-term cost savings, but these benefits are often not immediate and need to be evaluated appropriately (Jones et al., 2014). The cost-effectiveness assessment of CDSS involves assessing whether the health benefits provided are worth the costs incurred. The cost-effectiveness of CDSS can change based on many factors, including the clinical setting, the expertise of the system, and the patients served. According to a review by Moja, although some CDSS implementations have been found to be effective, others highlight the need for context-specific evaluations (Moja et al., 2014).

6. Future scope and opportunities

Semantic CDSS leverage advanced data representation technologies such as ontologies and semantic networks to improve the interpretability and accuracy of clinical data. Future progress in ontology development will enhance the ability of semantic CDSS to integrate diverse information. These improvements will enable accurate and comprehensive data representation, encouraging better collaboration across different medical systems. Ontology development will support more accurate and context-aware recommendations. Semantic CDSS will play a critical role as the healthcare industry keeps integrating data from various sources and AI. The future of semantic CDSS is depicted in Fig. 11.3. These systems are critical for connecting data from multiple sources, such as medical images, EHRs, genomics, and wearables.

Future semantic CDSS will facilitate the data exchange and the use of different platforms and organizations, thereby improving continuity of care. The combination of AI and ML learning with advanced technology will enhance the capabilities of CDSS. AI

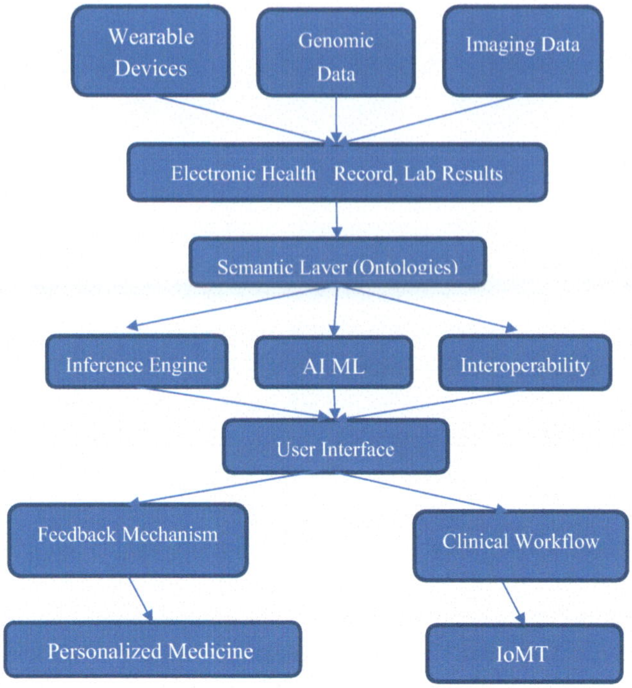

Figure 11.3 Future of semantic CDSS.

algorithms are used to analyze complex and large medical data, while advanced technology enables accurate and contextual interpretation of this data. This combination will result in better decision support and accuracy, along with the ability to provide personalized and predictive recommendations. AI and machine learning can also be used to accelerate the ontology learning and monitoring process. These ideas can help identify new concepts, relationships, and innovations in clinical knowledge to ensure that the use of semantic CDSS remains relevant and expanded. Automatic ontology management will reduce the burden on human experts and improve the scalability and adaptability of CDSS. The future of semantic CDSS will include greater integration of genomic and biomarker data, enabling more personalized and accurate medicine. Semantic CDSS can provide recommendations for disease prevention, diagnosis, and treatment by integrating genetic and other biomarker information. This approach will improve treatment and reduce adverse drug reactions. The future of semantic CDSS will include greater integration of genomic and biomarker data, enabling more personalized and accurate medicine. Semantic CDSS can provide recommendations for disease prevention, diagnosis, and treatment by integrating genetic and other biomarker information. This approach will improve treatment and reduce adverse drug reactions. The ability to continuously integrate information from various sources, including monitoring equipment and effective equipment, will increase the relevance and timeliness of medical decisions. Semantic CDSS will provide recommendations and information points related to the patient's health. The semantic CDSS of the future will be flexible and can be updated according to new information and changing medical procedures. These systems will update their knowledge bases and decision rules, allowing doctors to get the best, most accurate recommendations. This change is important for managing complexity and adapting quickly. NLP technology will enable machines to understand and process medical information and other text-based information, thus increasing the effectiveness of semantic CDSS. NLP can facilitate greater interaction between healthcare providers and CDSS by allowing users to use natural language for querying the system and receive valuable information and recommendations. Future semantic CDSS will feature more intuitive and user-friendly interfaces designed to integrate seamlessly into clinical workflows. These interactions will present information clearly and efficiently, reducing artificial intelligence and improving decision-making. As semantic CDSS becomes more complex, clarity and interpretation will become more important. Future guidelines should provide a clear explanation of their recommendations so that clinicians can understand and trust the decision-making process. This will include developing a process to visualize and communicate the logic behind CDSS results. A strong regulatory framework is necessary to ensure the safety, efficiency, and fair use of Semantic CDSS. This process should address issues related to data privacy, algorithmic bias, and the validity of clinical recommendations. Good governance will build trust in CDSS and support its leadership. Semantic CDSS will improve clinical outcomes by providing more accurate, personalized, and decision-supporting data. These systems will help reduce diagnoses, improve treatment outcomes, and improve patient safety. Semantic CDSS will have a significant impact on healthcare in the future, resulting in better health outcomes and improved quality. Semantic CDSS has the potential to improve the cost of healthcare by optimizing resource utilization, reducing the number of tests and treatments, and preventing adverse events. The ability to provide timely positive feedback will help improve clinical performance and reduction in the overall cost of care.

7. Conclusion

In this paper we have overviewed the history of development of CDSS. We discussed the importance of semantic CDSS. By using CDSS many benefits can be achieved, such as risk assessment, accurate diagnosis, cost savings, and improved health. Various features of the current semantic CDSS are discussed such as integration of data from EHR, lab reports, genomic data, imaging data, predictive analysis, wearable technologies. However, there are significant challenges in data quality, interactivity, knowledge representation, NLP, client acceptance, and management. These issues must be addressed to recognize its full potential. Continued research and collaboration among practitioners, professionals, and regulatory agencies will be necessary to address these issues and increase the effectiveness of CDSS in treatment. The future of semantic CDSS is full of exciting opportunities and potential for great impact. Advances in artificial intelligence, machine learning, data integration, and personalized medicine will increase the functionality and effectiveness of these systems. As semantic CDSS become more complex, they will enable more accurate, timely, and context-driven clinical decisions, thereby improving patient outcomes and healthcare performance. Addressing ethical and regulatory issues is critical to the responsible development and delivery of CDSS semantics. The continued development of these systems promises to revolutionize healthcare and usher in a new era of data-driven and intelligent medical decision making.

References

Akshay Krishna, R., Venkateswaran, D., Mala, T., Narasimhan, D., Mahalakshmi, 2024. Real-Time patient risk identification using deep learning. In: Futuristic Trends in Artificial Intelligence Volume 3 Book 9, IIP Series, 3, pp. 20–29. https://doi.org/10.58532/V3BIAI9P2CH2. e-ISBN: 978-93-6252-924-4.

Appari, A., Johnson, M.E., 2010. Information security and privacy in healthcare: current state of research. Int. J. Internet Enterp. Manag. 6 (4), 279–314.

Ash, J.S., Sittig, D.F., Poon, E.G., Guappone, K., Campbell, E., Dykstra, R.H., 2007. The extent and importance of unintended consequences related to computerized provider order entry. J. Am. Med. Inform. Assoc. 14 (4), 415–423.

Ashburner, M., Ball, C.A., Blake, J.A., et al., 2000. Gene ontology: tool for the unification of biology. Nat. Genet. 25 (1), 25–29.

Bates, D.W., Kuperman, G.J., Wang, S., et al., 2003. Ten commandments for effective clinical decision support: making the practice of evidence-based medicine a reality. J. Am. Med. Inform. Assoc. 10 (6), 523–530.

Baysari, M.T., Westbrook, J.I., Richardson, K.L., Day, R.O., 2018. The influence of computerized decision support on prescribing during ward-rounds: are the decision-makers targeted? J. Am. Med. Inform. Assoc. 18 (6), 754–759.

Bellazzi, R., Zupan, B., 2001. Predictive data mining in clinical medicine: current issues and guidelines. Int. J. Med. Inf. 77 (2), 81–97.

Benson, T., Grieve, G., 2016. Principles of Health Interoperability: HL7 and SNOMED. Springer.

Berner, E.S., Houston, T.K., Ray, M.N., et al., 2009. Improving ambulatory prescribing safety with a handheld decision support system: a randomized controlled trial. J. Am. Med. Inform. Assoc. 13 (2), 171–179.

Black, A.D., Car, J., Pagliari, C., et al., 2011. The impact of eHealth on the quality and safety of health care: a systematic overview. PLoS Med. 8 (1), e1000387.

Chen, J.H., Wang, J.C., Bates, D.W., 2020. Predictive models in clinical decision support: applications in practice. J. Am. Med. Inform. Assoc. 27 (8), 1150–1155.

Chowdhury, M., Sadek, A., Ma, Y., Kanhere, N., Bhavsar, P., 2006. Applications of artificial intelligence paradigms to decision support in real-time traffic management. Transp. Res. Rec. 1968 (1), 92–98.

Demner-Fushman, D., Chapman, W.W., McDonald, C.J., 2009. What can natural language processing do for clinical decision support? J. Biomed. Inf. 42 (5), 760–772.

Esteva, A., Robicquet, A., Ramsundar, B., Kuleshov, V., DePristo, M., Chou, K., Cui, C., Corrado, G.S., Thrun, S., Dean, J., 2019. A guide to deep learning in healthcare. Nat. Med. 25 (1), 24–29.

Ferrucci, D., Levas, A., Bagchi, S., Gondek, D., Mueller, E.T., 2013. Watson: beyond Jeopardy. Artif. Intell. 199–200, 93–105. https://doi.org/10.1016/j.artint.2012.06.009.

Froomkin, A.M., Kerr, I., Pineau, J., 2019. When AIs outperform doctors: confronting the challenges of a tort-induced over-reliance on machine learning. Ariz. Law Rev. 61, 33.

Garg, A.X., Adhikari, N.K., McDonald, H., et al., 2005. Effects of computerized clinical decision support systems on practitioner performance and patient outcomes: a systematic review. JAMA 293 (10), 1223–1238.

Goddard, K., 2012. Automation Bias and Prescribing Decision Support–Rates, Mediators and Mitigators. City University London. Doctoral dissertation.

Goldstein, M.K., Coleman, R.W., Tu, S.W., et al., 2012. Translating evidence-based clinical knowledge into practice. J. Am. Geriatr. Soc. 50 (5), 1115–1119.

Goodman, K.W., 2016. Ethical and legal issues in decision support. In: Clinical Decision Support Systems: Theory and Practice, pp. 131–146.

Gordan, L., Kovtun, V., Price, M., Doan, R., Lu, L., Reilly, E., , ... Mertens, W.C., 2019. A clinical decision support system–IBM Watson oncology and breast cancer treatment recommendations: agreement with an expert multidisciplinary tumor board. Ann. Oncol. 30 (7), 1085–1091.

Heckerman, D., Geiger, D., Chickering, D.M., 1992. Learning Bayesian networks: the combination of knowledge and statistical data. Mach. Learn. 20 (3), 197–243.

Hersh, W.R., 2007. Health informatics: information technology in healthcare. J. Am. Med. Inform. Assoc. 12 (4), 251–258.

Huang, J., Xiao, L., Yang, J., Cheng, M., 2021. Towards a semantic-enabled agent-oriented distributed clinical decision model. In: 2021 IEEE International Conference on Consumer Electronics and Computer Engineering (ICCECE). IEEE, pp. 500–510.

Isaac, T., Zheng, J., Jha, A.K., Stedman, M., 2012. Use of UpToDate and outcomes in US hospitals. J. Hosp. Med. 7 (2), 85–90.

Jaspers, M.W.M., Smeulers, M., Vermeulen, H., et al., 2011. Effects of clinical decision-support systems on practitioner performance and patient outcomes: a synthesis of high-quality systematic review findings. J. Am. Med. Inform. Assoc. 18 (3), 327–334.

Jones, L.R., Wilson, I., 2023. Supporting clinical decision making: semantics based classification of medical referral letters. Data Sci. J. Comput. Appl. Inform. 7 (1), 24–34.

Jones, S.S., Rudin, R.S., Perry, T., Shekelle, P.G., 2014. Health information technology: an updated systematic review with a focus on meaningful use. Ann. Intern. Med. 160 (1), 48–54. https://doi.org/10.7326/M13-1531.

Kahn, M.G., Brown, J.S., Chun, A.T., Davidson, B.N., Meeker, D., Ryan, P.B., Schilling, L.M., Weiskopf, N.G., Williams, A.E., Zozus, M.N., 2016. Transparent reporting of data quality in distributed data networks. EGEMS (Washington, DC) 4 (1), 18.

Kawamoto, K., Houlihan, C.A., Balas, E.A., et al., 2005a. Improving clinical practice using clinical decision support systems: a systematic review of trials to identify features critical to success. BMJ 330 (7494), 765.

Kawamoto, K., Houlihan, C.A., Balas, E.A., Lobach, D.F., 2005b. Improving clinical practice using clinical decision support systems: a systematic review of trials to identify features critical to success. BMJ 330 (7494), 765.

Kruse, C.S., Bolton, K., Freriks, G., 2017. The effect of patient portals on quality outcomes and its implications to meaningful use: a systematic review. J. Med. Internet Res. 17 (2), e44.

Liao, K.P., Cai, T., Savova, G.K., Murphy, S.N., Karlson, E.W., Ananthakrishnan, A.N., 2015. Development of phenotype algorithms using electronic medical records and incorporating natural language processing. BMJ 350, h1885.

Lindell, O.P., Karlsson, L.O., Nilsson, S., Charitakis, E., Hagström, E., Muhr, T., , ... Janzon, M., 2022. Clinical decision support for familial hypercholesterolemia (CDS-FH): rationale and design of a cluster randomized trial in primary care. Am. Heart J. 247, 132–148.

Lucas, P., van der Gaag, L., Abu-Hanna, A., 2004. Bayesian networks in biomedicine and healthcare. Artif. Intell. Med. 30 (3), 201–214.

Mandel, J.C., Kreda, D.A., Mandl, K.D., Kohane, I.S., Ramoni, R.B., 2016. SMART on FHIR: a standards-based, interoperable apps platform for electronic health records. J. Am. Med. Inform. Assoc. 23 (5), 899–908.

Martínez-Romero, M., Jonquet, C., O'connor, M.J., Graybeal, J., Pazos, A., Musen, M.A., 2017. NCBO Ontology Recommender 2.0: an enhanced approach for biomedical ontology recommendation. J. Biomed. Semant. 8, 1–22.

Meystre, S.M., Savova, G.K., Kipper-Schuler, K.C., Hurdle, J.F., 2010. Extracting information from textual documents in the electronic health record: a review of recent research. Yearb. Med. Inform. 19 (1), 128–144.

Middleton, B., Bloomrosen, M., Dente, M.A., et al., 2013. Enhancing patient safety and quality of care by improving the usability of electronic health record systems: recommendations from AMIA. J. Am. Med. Inform. Assoc. 20 (e1), e2–e8.

Mili, H., Gholami, J., Zowghi, D., Mahanti, A., 2018. Software tools for requirements integration: a systematic mapping study. Requir. Eng. 23 (4), 495–518.

Miller, R.A., Masarie, F., 1990. Quick Medical Reference (QMR): a microcomputer-based diagnostic decision-support system for general internal medicine. In: Proceedings of the Fourteenth Annual Symposium on Computer Applications in Medical Care. IEEE Computer Society Press, Los Alamitos, CA, pp. 986–998.

Miotto, R., Li, L., Kidd, B.A., Dudley, J.T., 2016. Deep patient: an unsupervised representation to predict the future of patients from the electronic health records. Sci. Rep. 6, 26094.

Moja, L., Alberti, R., Sani, D., Pistotti, V., D'Andrea, E., 2014. Clinical decision support systems for prescribing: a systematic review. BMC Med. Inf. Decis. Making 14, 91. https://doi.org/10.1186/1472-6947-14-91.

Murphy, D.R., Meyer, A.N.D., Bhise, V., Russo, E., Sittig, D.F., Singh, H., 2014. Computerized clinician decision support: an unmet challenge for information technology in patient safety. BMJ Qual. Saf. 23 (3), 238–246.

Musen, M.A., 2015. The Protégé project: a look back and a look forward. AI Matters 1 (4), 4–12.

Musen, M.A., Middleton, B., Greenes, R.A., 2014. Clinical decision-support systems. In: Biomedical Informatics. Springer, London, pp. 643–674.

Phansalkar, S., van der Sijs, H., Tucker, A.D., et al., 2013. Drug–drug interactions that should be non-interruptive in order to reduce alert fatigue in electronic health records. J. Am. Med. Inform. Assoc. 20 (3), 489–493.

Poly, T.N., Islam, M.M., Muhtar, M.S., Yang, H.C., Nguyen, P.A., Li, Y.C., 2020. Machine learning approach to reduce alert fatigue using a disease medication–related clinical decision support system: model development and validation. JMIR Med. Inform. 8 (11), e19489.

Powles, J., Hodson, H., 2017. Google DeepMind and healthcare in an age of algorithms. Health Technol. 7 (4), 351–367.

Raghupathi, W., Raghupathi, V., 2014. Big data analytics in healthcare: promise and potential. Health Inf. Sci. Syst. 2, 3.

Ramanathan, R., Di, Y., Jones, P., 2013. Privacy and security solutions for health care data. Commun. ACM 56 (11), 42–47.

Rector, A.L., Rogers, J., Zanstra, P.E., 1997. OpenGALEN: open source medical terminology and tools. In: Proceedings of the AMIA Symposium, 1012.

Rindfleisch, T.C., 1997. Privacy, information technology, and health care. Commun. ACM 40 (8), 92–100.

De Roulet, D., Seo, V., Sanders, L., Hopkins, A., 2020. Challenges and opportunities for clinical decision support in health care. Biomed. Inf. Insights 12, 117822261988847.

Shortliffe, E.H., Cimino, J.J., 2006. Biomedical Informatics: Computer Applications in Health Care and Biomedicine. Springer Science & Business Media.

Silverman, R.D., Skinner, W., 2017. Liability for use of clinical decision support systems. BMC Med. Inf. Decis. Making 17 (1), 1–6.

Simon, S.R., Evans, J.S., Benjamin, A., Delano, D., Bates, D.W., 2009. Patients' attitudes toward electronic health information exchange: qualitative study. J. Med. Internet Res. 11 (3), e30.

Sittig, D.F., Wright, A., Osheroff, J.A., et al., 2008a. Grand challenges in clinical decision support. J. Biomed. Inf. 41 (2), 387–392.

Sittig, D.F., Wright, A., Osheroff, J.A., Middleton, B., Teich, J.M., Ash, J.S., , … Bates, D.W., 2008b. Grand challenges in clinical decision support. J. Biomed. Inf. 41 (2), 387–392.

Topol, E.J., 2019. High-performance medicine: the convergence of human and artificial intelligence. Nat. Med. 25 (1), 44–56.

Van Melle, W., 1978. MYCIN: a knowledge-based consultation program for infectious disease diagnosis. Int. J. Man Mach. Stud. 10 (3), 313–322.
van der Sijs, H., Aarts, J., Vulto, A., et al., 2006. Overriding of drug safety alerts in computerized physician order entry. J. Am. Med. Inform. Assoc. 13 (2), 138–147.
Ventola, C.L., 2014. Mobile devices and apps for health care professionals: uses and benefits. Pharmacy and Therapeutics 39 (5), 356–364.
Wang, A.Y., Sable, J.H., Spackman, K.A., 2002. The SNOMED clinical terms development process: refinement and analysis of content. In: Proceedings of the AMIA Symposium. American Medical Informatics Association, p. 845.
Wright, A., Hickman, T.T., McEvoy, D., Aaron, S., Ai, A., Andersen, J.M., , … Bates, D.W., 2014. Analysis of clinical decision support system malfunctions: a case series and survey. J. Am. Med. Inform. Assoc. 23 (5), 1068–1076.

PART IV

Clinical intelligence: Domain-specific applications

CHAPTER 12

Multimodal semantic integration in radiology: From pixels to clinical knowledge graphs

Goguldeep Veluswamy and Satya Narayana Patro
Department of Radiology, University of Arkansas for Medical Sciences (UAMS), Little Rock, AR, United States

1. Introduction

1.1 The role of multimodal approaches in contemporary radiology

Radiology has traditionally been an image-centric discipline, with diagnostic information primarily extracted through visual interpretation of medical images. In contrast, contemporary clinical practice produces large volumes of supplementary information, containing electronic health records, clinical narratives, laboratory findings, genomic profiles, and data from wearable sensors (Amal et al., 2022). Accurate clinical diagnosis and prognosis depend on integrating these various data types, but current radiological workflows often do not include this extensive information (Vadlapudi and Kaladhar, 2025).

Integrating multiple data modalities into a unified clinical representation provides considerable benefits over unimodal approaches. Multimodal integration improves diagnostic accuracy, strengthens predictive modeling, and supports more personalized clinical decisions (Wang et al., 2025). For example, in cardiovascular disease prediction, integrating imaging biomarkers with clinical data and genomic information significantly improves risk assessment compared to using a single modality (Zhou and Wang, 2025).

1.2 The role of knowledge graphs in clinical data integration

Knowledge graphs, which emerge from combining multimodal data fusion and semantic integration, are structured representations of entities such as diseases, symptoms, imaging findings, laboratory values, and patient phenotypes, along with their relationships, including clinical associations, causal links, and diagnostic patterns. They convert raw data into valuable clinical insights by presenting domain knowledge in a clear, machine-readable format.

In radiology, knowledge graphs help transform pixelated images and diverse clinical data into meaningful semantic maps that support clinical reasoning, decision-making, and discovery. This shift from "pixels to clinical knowledge graphs" represents a fundamental change in how radiological information is processed and utilized.

Intelligent Semantic Analysis for Healthcare
ISBN 978-0-443-33512-9,
https://doi.org/10.1016/B978-0-443-33512-9.00019-6

2. Foundations of multimodal data integration

2.1 Data modalities in clinical radiology

Modern radiological practice integrates diverse data types:

Imaging Modalities: Radiological images, including CT, MRI, X-ray, angiography, ultrasound, and PET, form the traditional foundation of radiological practice by providing visual information about anatomical structures and pathological processes (Najjar, 2023). These 2D and 3D data contain rich spatial and temporal information.

Clinical Text: Free-text radiology reports, clinical notes, and patient narratives provide semantic information that cannot be acquired by images alone (Tayefi et al., 2021). Natural language processing techniques facilitate the extraction of clinical information and relationships from these unstructured documents.

Structured Clinical Data::Electronic health records (EHRs) include coded diagnoses, procedures, medications, and laboratory results, offering temporal and quantitative clinical context (Amal et al., 2022). Integrating structured data enables longitudinal analysis and comprehensive patient characterization.

Genomic and Molecular Data: Genomic profiles, biomarkers, and molecular characteristics are increasingly integrated with imaging to stratify patients and predict treatment responses (Acosta et al., 2022). Multimodal fusion of imaging and genomics has shown increased performance for survival prediction and treatment planning compared with unimodal approaches.

Physiological Signals: Vital signs, wearable sensor data, and biomedical signals (ECG, EEG) provide real-time, continuous monitoring of patient condition.

2.2 Semantic integration approaches

Semantic integration closes the gap between raw heterogeneous data and unified clinical understanding. Three primary levels of fusion are commonly employed (Liu and Ye, 2025):

Early Fusion (Data-Level): Raw data from different modalities are combined before processing. It enables joint representation yet can face issues with heterogeneous features and varying quality.

Intermediate Fusion (Feature-Level): Features extracted from individual modalities are combined at intermediate representation layers. This approach enables modality-specific processing before integration and has proven effective through various applications.

Late Fusion (Decision-Level): Individual modality-specific models generate independent predictions, which are then aggregated via collaborative methods, voting, or probability-based combination (Liu and Ye, 2025). Late fusion enables handling of missing modalities and provides modality-specific interpretability (Wang et al., 2025).

Modern approaches increasingly employ hybrid fusion strategies that combine multiple fusion levels to leverage their complementary advantages.

3. Multimodal fusion technologies in radiology

3.1 Transformer-based multimodal integration

Transformer architectures have become powerful for multimodal integration because they can model complex intermodal dependencies and long-range relationships. Vision-language models that combine vision transformers with language models enable unified encoding of visual and textual information. Attention methods, especially intermodal attention, enable dynamic weighting of modalities based on task context and data quality (Liu et al., 2025). The MDFormer framework demonstrates that using intermodal attention fusion mechanisms with transformers effectively combines medical imaging, clinical data, and vital signs, maintaining robustness even when some modalities are missing (Liu et al., 2025).

3.2 Graph neural networks for multimodal relationships

Graph neural networks (GNNs) provide a natural representation for modeling complex relationships within and across modalities. Unlike Convolutional Neural Networks (CNNs) and transformer architectures, which process data as sequences or grids, GNNs explicitly represent entities such as patients, diseases, and imaging findings as nodes and relationships as edges.

GNNs are especially well-suited to combining knowledge from biomedical knowledge bases with patient-specific data, enabling both knowledge-driven and data-centered reasoning.

3.3 Large language models and vision-language integration

Recent advancements in large language models (LLMs) and multimodal large language models (MLLMs) have transformed the domain of medical data integration (Lin and Kuo, 2025). These models can integrate text with medical scans for various radiologic tasks, including diagnosis, visual question answering, report generation, and rationale diagnosis.

Vision-language models support semantic search in radiology reports by learning joint embeddings of images and text that preserve clinical meaning. Multimodal retrieval systems can greatly improve LLMs' capacity to provide accurate diagnoses from radiology reports.

4. Knowledge graphs in radiology and healthcare

4.1 Construction and representation

Knowledge graphs transform clinical data into machine-interpretable semantic networks. The basic components are:

Entities: Biomedical concepts include diseases, imaging findings, anatomical structures, laboratory tests, medications, and patient phenotypes. The Unified Medical

Language System (UMLS) provides standardized vocabularies for representing these entities with consistent identifiers throughout healthcare systems.

Relationships: Semantic connections between entities, such as "shows finding," "causes disease," "indicated by symptom," and "treated by medication," can be categorized to represent different types of associations, including diagnostic, causal, or procedural.

Properties: Attributes of entities include severity, temporal data, confidence levels, and evidence strength.

Building a knowledge graph involves combining various information sources such as biomedical literature, clinical guidelines, structured EHR data, and natural language processing of free text. Large language models have become effective tools for automating this process (Lin and Kuo, 2025). In specialized fields, these models have been shown to perform better in capturing medical domain knowledge.

4.2 Semantic standardization and interoperability

A key challenge in building knowledge graphs is semantic heterogeneity, as various healthcare institutions employ different coding systems, terminologies, and data schemas. The Observational Medical Outcomes Partnership (OMOP) Common Data Model helps standardize and harmonize clinical data across several institutions and regions.

For medical imaging specifically, extensions to common data models address the ongoing challenge that many of these models captured clinical data effectively but failed to integrate imaging information in a standardized manner (Amal et al., 2022).

Standardized schemas facilitate integration of knowledge graphs across various institutions and research projects.

A critical step in converting pixel-level outputs into clinically reliable knowledge is the explicit mapping of image-derived features to standardized ontologies such as RadLex and SNOMED CT. Deep learning systems process numerical pixel intensities and latent embeddings, which lack inherent clinical meaning. Linking these representations to formally defined ontology concepts, each with unique identifiers, hierarchical placement, and structured relationships, constrains model outputs semantically rather than producing unconstrained narrative text.

RadLex provides radiology-specific terminology for anatomy, imaging observations, and procedural concepts, whereas SNOMED CT encodes broader clinical findings and their defined relationships, such as is-a and finding site. Mapping detected imaging features, such as "ground-glass opacity in the right upper lobe," to ontology-backed concepts transforms probabilistic predictions into standardized semantic entities. This process establishes a shared language across institutions, datasets, and AI systems.

Ontology grounding serves as a safeguard against AI hallucinations, although it does not eliminate model hallucinations entirely. Generative and multimodal models may generate plausible yet unsupported findings, especially under domain shift or incomplete input.

Restricting outputs to a predefined concept space reduces the fabrication of nonexistent entities. Additionally, hierarchical and relational structures enable rule-based or constraint-based semantic validation, such as verifying that a diagnosis is anatomically compatible with the detected imaging region. Integrating knowledge graphs facilitates cross-modal consistency checks among imaging findings, clinical metadata, and established disease–phenotype relationships, further minimizing unsupported inference.

Although ontologies are not exhaustive and require ongoing curation, integrating them embeds symbolic structure within data-driven models. This hybrid approach, which combines learned representations with formal semantic constraints, enhances interoperability, longitudinal consistency, and factual reliability. Thus, mapping pixels to ontologies establishes a common clinical language that promotes transparency and reduces hallucinations in multimodal radiologic AI systems.

The practical implications of ontology-grounded mapping in Neuroradiology are illustrated in the example of the semantic pipeline below.

Table 12.1 shows how a contrast-enhancing intra-axial lesion is identified at the pixel level, translated into structured imaging descriptors, mapped to RadLex and SNOMED CT concepts, and encoded in a clinical knowledge graph. This ontology-based approach allows integration with established glioma classification and treatment guidelines, ensuring transparent and reliable decision support.

4.3 Clinical applications of knowledge graphs

Diagnostic Decision Support: Integrating a knowledge graph with large language models can strengthen diagnostic predictions from electronic health records. By retrieving contextually relevant patient-specific information, these systems achieve higher diagnostic accuracy than standard models.

Risk Prediction: Knowledge graph methods present how incorporating external knowledge can improve EHR-based predictions by constructing patient-specific graphs from clinical data and biomedical information. This strategy greatly improves accuracy across key clinical prediction tasks, including mortality, readmission, length of stay, and drug recommendations.

Clinical Reasoning: Knowledge graphs facilitate transparent and explainable clinical reasoning by clearly highlighting associations.

Multicenter Data Integration: Knowledge graph systems enable the privacy-preserving integration of segmented patient data across various institutions.

5. Semantic correlation and cross-modal representation learning

5.1 Vision-language pre-training

Vision-language pre-training aligns image and textual representations via self-supervised contrastive learning. This method skillfully utilizes medical image-text datasets, even

Table 12.1 Semantic integration workflow for brain glioma imaging analysis.

Stage	Example: Enhancing brain glioma	Role in semantic integration
1. Pixel detection	Irregular ring-enhancing lesion with central necrosis on post-contrast T1; peritumoral FLAIR hyperintensity	Deep model identifies the lesion and extracts radiomic features
2. Structured finding	"Heterogeneously enhancing intra-axial mass with necrosis and vasogenic edema"	Converts latent features into structured descriptors
3. Ontology mapping	RadLex concept: Ring enhancement; SNOMED CT concept: Malignant neoplasm of brain; anatomy: Frontal lobe structure	Anchors findings to standardized identifiers
4. Knowledge graph encoding	Nodes: Tumor, enhancement pattern, edema; edges: has morphology, located in, associated with	Enables computable relationships and cross-modal reasoning
5. Guideline linkage	Association with NCCN/WHO glioma framework	Triggers structured diagnostic and management pathways
6. Decision support	"Imaging features concerning for high-grade glioma. Recommend MRI perfusion, spectroscopy, and neurosurgical consultation."	Produces explainable, ontology-traceable recommendations

when they are much smaller than those in general domains. Focused pre-training allows models to outperform once trained on larger datasets. Fine-tuned vision-language models and domain-specific pre-training support zero-shot and few-shot medical image classification, greatly reducing dependence on labeled data (Chandrakala and Fatima, 2025).

MedViLaM demonstrates that a single model can handle multiple medical data types and tasks, achieving strong zero-shot generalization and transfer learning (Chandrakala and Fatima, 2025).

5.2 Region-grounded reasoning

A key challenge in vision-language models is maintaining a clear connection between model outputs and specific image regions. When a model generates a radiology report, clinicians need to know which image regions correspond to each clinical finding.

Detailed alignment across various levels enhances both diagnostic accuracy and interpretability.

This region-based approach is especially important for providing clear visual explanations of model decisions. In chest radiography, visual XAI methods such as Grad-CAM, which denote particular image regions, offer greater coherence and clinical trust than model-agnostic methods lacking spatial grounding (Ihongbe et al., 2024).

5.3 Handling missing modalities

In actual clinical practice, incomplete multimodal data are often encountered due to patient-centered factors (e.g., contraindications to certain imaging modalities), equipment constraints, or inadequate data collection (Liu et al., 2025). Robust multimodal systems must gracefully degrade when modalities are unavailable.

MDFormer meets this challenge with a two-stage masked-enhancement classification and contrastive learning framework, boosting the model's robustness to missing modalities. Remarkably, even when 50% of the modalities are absent, MDFormer retains 85% of the performance it achieves with full data (Liu et al., 2025).

The missing-modality fusion architecture performs better and stays robust with missing data by using multivariate loss functions during training that specifically prepare the model for inference with incomplete data (Wang et al., 2025).

6. Applications of multimodal semantic integration

6.1 Radiology report generation

Automated radiology report generation is a practical application where multimodal semantic integration directly affects clinical workflows. Traditional approaches focused solely on image analysis, whereas modern systems integrate clinical context.

Recent innovations enable the generation of radiology reports for medical images through employing transformer architectures combined with multimodal fusion modules and hierarchical memory, which support the integration of longitudinal multimodal data.

Research shows that retrieval-augmented generation methods, which use multimodal fusion, can make automatically generated reports more accurate and clinically relevant. These methods use reference materials to identify strong examples and incorporate factual knowledge during the writing process.

6.2 Disease diagnosis and outcome prediction

Multimodal fusion allows more precise disease characterization and outcome prediction through integrating complementary information from varied data sources.

Survival Prediction: Hierarchical techniques that integrate multiple imaging modalities using attention mechanisms for survival prediction in cancer patients demonstrate superior performance compared to unimodal approaches. By learning hierarchical visual concepts and modeling multimodal interactions, these systems achieve better predictive outcomes (Yang et al., 2025).

Extended approaches use clinical reports as linguistic surrogates for absent molecular information, thereby maintaining high predictive performance even in the absence of genomic data (Yang et al., 2025).

Early Disease Detection: Multimodal fusion provides considerable benefits for early detection in various disease domains. For example, a glaucoma detection system that utilizes quality-aware federated learning with multimodal fusion achieved 98.6% accuracy while maintaining privacy through secure aggregation mechanisms (Jabbar et al., 2025).

6.2.1 Radiogenomics: The semantic bridge between imaging phenotypes and molecular genotypes

Radiogenomics exemplifies multimodal semantic integration in precision medicine by linking imaging phenotypes to molecular and genomic characteristics. In gliomas and other cancers, molecular alterations such as IDH mutation, 1p/19q co-deletion, MGMT promoter methylation, and EGFR amplification increasingly guide diagnosis, prognosis, and therapy selection. However, molecular testing can be delayed, invasive, or unavailable. Radiogenomic modeling seeks to infer these genotypes from imaging features.

6.2.2 Imaging phenotypes as structured semantic entities

Radiologic traits such as contrast enhancement, necrosis, tumor margins, perfusion, diffusion restriction, and infiltration can be quantified using radiomics and deep learning. Mapping these phenotypes to standardized ontological descriptors creates computable entities within a knowledge graph.

For example, a "ring-enhancing necrotic mass with elevated relative cerebral blood volume" can be encoded as semantic nodes for enhancement morphology, necrosis, and perfusion metrics. These imaging nodes can then link to genomic nodes such as "IDH-wildtype" or "TERT promoter mutation" based on validated statistical associations.

6.2.3 Knowledge graphs as radiogenomic integrators

Knowledge graphs offer an effective architecture for radiogenomic reasoning. Rather than treating imaging and genomic data as parallel feature vectors, graphs explicitly model their relationships. Edges can represent known biological associations, such as "associated with higher grade" or "predicts treatment resistance," or probabilistic correlations from multimodal learning.

This graph-based framework supports:

- Cross-modal inference when genomic data are absent
- Transparent reasoning pathways linking phenotype to genotype
- Integration of WHO molecular classification frameworks

- Longitudinal updating as new genomic knowledge emerges

Embedding WHO molecular classification entities within the same knowledge graph structure as imaging findings allows radiogenomics to function as an operational bridge, not merely a post hoc correlation.

6.2.4 Clinical impact: Toward personalized neuro-oncology

In neuro-oncology, radiogenomic integration enables noninvasive molecular stratification before surgery, supports biopsy targeting, and refines prognostic modeling. Predicting IDH mutation status or MGMT methylation from imaging can guide surgical decisions, adjuvant therapy, and clinical trial enrollment.

More broadly, radiogenomic semantic integration advances personalized medicine by transforming imaging from a descriptive tool into a surrogate biomarker platform. As multimodal foundation models mature, combining imaging phenotypes with genomic signatures will increasingly guide therapy selection, outcome prediction, and adaptive treatment planning.

6.3 Medical imaging narrative generation

Domain-adaptive methods for report generation address the challenge of producing clinically consistent narratives by constructing detailed representations that differentiate disease-specific attributes from normal findings. This methodology captures the semantic distinction between pathological and normal states, thus enhancing generalization across different datasets.

6.3.1 The role of foundation models in semantic discovery

Recent advances in multimodal foundation models, trained on millions of imaging studies with corresponding reports, clinical variables, and molecular data, have expanded semantic integration beyond predefined ontologies. Unlike task-specific models, foundation models learn latent representations that capture complex statistical patterns across populations.

When integrated into a knowledge graph, these models can identify previously unrecognized associations. For example, subtle imaging phenotypes may cluster with specific molecular alterations or treatment responses, even if these relationships are not captured in current classification systems. In this way, foundation models can suggest candidate associations for semantic discovery by proposing novel connections within the graph, rather than only populating known nodes.

Importantly, discovery remains hypothesis-generating rather than definitive. Newly inferred associations require clinical validation and biological plausibility assessment before integration into practice. However, combining large-scale representation learning with structured graph reasoning creates a dynamic knowledge ecosystem that evolves with emerging clinical evidence, rather than remaining limited to static textbook relationships.

6.3.2 Handling missing data and uncertainty in acute care

Acute care environments pose a fundamental challenge for multimodal systems due to incomplete information. For example, a patient arriving at the emergency department with an altered mental status may lack prior imaging, genomic data, or a reliable history. Semantic systems must therefore reason under uncertainty.

Architectures such as multimodal transformers can be designed to handle missing data through masking and training-time regularization strategies. Instead of requiring all inputs at once, these systems learn cross-modal dependencies and dynamically weight available information. When genomic or longitudinal data are missing, the model relies more on imaging and immediate clinical variables. The knowledge graph can be incrementally updated as new data becomes available.

In a graph-based framework, uncertainty can be explicitly represented. Nodes may carry probabilistic weights to reflect confidence levels, and edges can encode conditional likelihoods instead of deterministic associations. This enables the system to generate differential diagnoses ranked by confidence rather than relying on binary predictions.

In acute stroke, for example, imaging-derived perfusion deficits may initially guide reasoning. As laboratory results or prior medical history become available through FHIR integration, the graph recalibrates and refines etiologic inference. This incremental semantic updating mirrors clinical reasoning and supports decision-making without overconfidence in the face of incomplete data.

6.3.3 Bridging research and hospital practice

6.3.3.1 Bridging research and clinical practice: Standards as semantic custodians

A key challenge in multimodal radiologic AI is translating research innovation into hospital practice. While many models perform well on curated datasets, they often do not generalize to diverse clinical settings. Bridging this gap requires not only robust models but also adherence to interoperable data standards that maintain semantic integrity across institutions.

6.3.3.2 DICOM as the imaging information custodian

Digital Imaging and Communications in Medicine (DICOM) is the primary standard for imaging acquisition, storage, and transmission. In addition to pixel data, DICOM encodes essential metadata such as scanner parameters, acquisition protocols, anatomical descriptors, temporal information, and structured reporting elements. These metadata fields preserve imaging context, ensuring that findings remain consistent across institutions.

When multimodal systems map imaging findings to ontologies like RadLex and SNOMED CT, embedding these mappings in DICOM Structured Reports (DICOM-SR) preserves standardized identifiers with the image. This approach allows ontology-linked findings to accompany the imaging study, supporting the construction of a knowledge graph that remains semantically consistent across sites.

6.3.3.3 HL7 FHIR as the clinical interoperability backbone

While DICOM manages imaging, HL7 Fast Healthcare Interoperability Resources (FHIR) standardizes the exchange of structured clinical data such as laboratory values, genomic reports, medications, and diagnoses. FHIR resources represent clinical entities in a modular, machine-readable format, enabling secure and scalable data exchange across electronic health record systems.

When knowledge graph nodes, such as "enhancing intra-axial mass" or "IDH-mutant astrocytoma," are linked to FHIR-based clinical data, multimodal reasoning becomes portable. A knowledge graph built in one hospital retains its meaning in another because both imaging and clinical data follow shared structural standards. In this framework, DICOM and FHIR serve as dual custodians: DICOM protects imaging semantics, while FHIR maintains longitudinal clinical context.

6.3.3.4 Toward operational Semantic continuity

By grounding multimodal AI systems in DICOM and FHIR standards, institutions ensure semantic continuity across research, validation, and deployment. Knowledge graphs become transferable infrastructure rather than institution-specific artifacts. This standards-based approach turns multimodal semantic integration from a research prototype into a scalable clinical architecture.

6.3.3.5 Clinical workflow integration and the human-in-the-loop

To advance beyond research settings, multimodal semantic systems must integrate seamlessly into radiologists' daily workflows. The effectiveness of a knowledge graph–driven architecture depends not only on computational sophistication but also on its ability to enhance human expertise in real time.

6.3.3.6 Decision Support within the PACS environment

In practice, the knowledge graph functions as an embedded decision-support engine within the PACS and reporting ecosystem, rather than as a separate analytical layer. As imaging studies are opened, deep learning models process pixel-level data in the background and map findings to standardized ontology concepts. These entities populate a graph instance linked to relevant clinical and genomic data retrieved through FHIR-based interfaces:

- During report generation, the system may:
- Suggest structured descriptors consistent with detected findings
- Highlight prior imaging correlations or interval changes
- Surface guideline-based recommendations triggered by ontology-linked nodes
- Alert the radiologist to discordance between imaging phenotype and available molecular data

For example, if a radiologist describes a heterogeneously enhancing intra-axial mass with necrosis, the system may suggest high-grade glioma descriptors and prompt consideration of advanced MRI sequences or neurosurgical consultation. If genomic data indicate an IDH mutation, the system may contextualize imaging aggressiveness within that molecular framework.

The radiologist remains the final decision-maker. The system provides structured suggestions, which the clinician reviews, modifies, or accepts. This approach preserves professional autonomy while enhancing cognitive efficiency.

6.3.3.7 The human-in-the-loop model

Multimodal knowledge systems serve as cognitive amplifiers rather than replacements for radiologic judgment. The "human-in-the-loop" paradigm ensures that:

- AI outputs are continuously validated by domain expertise
- Structured corrections refine downstream model performance
- Institutional practice patterns shape localized reasoning pathways

Each interaction serves as feedback, improving model calibration and semantic precision over time. This two-way exchange transforms static AI tools into adaptive clinical collaborators.

6.3.3.8 Trust and multimodal explainability

Trust is essential for adoption. In multimodal systems, explainability must go beyond visual saliency maps to include cross-domain reasoning.

At the imaging level, techniques such as attention mapping or Grad-CAM show which lesion regions contributed to a classification. When decisions include genomic markers or clinical variables, explanations must also operate at the graph level. Knowledge graphs offer structural transparency. Recommendations can be traced through explicit nodes and edges:

Imaging phenotype → Ontology mapping → Linked molecular marker → Guideline association → Proposed action

For example, a system suggesting "features concerning for IDH-wildtype high-grade glioma" should be able to display:

- The imaging traits contributing to aggressiveness (e.g., necrosis, elevated perfusion)
- The absence or presence of specific molecular markers
- The guideline or classification framework informing the recommendation

This layered explanation, combining pixel-level attention with semantic path tracing, creates a comprehensive interpretability model. Radiologists can review both where the model focused and why specific cross-modal associations were activated.

6.3.3.9 Building durable trust

Sustained trust relies on three principles:

- Transparency of reasoning pathways
- Consistency across cases and institutions
- Preservation of clinician authority

When DICOM and FHIR standards preserve semantic fidelity and ontology-based knowledge graphs enable traceable reasoning, AI systems become accountable clinical partners rather than opaque predictors. In this framework, multimodal integration enhances clinical judgment rather than obscuring it.

7. Challenges in multimodal semantic integration

7.1 Data disparity and integration

Multimodal biomedical data demonstrate significant heterogeneity in structure (structured vs. unstructured), distribution (imaging pixels vs. categorical variables), modality-specific scales and units, and data quality (Vadlapudi and Kaladhar, 2025). Such heterogeneity poses challenges for effective representation learning and data fusion.

Data modalities are derived from diverse sources, each applying distinct coding systems, naming conventions, and quality standards. Achieving semantic interoperability necessitates explicit mapping and harmonization, which is a resource-intensive process requiring collaboration between domain experts and data scientists.

Evidential fusion frameworks address this heterogeneity by modeling both data uncertainty and model-related uncertainty, while simultaneously learning modality-level reliability to alleviate the impact of noisy data sources (Mobadersany et al., 2018).

7.2 Data privacy and security

The integration of multimodal data from multiple institutions presents major privacy challenges. Patient identifiers, imaging data, genomic information, and clinical records constitute sensitive protected health information as defined by regulations such as HIPAA and GDPR.

Federated learning methods facilitate collaborative model training across institutions without the need to centralize sensitive data. Quality-aware vertical federated learning, combined with homomorphic encryption and differential privacy, can maintain high diagnostic accuracy while protecting patient confidentiality (Jabbar et al., 2025).

Blockchain-based methods support the safe aggregation of intermediate reasoning results across institutions, thereby enabling collaborative clinical reasoning while ensuring that original patient data remains localized.

7.3 Interpretability and explainability

A key challenge in multimodal systems is clarifying how various modalities influence clinical predictions. Although explanations for unimodal image-based methods, like Grad-CAM highlights, are fairly straightforward, providing explanations for multimodal fusion decisions remains largely unresolved.

Multimodal transformers that use attention mechanisms can reveal how different modalities interact, indicating which image regions and textual concepts had the most influence (Ahmed et al., 2026). However, it remains difficult to ensure that the attention factors accurately represent clinically significant connections (Hafeez et al., 2025).

7.4 Validation and clinical integration

The translation of multimodal systems from research prototypes to clinical tools demands strict prospective validation across diverse populations and institutions. Retrospective validation utilizing selected cohorts frequently overestimates real-world performance.

Effective clinical integration calls for careful workflow design to ensure that multimodal decision-support tools enhance radiologist performance rather than introduce additional burdens, such as interruptions or information overload.

7.4.1 Emerging directions and future perspectives

7.4.1.1 Foundation models for radiological intelligence

Large-scale foundation models pre-trained on extensive multimodal clinical datasets offer significant potential for generalizability. Recent studies indicate that models trained on millions of images and thousands of 3D scans achieve cutting-edge performance, underscoring the ongoing need to broaden and improve medical datasets and benchmarks.

Foundation models capable of flexibly interpreting diverse modality combinations and generating expressive outputs, such as free-text explanations and image annotations, represent a major shift toward generalist medical artificial intelligence.

7.4.1.2 Knowledge-integrated AI

The integration of explicit medical knowledge, encoded in knowledge graphs, with learned representations from data constitutes a critical research direction. Recent approaches demonstrate that grounding model projections in structured medical knowledge enhances both diagnostic accuracy and interpretability.

Knowledge graph construction driven by large language models allows scalable and automated discovery of clinical relationships from unstructured text, thereby substantially accelerating knowledge acquisition.

7.5 Personalized medicine through multimodal integration

By combining imaging, genomics, clinical data, and molecular biomarkers, multimodal systems promote more accurate patient stratification and treatment planning (Amal et al., 2022). Going forward, these systems are expected to aid in predicting customized treatment responses and optimizing therapy choices.

7.6 Real-time and edge deployment

Current multimodal systems often demand considerable computational resources, which restrict their deployment at the point of care. Future efforts aim to develop efficient models suited for real-time processing and edge deployment.

Techniques such as knowledge distillation, quantization, and pruning help reduce model size and computational requirements while preserving diagnostic accuracy.

7.6.1 Illustrative case study: Longitudinal semantic growth in glioma care

Consider a patient presenting with new-onset seizures. Initial MRI shows a nonenhancing T2/FLAIR hyperintense lesion in the left frontal lobe. At this stage, the knowledge graph includes nodes for imaging phenotype (nonenhancing intra-axial lesion, cortical involvement), clinical presentation (seizure), and preliminary differential diagnoses.

After surgical biopsy, the pathology reveals IDH-mutant astrocytoma. Molecular markers, such as IDH mutation status and ATRX loss, are added as genomic nodes linked to WHO classification entities. The graph now encodes both phenotype and genotype, supporting prognostic modeling and therapy planning.

Postoperative imaging adds extent-of-resection metrics, and longitudinal follow-up introduces treatment variables (radiation, temozolomide) and response assessments. If recurrence occurs with new contrast enhancement and elevated perfusion, additional nodes indicate malignant transformation. These updates alter predictive pathways in the graph, thereby modifying survival estimates and therapeutic recommendations.

Over time, the patient-specific knowledge graph evolves from a simple imaging-based structure to a richly interconnected semantic network that spans radiology, genomics, pathology, and treatment response. Each clinical event expands and refines the graph, demonstrating how multimodal semantic systems support precision neuro-oncology throughout the disease trajectory.

8. Conclusion

Radiology is evolving beyond image analysis to incorporate integrated, graph-based clinical reasoning. Multimodal semantic systems combine imaging data with genomics, laboratory results, pathology, and clinical histories, converting diverse data into structured, computable knowledge. Built on interoperable standards such as DICOM and HL7 FHIR, this approach preserves semantic meaning across institutions and enables research innovations to become lasting clinical infrastructure.

Knowledge graphs organize this transformation by mapping imaging phenotypes to standardized ontologies and linking them to molecular genotypes, treatment variables, and outcomes. This positions radiology as a central component of precision medicine. Radiogenomics demonstrates this integration, connecting noninvasive imaging features to molecular classification systems to inform prognosis, guide biopsy targeting, and shape treatment strategies. These systems are shifting from retrospective correlation tools to prospective decision-support platforms embedded in daily workflows.

Multimodal foundation models further expand this landscape. Trained on large, diverse datasets, these models can identify associations beyond established textbook relationships. When combined with structured graph reasoning, these findings become traceable and testable, supporting hypothesis generation and maintaining clinical accountability. At the same time, semantic architecture that handles missing data and probabilistic reasoning enables AI systems to function effectively in acute care, where uncertainty and incomplete information are common.

The human-in-the-loop paradigm is equally important. Integrated into PACS and reporting environments, multimodal systems act as cognitive amplifiers by suggesting structured descriptors, highlighting phenotype–genotype concordance or discordance, and updating dynamically as new clinical data become available. Explainability operates at multiple levels: visual attention mechanisms clarify the contributions of imaging, and graph-based path tracing reveals cross-modal reasoning. This transparency reinforces trust and maintains radiologist authority.

The longitudinal glioma example demonstrates the broader vision: a patient-specific knowledge graph that evolves with each diagnostic, molecular, and therapeutic milestone. In this model, radiology becomes an active, iterative contributor to a dynamic semantic ecosystem of care, rather than a static interpretive endpoint.

Ultimately, the convergence of standardized data exchange, knowledge graph architectures, foundation models, and human-centered workflow design is redefining radiology's role. The field is shifting from isolated image interpretation to orchestrating integrated clinical intelligence. As these systems mature, radiologists remain central, supported by tools that connect pixels to pathways, phenotypes to genotypes, and individual patients to evolving medical knowledge.

References

Acosta, J.N., Falcone, G.J., Rajpurkar, P., Topol, E.J., 2022. Multimodal biomedical AI. Nat. Med. 28 (9), 1773–1784. https://doi.org/10.1038/s41591-022-01981-2.

Ahmed, F., Naz, N.S., Khan, S., Rehman, A.U., Ismael, W.M., Khan, M.A., 2026. Explainable artificial intelligence (XAI) in medical imaging: a systematic review of techniques, applications, and challenges. BMC Med. Imag. 26 (1). https://doi.org/10.1186/s12880-025-02118-w.

Amal, S., Safarnejad, L., Omiye, J.A., Ghanzouri, I., Cabot, J.H., Ross, E.G., 2022. Use of multi-modal data and machine learning to improve cardiovascular disease care. Front. Cardiovasc. Med. 9. https://doi.org/10.3389/fcvm.2022.840262.

Chandrakala, K., Fatima, S., 2025. Advances in natural language processing for radiology: state-of-the-art techniques, applications, and open challenges. Int. J. Adv. Comput. Sci. Appl. 16 (11). https://doi.org/10.14569/ijacsa.2025.0161145.

Hafeez, Y., Memon, K., Al-Quraishi, M.S., Yahya, N., Elferik, S., Ali, S.S.A., 2025. Explainable AI in diagnostic radiology for neurological disorders: a systematic review, and what doctors think about it. Diagnostics 15 (2). https://doi.org/10.3390/diagnostics15020168.

Ihongbe, I.E., Fouad, S., Mahmoud, T.F., Rajasekaran, A., Bhatia, B., 2024. Evaluating explainable artificial intelligence (XAI) techniques in chest radiology imaging through a human-centered lens. PLoS One 19 (10). https://doi.org/10.1371/journal.pone.0308758.

Jabbar, A., Huang, J., Jabbar, M.K., Ali, A., 2025. Adaptive multimodal fusion in vertical federated learning for decentralized glaucoma screening. Brain Sci. 15 (9). https://doi.org/10.3390/brainsci15090990.

Lin, C., Kuo, C.-F., 2025. Roles and potential of Large language models in healthcare: a comprehensive review. Biomed. J. 48 (5), 100868. https://doi.org/10.1016/j.bj.2025.100868.

Liu, C., Ye, F., 2025. A review of multimodal medical data fusion techniques for personalized medicine. In: Proceedings of The 4th International Conference on Biomedical and Intelligent Systems, IC-BIS 2025. Association for Computing Machinery, Inc, China, pp. 338–347. https://doi.org/10.1145/3745034.3745088. http://dl.acm.org/citation.cfm?id=3745034.

Liu, X., Pan, F., Song, H., Cao, S., Li, C., Li, T., 2025. MDFormer: transformer-based multimodal fusion for robust chest disease diagnosis. Electronics 14 (10), 1926. https://doi.org/10.3390/electronics14101926.

Mobadersany, P., Yousefi, S., Amgad, M., Gutman, D.A., Barnholtz-Sloan, J.S., Velázquez Vega, J.E., Brat, D.J., Cooper, L.A.D., 2018. Predicting cancer outcomes from histology and genomics using convolutional networks. Proc. Natl. Acad. Sci. U.S.A. 115 (13), E2970. https://doi.org/10.1073/pnas.1717139115.

Najjar, R., 2023. Redefining radiology: a review of artificial intelligence integration in medical imaging. Diagnostics 13 (17), 2760. https://doi.org/10.3390/diagnostics13172760.

Tayefi, M., Ngo, P., Chomutare, T., Dalianis, H., Salvi, E., Budrionis, A., Godtliebsen, F., 2021. Challenges and opportunities beyond structured data in analysis of electronic health records. WIREs Comput. Stat. 13 (6). https://doi.org/10.1002/wics.1549.

Vadlapudi, V., Kaladhar, D.S.V.G.K., 2025. Artificial intelligence in medical imaging: bridging innovation, ethics, and clinical impact. Int. J. Adv. Med. 12 (6), 621–626. https://doi.org/10.18203/2349-3933.ijam20253360.

Wang, M., Fan, S., Li, Y., Xie, Z., Chen, H., 2025. Missing-modality enabled multi-modal fusion architecture for medical data. J. Biomed. Inf. 164, 104796. https://doi.org/10.1016/j.jbi.2025.104796.

Yang, J., Hu, J., He, X., 2025. HMSP: hierarchical multimodal semantic fusion of pathology-text-genomic representations for robust survival prediction. In: 2025 IEEE International Conference on Medical Artificial Intelligence (MedAI).

Zhou, Z., Wang, J., 2025. A narrative review of multimodal data fusion strategies for precision risk prediction in coronary artery disease: advances, challenges, and future informatics directions. Rambam Maimonides Med. J. 16 (4), e0023. https://doi.org/10.5041/rmmj.10558.

CHAPTER 13

A framework for brain tumor (Glioma) analysis using SVM and graph convolutional networks

Geetika Dikshit[1], Shivendra Dubey[2] and Rahul Shrivastava[3]
[1]Maulana Azad National Institute of Technology, Bhopal, Madhya Pradesh, India; [2]Department of Artificial Intelligence and Machine Learning, Manipal University, Jaipur, Rajasthan, India; [3]SRM Institute of Science and Technology, Tiruchirappalli, Tamil Nadu, India

1. Introduction

In the year 2020, brain tumors were identified as the 10th most prevalent type of tumor within the Indian population. According to research by the International Association of Cancer Registries (IARC), the yearly incidence of brain tumors in India exceeds 28,000 cases, with a corresponding mortality rate of over 24,000 individuals per year.

Medical imaging methods such as magnetic resonance imaging (MRI), electrogastrography (EGG), positron emission tomography (PET), fMRI (functional magnetic resonance imaging) and others can be used to scan the human brain. The most often used acquisition among all of these for research on brain analysis is MRI data. Additionally, there are many MRI data collection methods, including functional MRI (fMRI) and diffusion tensor imaging (DTI), from which functional and anatomical brain networks can be built (Cui et al., 2023). In particular, the connection in structural brain networks depicts the physical connectivity between gray matter regions, whereas connectivity in functional brain networks describes correlation between time-series signals of the regions of the human brain.

Modern neuroscience seeks to understand the brain's complicated function. For modeling the brain as a graph, we need to implement high end mathematical constructs describing the connections or interactions (edges) between distinct items (nodes). These graphs use brain areas of interest (ROIs) as nodes and functional connectivity between ROIs as edges, estimated as pairwise correlations of fMRI time series.

Previous research on brain network analysis has focused primarily on the study of shallow models that are based on graph theory and tensor factorization. These models are primarily concerned with the proposal of neurobiologically insightful graph measures and approaches from the node, motif and graph level in order to identify central network elements and detect network communities or modules Fig. 13.1.

An atlas is used to parcellate the fMRI pictures, and the results are then converted to graphs. After that, the graphs are sent to the suggested BrainGNN framework, which

Intelligent Semantic Analysis for Healthcare
ISBN 978-0-443-33512-9,
https://doi.org/10.1016/B978-0-443-33512-9.00018-4

then provides the prediction of certain tasks. It chooses prominent brain areas that are relevant to the predicting outcomes and grouping brain areas into interconnected communities associated with prediction. Both of these processes take place simultaneously (Li et al., 2021). This framework works with.niiimage format which stands for Neuroimaging Informatics Technology initiative. This is the latest format to analyze the various brain related diseases.

It is commonly used to store magnetic resonance imaging (MRI) data. It uses nibabel library for implementing the training/testing of various images of brain tumor and its various types—glioma, meningioma, no-tumor, pituitary. For checking the result, it uses Dataset_Brain_tumor_MRI_images dataset from Kaggle which contains approx. 400 images for analyzing the testing, training and predictions checking the input of images in.nii format as well as.jpg format.

The signs and symptoms of a brain tumor might change depending on the size of the tumor, where it is located, and how quickly it is growing. The most common symptoms include headaches, seizures, changes in vision, issues with cognitive function and memory, changes in mood or personality, nausea or vomiting, and problems with balance or coordination.

In the event that a brain tumor is suspected, diagnostic procedures including imaging examinations (such as MRI or CT scans) and a biopsy may be carried out in order to identify the tumor's subtype and severity level. Surgery, radiation therapy and chemotherapy are all possible treatment choices for patients who have brain tumors. The particular course of treatment will be determined by a number of criteria, such as the nature, size, and location of the tumor, amongst others (Gao et al., 2019).

In this paper we have used various concepts of Functional correlation matrix and functional graph. Calculating functional correlation matrices can be done using a variety of ways, such as nonlinear correlation measures, kernel-based methods or even by directly predicting functional correlations between variables using methods such as

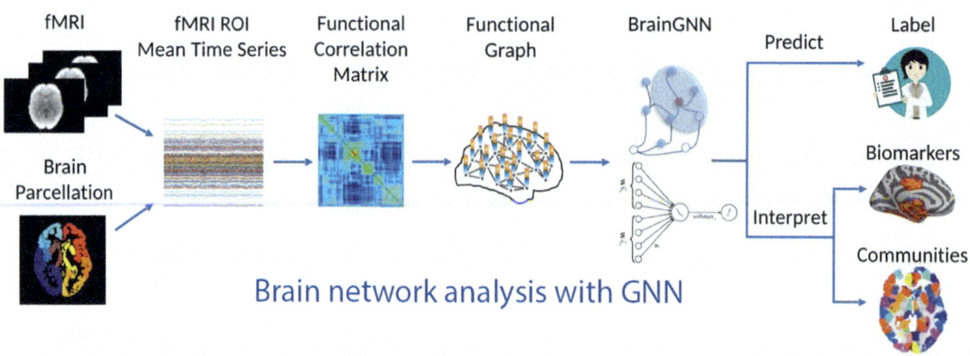

Figure 13.1 Architecture of my deep learning model to classify the graph neural network of brain network analysis framework.

functional regression or dynamic modeling. All of these methods can be found in the literature.

Functional correlation can be used to measure the correlation and dependence between two variables. This technique is a form of dimensionality reduction known as functional correlation. Quantification of the relationship between two functional variables has been approached from a number of different angles. The concept of a functional correlation matrix is one that is utilized in a variety of industries, including statistics and neuroscience, amongst others. It is quite similar to a correlation matrix but its primary purpose is to describe the connections between functional data and time series data.

We can also discuss about a framework called BrainGNN that uses graph neural networks (GNNs) to analyze functional magnetic resonance imaging (fMRI) and find neurological biomarkers. Architecture of BrainGNN defined below can help to predict and interpret the Brain Tumor. Also, its output will help in labels, biomarkers and communities.

Main findings are as follows:

(1) A thorough set of recommendations for creating high-performing GNNs;
(2) Despite the fact that the best GNN designs for various tasks differ greatly, the GNN task space enables the transfer of the best designs between tasks;
(3) Implemented CNN, SVM, Deep learning concepts and mathematical calculations to get a 97.3% result.

2. Literature survey

A literature assessment on brain tumor entails the examination's past work and consolidation of current research. It also entails the publications pertaining to different facets of brain tumors. With the help of this I am able to identify the past work related to brain tumor and future scope of the same.

2.1 Identical work

Searching the Graph Architecture: There has been some experimentation on using architecture search methods with GNNs (Shchur et al., 2018). However, these studies only evaluate the designs on a limited number of node classification tasks and only focus on the design within each GNN layer rather than a broad GNN design space.

Evaluation of models built using GNN: Several publications explain different methods that can be used to make fair comparisons between GNN models. However, existing models only take into account a few particular GNN designs (for example, GCN, GAT, and GraphSAGE) (You et al., 2019), whereas my method investigates the broad design space of GNNs in great depth.

Different models for graph learning: My primary emphasis is placed on GNNs that communicate by means of messages because this type of GNN has been shown to be the most effective and efficient across a wide range of GNN tasks. There are also other architectures for graph learning models (Wong et al., 2018), but their design spaces are distinct from those of GNNs and are less modularized.

Architecture Search That Can Be Transferred: In the area of computer vision tasks, the concept of transferring the results of an architectural search from one task to another has been the subject of research. The design of meta-level architecture has also been researched in (Nazir et al., 2021), under the presumption that various tasks follow the same distribution (for example, versions of the ImageNet dataset). These approaches frequently make the premise that a single neural architecture may perform well on all tasks. This is a strategy that works well for tasks that have a relatively minimal range of possible outcomes. This assumption is no longer valid, however, because there are so many different graph learning tasks to complete.

2.2 Literature review

The diagnosis of brain tumors through the use of MRI image categorization has received a significant amount of support from researchers over the past few decades. Now many FMRI images are being used for understanding, diagnosis, and treatment of various types of brain tumors. The researchers have employed a wide range of methods in order to improve the identification of brain tumors for the purposes of diagnosing and classifying them, some of the techniques which were used in past times are as follows:

- Biswas and Islam developed a tumor categorization by utilizing K-means, ANN, and Principal Component Algorithms in addition to other machine learning approaches (PCA). In addition, they used PCA. When it was time to pre-process the data from the images of brain tumors, they turned to the k-means technique. The results of this investigation showed an accuracy of 95.4%, a sensitivity of 94.58%, and a specificity of 97.83% (Biswas & Islam, 2021). After a large number of studies utilizing transfer learning but without significant dataset growth, researchers explored brain tumor classification (BTC) using convolutional neural network (CNN) images to improve classification accuracy and feature extraction (Sejuti and Islam, 2021). The authors of this study gave a comprehensive analysis, not only of previously finished surveys, but also of BTC's most recent deep learning approaches.
- The researchers developed a model to interpret magnetic resonance imaging (MRI) scans by utilizing mathematical equations and matrix operations within an Artificial Convolutional Neural Network (CNN), enabling accurate feature extraction and brain tumor classification. The model was created to analyze magnetic resonance images. In this investigation, the proposed model has the capacity to grow size by a factor of 14 through the use of data augmentation. The existence of a tumor could be

predicted by the model with an accuracy of 96.7% in validation data and an accuracy of about 88.25% in test cases (Noreen et al., 2020).

- According to proposed approach that was created will reliably identify the tumor regardless of where it is located in the brain, including the temporal lobe (which corresponds with the level of the eye). The accuracy rate of the system is 94%, and it achieves a diagnosis score of 81.25% on tumor scans while achieving a score of 100% on scans that do not include tumors.
- According to Deepak and Ameer's research (2019), transfer learning looks like a promising strategy for cutting down on the number of medical scans that are needed. The area under the curve (AUC), accuracy, recall, the F-score and specificity are all components of this study.
- Accurate and generalized segmentation of brain tumors continues to be a significant challenge because of the complex characteristics of brain tumors observed in magnetic resonance imaging (MRI), including variations in tumor size, shape, location, intensity, and texture. These factors make it difficult to achieve consistent and reliable segmentation across different patients and imaging conditions. They developed an innovative method for the segmentation of brain tumors that was based on the segregated local square properties. This would allow them to circumvent the issue (Desikan et al., 2006).

3. Proposed methodology

To achieve optimal classification results with little mathematics resources and maximum accuracy, a proposed model is exhibited which uses the SVM, CNN, Decision Tree, KNN and few mathematical calculations to get the proper result. The proposed model uses multiple effected human brain tumor images as an input and output value. This shall enable getting accurate results, recall values, F1 score and precision values after implementing various calculations (Deco et al., 2011).

Therefore, the task of identifying an appropriate and pertinent classification is carried out. The following is a representation of the proposed approach and the procedures that are followed:

1. Searching for the dataset obtained from many sources that contains the Brain Tumor units' sample dataset, in the format of FMRI images.
2. Images reshaped from 3D to 2D for feature extraction. Here a conversion of data is performed which strengthens the complete Brain tumor dataset.
3. Identification of images on the basis of five types of different Brain Tumors like Low-Grade Pilocytic Astrocytoma, Diffuse Astrocytoma, High-Grade Anaplastic Astrocytoma, Glioblastoma Multiforme and no tumor. Here, the data is divided into 0 level to 4 level according to the brain tumor identification.

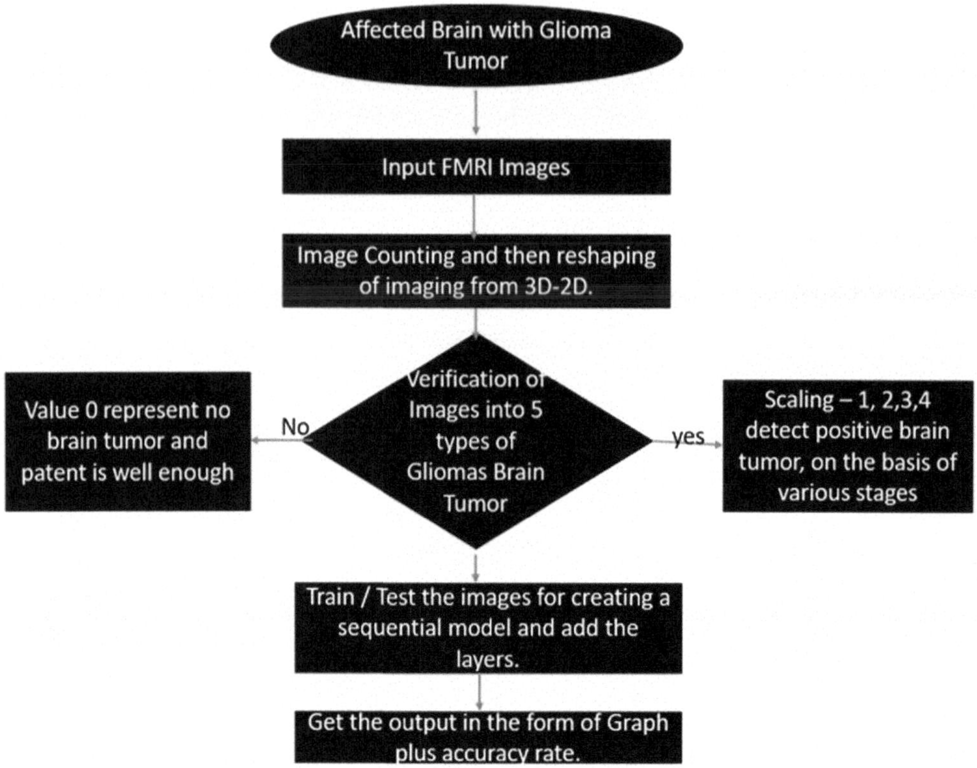

Figure 13.2 Brain tumor classification model.

4 In this section, train/test the images in the 8:2 ratio to get proper accuracy. Created a sequential model & added the layers. Total number of images used in this paper is 315.
5. Validation and performance are evaluated via applying a final dense layer to Keras—Model sequential.
6. Ultimately, the computation parameters were calculated and then compared to the current approach in order to demonstrate the effectiveness of the suggested model Fig. 13.2.

The process of establishing a system for detecting glioma brain tumors encompasses various stages, such as gathering data, preparing it for analysis, choosing an appropriate model, training the model, assessing its performance, and implementing it.

The above presented method is a suggested approach for constructing a machine learning-based system for detecting glioma brain tumors.

4. Computational modules

In this Section I would like to introduce instances of general pipeline design for Brain Tumor model named as G-T4. Under this introduce three sub-components of propagation modules: Computational Modules, Processing of FMRI and Segregation of Coordinates, analysis of output using CNN, SVM and Deep learning mechanisms. Output generated from this architecture shall be helpful in medical field for effective surgeries of Brain Tumor Patients.

In this module able to perform algorithm implementation of FMRI images. Used 310 images for training and testing point of view. We am using various algorithms for analyzing the result—CNN, SVM and Deep learning concepts plus utilized relevant mathematical calculations as well.

Output of this section generates 97.3% accuracy rate for my FMRI dataset. At the time of analyzing the dataset, need to check images of brain tumor patients one by one. Identified that the brain tumor Gliomas are generally found at the right side of human brain with four stages.

4.1 Low-grade pilocytic astrocytoma

Grade I gliomas are the ones that cause the least amount of damage. Pilocytic astrocytoma are often tumors that have a slow growth rate, are well-differentiated and have a substantially better prognosis than other types of brain tumors. They are connected with separate cystic and solid components and are frequently found in youngsters as well as adults.

4.2 Diffuse astrocytoma (low-grade)

Grade II gliomas are likewise characterized by relatively moderate rates of growth; nevertheless, they have a greater propensity to invade neighboring brain tissues as compared to Grade I tumors. Adults are most likely to develop diffuse astrocytoma, which differs from lower-grade tumor in a way that their borders are not clearly defined. They have the potential to advance to further difficult levels in the future.

4.3 High-grade anaplastic astrocytoma

Grade III gliomas are more aggressive and grow at a faster rate than grade II and grade I tumor. Anaplastic astrocytoma is distinguished from other types of astrocytoma by the presence of cellular abnormalities and elevated mitotic activity. When compared to tumor of lower grade, they are more likely to penetrate the surrounding tissue.

4.4 Glioblastoma multiforme (high-grade)

Grade IV gliomas, more especially glioblastoma multiforme (GBM), are the type of glioma that are the most aggressive and malignant. They expand quickly, invade the surrounding brain tissue, and are linked to high levels of angiogenesis, which refers to the production of new blood vessels. The most prevalent and deadly kind of brain cancer is glioblastoma multiforme (Noreen et al., 2020).

Gliomas are graded according to a variety of microscopic features, including the appearance of the cells, the rate at which they divide, the presence of aberrant blood vessels and a number of other characteristics. It is essential to keep in mind that gliomas can be difficult to treat and that the patient's prognosis and treatment will be determined by a number of circumstances (Duong et al., 2019). These considerations include the grade of the tumor, its location, its size, and the patient's overall health.

4.5 Processing of FMRI

In order to derive relevant information from the raw measurements, the processing of fMRI data requires multiple processes to be carried out. The following is an outline of the several processing processes that are often engaged in fMRI analysis.

4.5.1 Preprocessing

- **Adjustments to the Slice Dimensions:** Correct for differences in image acquisition dimensions reduction, converted 3D FMRI images into 1D images.
- **The smoothing of space:** It is recommended to use a spatial filter so that the signal-to-noise ratio can be improved.

4.5.2 Post-processing and visualization

- **Superimposing on Top of Structural Images:** Visualization is improved by superimposing activation maps on top of anatomical images.
- **An Analysis of the Region of Interest (ROI):** Take the average signal or data from a region that has been predefined.
- **An Analysis of the Functional Connectivity:** Examine the ways in which distinct parts of the brain exhibit patterns of connected activity.
- **Network Analysis:** Investigate the functional connections in the brain using graph theory methods.

The most challenging task in Processing of FMRI images was to identify the correct brain tumor stages, to get the 97.3% result. Final output can be used to predict the surgeries more effectively, accurately and thereby reduce the death rate in India Fig. 13.3.

This model currently works for one type of human brain tumor, in future I will work on exact location identification of brain tumor with the help of dimension analysis using Graphical neural network architecture.

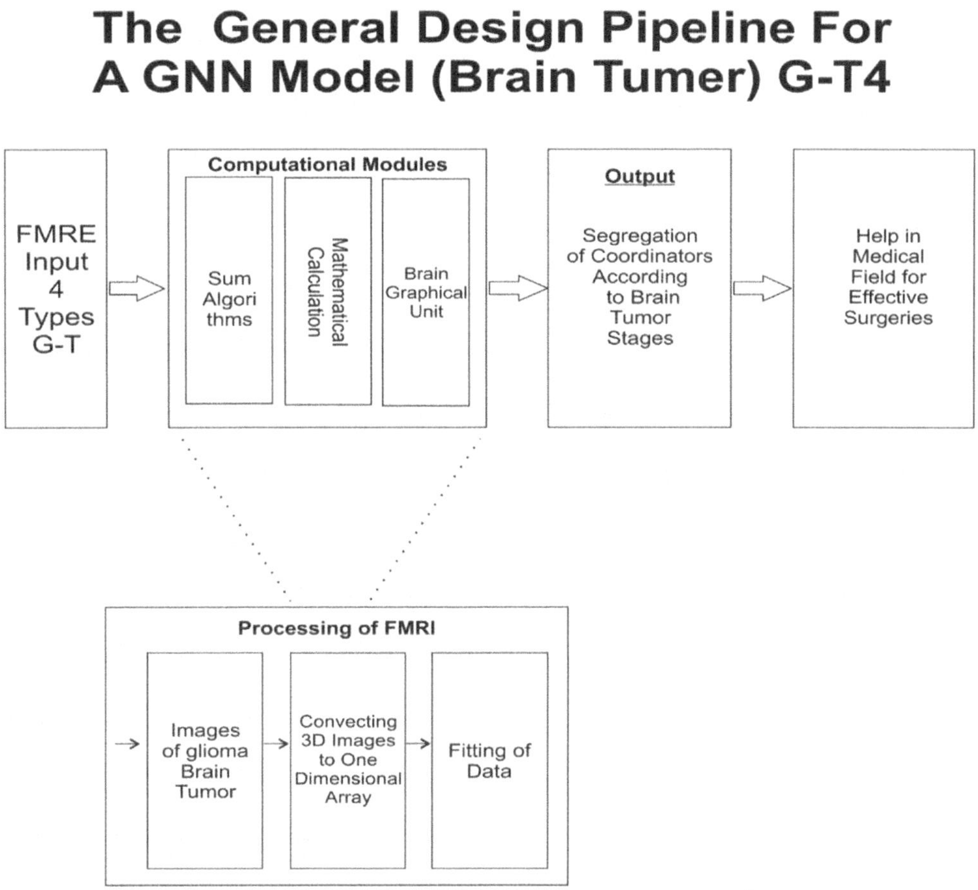

Figure 13.3 Model representing the working process to analysis the brain tumer - gliomas.

5. Experimental details

Under this section, we have taken various FMRI Brain tumor images as input and performed comparative studies to analyze the various stages of it. After analyzing the stages, passed the input to Brain Graphical Neural Network, which is represented by Brain G1, G2 in below diagram. Next, passed the results to various hidden layers which are denoted by Z series and also assigned the weights wij to end points for getting better results.

Out of this get explanation generation and at the end get the marking location or the approximate location of Glioma brain tumor. This process is applied in all 310 FMRI images to get 97.3% correct result. For getting exact marking location we have to perform Brain NN nth times, which is arrowed properly in below diagram (Wu et al., 2020) Fig. 13.4.

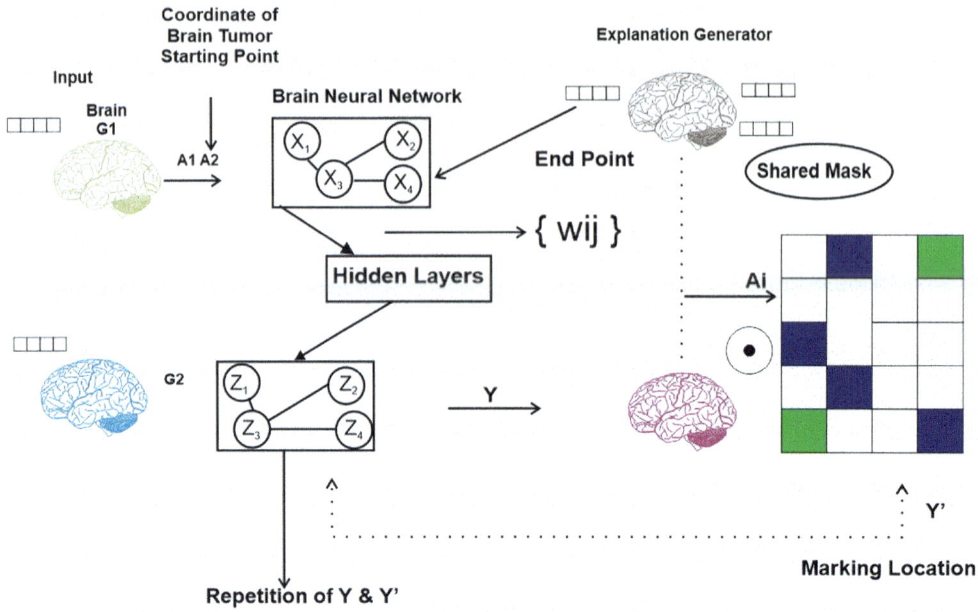

Figure 13.4 Model represents the mathematical calculation for representing the exact location of brain tumor position using explanation generation.

Mathematical equation for above diagram is as follows:

$$\mathbf{E_m} = -\sum_{c=1}^{G} 1(\mathbf{y} = \mathbf{c})\log \mathbf{P}_{\varphi}\left(\mathbf{y}' = \frac{\mathbf{y}}{\mathbf{G}} = \mathbf{H}'\right) \tag{13.1}$$

In the above equation "y" represents the repetition process of getting hidden layer values. With the help of variable "z" we get marking location of effected cells due to brain tumor. Variable "H" represents the total weight of hidden layer. Log P is the computation prediction of logarithm of partition coefficient. Variable "G" depicts various affected brains due to different types of brain tumors and E_m (Mutual Information Loss) ishelpful for determining whether or not two random variables are statistically independent of one another.

Using a masked matrix, one can hide any discrepancies that exist between the actual values and the predicted values known as masked prediction loss $L_{p.}$ Masked prediction loss is the sum of the above mutual information loss E_m and the supervised disease prediction loss $L_{p.}$

$$\mathbf{L_{p'}} = E_m + L_{p'} \tag{13.2}$$

In order to get a compact explanation mask, first apply a sparsity loss P_s that is defined as the sum of all elements of the mask parameters and that imposes a regularization on the edge size of G'. Next, we apply an element-wise entropy loss P_e to each of the mask's elements. As defined by P_e

$$\mathbf{P_e} = -(\mathbf{N}\,\mathbf{log}(\mathbf{N}) + (1-\mathbf{N})\mathbf{log}(1-\mathbf{N})) \tag{13.3}$$

To encourage discreteness in mask weight values, a regularization strategy was employed during model training. As a consequence of this, my explanation generator will provide a globally shared edge mask "N". This edge mask "N" will be able to highlight disease-specific significant brain network connections and it will also be able to be applied to all testing graphs in order to investigate disease-specific neurological biomarkers as well as salient Regions of Interest (ROIs) (Msonda et al., 2020).

My final training objective is to get the sum of all generator values, for which the equations are mentioned below:

$$\mathrm{L} = \mathbf{L}_{\mathbf{P'}} + \mathrm{P_s} + \mathrm{P_e} \tag{13.4}$$

In order to accomplish this, suggest the learning of a globally shared edge mask $N \in R^{n \times n}$ and apply it on the individual brain. N is trained by maximizing the reciprocal information between BNN (brain Neural Network) projection on the original graph G and the masked graph G'. where $\mathbf{W}' = \mathbf{W} \odot \boldsymbol{\sigma}\ (\mathbf{N})$. $\odot$ Denotes element-wise multiplication and $\boldsymbol{\sigma}$ the term "denotes" is used to indicate that the sigmoid function is being used to map the mask to (Cui et al., 2023)$^{\mathbf{nxn}}$.

Graph convolutional layer is used to probe the graph structure by using edge features, which contains important information about graphs. For example, the weights of the edges in brain fMRI graphs can represent the relationship between different ROIs.

Following details define below

$$h_i^{(1)} \quad \varepsilon R^{d^{(1)}}$$

as the features for the ith node in the lth layer, where $d^{(l)}$ is the dimension of the lth layer features. To figure out the spread model for the forward pass update of the node representation

$$\boldsymbol{h_i}^{(l+1)} = \mathbf{relu}\left(\boldsymbol{W_i}^{(l)}\boldsymbol{h_i}^{(l)} + \sum\nolimits_{\mathrm{J}\varepsilon\mathrm{N}^{(i)}(i)} \boldsymbol{E_{ij}}^{(l)}\boldsymbol{W}^{(l)}\boldsymbol{h_j}^{(l)}\right) \tag{13.5}$$

When the notation $N(l)$ (i) denotes the collection of indices corresponding to the neighboring nodes of the node vi.

$Eij(l)$ denotes the characteristics associated with the edge connecting vi to vj.

The symbol $W_i(l)$ represents the parameter of the model that needs to be trained.

The initial layer is applied to the original graph., that is, $h_i^{(0)} = h_i$, $E_{ij}^{(0)} = E_{ij}$. In order to prevent the amplification of output characteristics, it is necessary to normalize the edge features, similar to the approach used in Graph Attention Networks (GAT). and GNN (Noreen et al., 2020). Normalize the weights using the aggregation process.

$$\boldsymbol{E}_{ij}{}^{(l)} = \boldsymbol{E}_{ij}{}^{(l)} / \sum \mathrm{j}\boldsymbol{\varepsilon}\ \boldsymbol{N}_{(i)}^{(l)} \boldsymbol{E}_{ij}{}^{(l)} \tag{13.6}$$

A node pooling layer cuts down on the size of the graph by either putting the nodes together or cutting the original graph G into a subgraph Gs that only has the most important nodes. I will focus on the cutting method because it is easier to understand and can help find biomarkers (Pasa et al., 2019).

The node feature vectors are summed up by a reading layer $[h_i{}^{(1)}]$ into a single vector $z^{(l)}$ which is then put into a classifier to sort the graph into groups.

6. Classification performance and result

During the analysis of result of Gliomas brain tumor, we got the following results from previous papers as mentioned below. With the help of below table, we are able to analyses the working methods of various machine learning algorithms under varied sized datasets and different percentage of accuracy in result output (Sun et al., 2020) Table 13.1.

We have achieved the maximum result while analyzing Brain Tumor containing 224 MRI images Benign and 472 malignant tumors. The countable value was 99.04% using Convolutional Neural Network (AlexNet). This model is the proposed system, but in my research, I am able to get 97.3% result using various machine learning algorithms and mathematical calculations.

For finding the results of various stages of Gliomas Brain Tumor and exact location in human brain is implemented various Machine learning algorithms, which are mentioned in the below diagram.

While implementing decision tree classifier, have taken the maximum depth of four layers. In KNeighbors classifier seven neighbors are taken and Support vector machine shows 97.3% accuracy Fig. 13.5.

Performed CNN in shape of 201,512,512,1 dimensions, under this have taken 201 trained number of images and 99 test number of images. We found 97.3% accurate result in 20 epochs.

Below table classified few machine learning methods on the basis of following parameters—Accuracy, Recall, F1 score, Precision and computation time Table 13.2.

The following parameters such as Accuracy, Precision, Recall, F1-score, and computation time are calculated based on proposed methodology with the help of False negative (FN), True Negative (TN), True Positive (TP), False Positive (FP) (Tian, 2020). The equations are given below:

$$\text{Accuracy} = (TP + TN) \ / \ (\mathrm{TP} + \mathrm{TN} + \mathrm{FP} + \mathrm{FN}) \tag{13.7}$$

Table 13.1 Analysis of brain tumor result using various machine learning algorithms.

Authors name	Methods used	Type of grading	Work of following dataset	Precision (%)
Cheng et al. Year - (2015)	Support vector machine and K-Nearest neighbor.	Multi-classification	The study utilized T1-weighted contrast-enhanced MRI to examine a total of 708 meningiomas, 1426 gliomas, and 930 pituitary tumors.	91.28
Paul et al. Year - (2017)	Convolutional neural network.	Multi-classification	T1-weighted contrast-enhanced MRI was performed on a total of 208 meningiomas, 492 gliomas, and 289 pituitary tumors.	91.43
Afshar et al. Year - (2019)	Convolutional neural network.	Multi-classification	T1-weighted contrast-enhanced magnetic resonance imaging (CE-MRI) was performed on a total of 708 meningiomas, 1426 gliomas, and 930 pituitary tumors.	90.89
Anaraki et al. Year - (2019)	Genetic algorithm—Convolutional neural network.	Multi-classification	A total of 708 meningiomas, 1426 gliomas, and 930 pituitary tumors were analyzed using T1-weighted contrast-enhanced MRI.	94.20
Bahadure et al. Year - (2017)	Burrows-Wheeler Transform + support vector machine.	Binary-classification	There are 67 normal and 134 aberrant T2-weighted brain pictures.	95.00
Ansari et al. Year - (2020)	Discrete wavelet transform ÷ Principal component analysis + general intelligence. Continuous integration + support vector machine	Binary-classification	A T1-weighted contrast-enhanced MRI was performed on 140 patients with tumor and 60 patients without tumor.	98.91

Continued

Table 13.1 Analysis of brain tumor result using various machine learning algorithms.—cont'd

Authors name	Methods used	Type of grading	Work of following dataset	Precision (%)
Sultan et al. Year - (2019)	Convolutional neural network.	Multi-classification	A total of 708 meningiomas, 1426 gliomas, and 930 pituitary tumors were evaluated using T1-weighted contrast-enhanced MRI.	96.13
Kumar et al.	Gray wolf optimizer + Machine - support vector machine	Multi-classification	A total of 248 meningiomas, 12 gliomas, and 55 pituitary tumors were identified using T1-weighted contrast-enhanced MRI.	95.23
Abiwinanda et al.	Convolutional neural network.	Multi-classification	A total of 708 meningiomas, 1426 gliomas, and 930 pituitary tumors were evaluated using T1-weighted contrast-enhanced MRI.	84.19
Proposed system	Parity time-convolutional neural network (AlexNet)	Binary-classification	A T1-weighted MRI scan revealed the presence of 224 benign tumors and 472 malignant tumors.	99.04

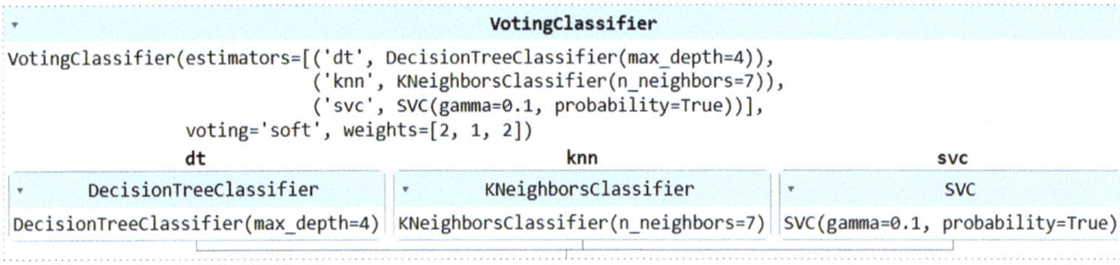

Figure 13.5 Used decision tree classifier, KNeighbors classifier and support vector machine to analyze the result.

Table 13.2 Classification method with performance parameter.

S No.	Method	Accuracy (%)	F1 score (%)	Precision (%)	Recall (%)	Computation time
1	Decision tree classification	71.85	68.20	70.89	69.30	2.56
2	K-Nearest neighbor	70.56	68.55	69.75	69.60	2.40
3	Support vector machine	98.45	97.56	98.40	97.98	1.55
4	Convolutional neural network	97.54	96.35	97.32	96.85	1.26

$$\textbf{F1-Score} = \mathbf{(2*Precision*Sensitivity)/(Precision + Sensitivity)} \quad (13.8)$$

$$\textbf{Precision} = \boldsymbol{TP}\mathbf{/(TP + FP)} \quad (13.9)$$

$$\textbf{Recall} = \boldsymbol{TP}\mathbf{/(TP + FN)} \quad (13.10)$$

If talk about Fig. 13.6, we applied CNNs algorithm using deep learning model primarily for tasks involving images and spatial data. They consist of multiple layers, including convolutional layers that automatically learn features from the input data.

During training, the CNN goes through a series of iterations called epochs. In each epoch, the entire training dataset is presented to the model in batches. For each batch, the model predicts using the current parameters and compares them to goal outputs (measures model prediction error).

While performing the training and validation loss of my dataset and we achieved the 97.3% result in 20 epochs. Applied deep neural network, for getting the accurate result.

Below diagram 7 is showing training and validation accuracy of CNN results. In this also we find out 97.3% accuracy in 19 epochs.

In Figs. 13.6 and 13.7, we most frequently encounter examples of the following two types of learning curves:

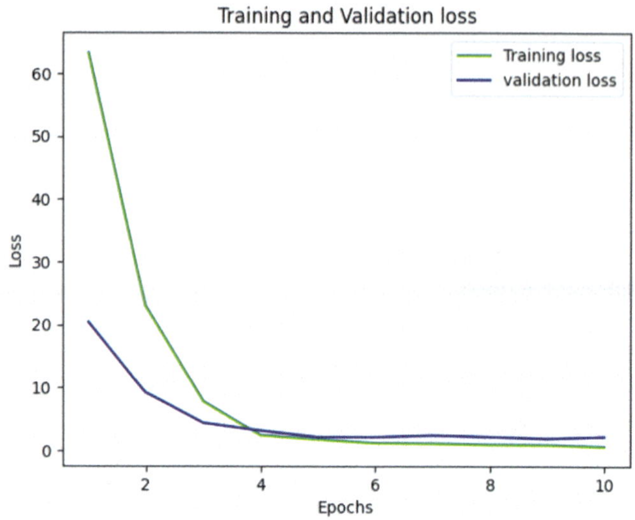

Figure 13.6 *The process of optimizing* - CNN training and validation loss of glioma human brain tumor results.

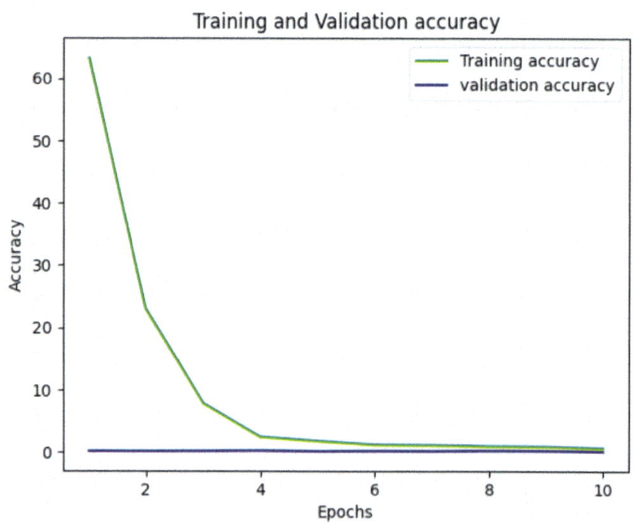

Figure 13.7 *Performance learning curves* - CNN training and validation accuracy of gliomas human brain tumor results.

The process of optimizing: Learning curves are created based on the metric that is being used to optimize the parameters of the model, such as the amount of loss or the mean square error (Utkarsh, 2021).

Performance Learning Curves: Performance learning curves are learning curves calculated on the metric by which the model will be assessed and selected, such as accuracy, precision, recall, or F1 score. Performance learning curves are used to evaluate and choose models.

Below you can see an example of machine translation which is showing (a **performance score**—It's utilized for the purpose of assessing the high quality and efficiency of a model) and above diagram is depicting the loss (**optimization score** - Parameter optimization is the act of determining the optimal values for a model's parameters which is weights and biases, in order to maximize its performance) for two different models used blue and green curve lines.

In below diagram, performed comparative study between CNN, SVM and DL algorithms. Under this process we are able to convert 3D fMRI images in to 2D format firstly. Next step, analyzed all human effected brains on the basis of four different stages of brain tumor. Level 0 represents pilocytic, level 1 represents diffuse astrocytoma, level 2 represents high grade anaplastic astrocytoma and level 3 represents glioblastoma multiforme (highest grade glioma).

As we go-through 310 different patients fMRI images of brains, in the below figure yellow color equates level 0 patients, green color equates level 1 patients, light blue color equates level 2 patients and lastly dark purple color equates last level 4 patients Fig. 13.8.

As we know Gliomas are usually found on top—right most side of the human brain. It also has four stages, out of which the last stage is the most dangerous. With the help of the above figure are able. To get the affected values of patients and it also helps doctors to operate the patient on time.

7. Conclusion & future work

In this paper we have analyzed the accuracy in detecting the Gliomas Brain Tumor. We get 97.3% result with the help of Machine learning, deep learning and mathematical

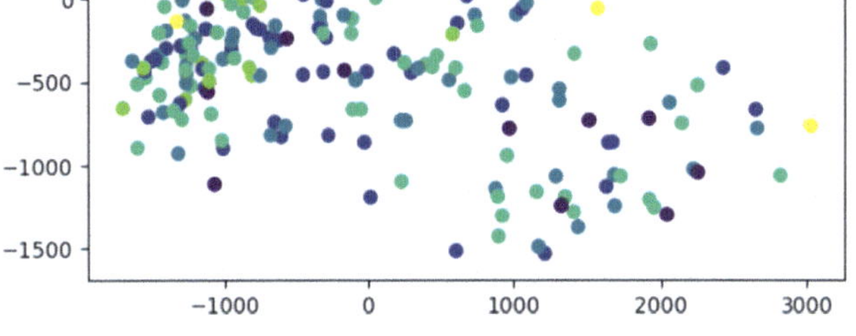

Figure 13.8 Results applied in 310 fMRI images to represent, different stages of gliomas brain tumor.

equations. Also identified the various stages of tumor and where exactly it is located in the human brain.

In my next research paper, we will try to find the exact location using dimension identification and coordinate localization interpretation using the advance concepts of neural networks and relevant mathematical equations.

It will be more challenging for me to work with.nii images and get more accurate results. The.niiimage format stands for Neuroimaging Informatics Technology initiative (Zhou, 2020). Is the small result of.nii images and indicates my future work area. Our research really helps the health care sector to cure the patient and reduce the death rate in India. In order to satisfy the requirements of an expanding population, decisions on the expansion of healthcare infrastructure and the distribution of resources might be influenced by research.

Statements and declarations

Funding - The author did not receive support from any organization for the submitted work.
Competing Interests and Funding - The author has no conflicts of interest to declare that are relevant to the content of this article.
Ethical approval - which has not been previously published elsewhere. The paper is not currently being considered for publication elsewhere. The paper reflects the author own research and analysis in a truthful and complete manner.

References

Cui, H., Dai, W., Zhu, Y., Kan, X., Gu, A.A.C., Lukemire, J., Zhan, L., He, L., Guo, Y., Yang, C., 2023. BrainGB: a benchmark for brain network analysis with graph neural networks. IEEE Trans. Med. Imaging. 42 (2), 493–506. https://doi.org/10.1109/tmi.2022.3218745.

Deco, G., Jirsa, V.K., McIntosh, A.R., 2011. Emerging concepts for the dynamical organization of resting-state activity in the brain. Nat. Rev. Neurosci. 12 (1), 43–56. https://doi.org/10.1038/nrn2961.

Desikan, R.S., Ségonne, F., Fischl, B., Quinn, B.T., Dickerson, B.C., Blacker, D., Buckner, R.L., Dale, A. M., Maguire, R.P., Hyman, B.T., Albert, M.S., Killiany, R.J., 2006. An automated labeling system for subdividing the human cerebral cortex on MRI scans into gyral based regions of interest. Neuroimage 31 (3), 968–980. https://doi.org/10.1016/j.neuroimage.2006.01.021.

Duong, C.T., Hoang, T.D., Dang, H.T.H., Nguyen, Q.V.H., Aberer, K., 2019. On node features for graph neural networks. arXiv, Undefined arXiv. https://doi.org/10.48550/arxiv.1911.08795.

Gao, Y., Yang, H., Zhang, P., Zhou, C., Hu, Y., 2019. GraphNAS: graph neural architecture search with reinforcement learning. arXiv, China arXiv. https://doi.org/10.48550/arxiv.1904.09981.

Li, X., Zhou, Y., Dvornek, N., Zhang, M., Gao, S., Zhuang, J., Scheinost, D., Staib, L.H., Ventola, P., Duncan, J.S., 2021. BrainGNN: interpretable brain graph neural network for fMRI analysis. Med. Image Anal. 74, 102233. https://doi.org/10.1016/j.media.2021.102233.

Msonda, P., Uymaz, S.A., Karaağaç, S.S., 2020. Spatial pyramid pooling in deep convolutional networks for automatic tuberculosis diagnosis. Trait. Du. Signal 37 (6), 1075–1084. https://doi.org/10.18280/TS.370620.

Nazir, M., Shakil, S., Khurshid, K., 2021. Role of deep learning in brain tumor detection and classification (2015 to 2020): a review. Comput. Med. Imag. Graph. 91, 101940. https://doi.org/10.1016/j.compmedimag.2021.101940.

Noreen, N., Palaniappan, S., Qayyum, A., Ahmad, I., Imran, M., Shoaib, M., 2020. A deep learning model based on concatenation approach for the diagnosis of brain tumor. IEEE Access 8, 55135–55144. https://doi.org/10.1109/ACCESS.2020.2978629.

Pasa, F., Golkov, V., Pfeiffer, F., Cremers, D., Pfeiffer, D., 2019. Efficient deep network architectures for fast chest X-ray tuberculosis screening and visualization. Sci. Rep. 9 (1). https://doi.org/10.1038/s41598-019-42557-4.

Sejuti, Z.A., Islam, M.S., 2021. Bangladesh an efficient method to classify brain tumor using CNN and SVM. In: International Conference on Robotics, Electrical and Signal Processing Techniques, pp. 644–648. https://doi.org/10.1109/ICREST51555.2021.9331060. In: https://ieeexplore.ieee.org/xpl/conhome/1830504/all-proceedings.

Shchur, O., Mumme, M., Bojchevski, A., Günnemann, S., 2018. Pitfalls of graph neural network evaluation. arXiv, Germany arXiv. https://doi.org/10.48550/arxiv.1811.05868.

Sun, Y., Xue, B., Zhang, M., Yen, G.G., 2020. Evolving deep convolutional neural networks for image classification. IEEE Trans. Evol. Comput. 24 (2), 394–407. https://doi.org/10.1109/tevc.2019.2916183.

Tian, Y., 2020. Artificial intelligence image recognition method based on convolutional neural network algorithm. IEEE Access (8), 125731–125744. https://doi.org/10.1109/access.2020.3006097.

Utkarsh, 2021. Detecting the disease before it's too late. J. Pharm. Res. Int. 82–89. https://doi.org/10.9734/jpri/2021/v33i19a31331.

Wong, C., Houlsby, N., Lu, Y., Gesmundo, A., 2018. Transfer learning with neural automl. In: Advances in Neural Information Processing Systems (NeurIPS).

Wu, W., Li, D., Du, J., Gao, X., Gu, W., Zhao, F., Feng, X., Yan, H., 2020. An intelligent diagnosis method of brain MRI tumor segmentation using deep convolutional neural network and SVM algorithm. Comput. Math. Methods Med. 2020, 1–10. https://doi.org/10.1155/2020/6789306.

You, J., Ying, R., Leskovec, J., 2019. Position-aware graph neural networks. Int. Conf. Mach. Learn. (ICML).

Zhou, D.-X., 2020. Universality of deep convolutional neural networks. Appl. Comput. Harmon. Anal. 48 (2), 787–794. https://doi.org/10.1016/j.acha.2019.06.004.

CHAPTER 14

Sentinel Eye: Hybrid AI for noninvasive anemia detection using eye scans, symptoms, and history

Atanu Jana[1], Snehal Panda[1], G Sekhar Reddy[2], Kunjabihari Swain[3], Sandipan Mallik[1] and Priyadarsan Patra[2]
[1]Department of Electronics and Communication Engineering, NIST University, Berhampur, Odisha, India; [2]Department of Computer Science and Engineering, NIST University, Berhampur, Odisha, India; [3]Department of Electrical Engineering, NIST University, Berhampur, Odisha, India

1. Clinical motivation and epidemiology

Anemia is a widespread global health issue. The World Health Organization estimates anemia affects roughly 1.6 billion people worldwide (Olawade et al., 2025), with prevalence particularly high among women of childbearing age and populations in low-resource settings. It manifests as reduced hemoglobin (Hb) and red blood cell counts, leading to symptoms like fatigue, pallor, and shortness of breath. Traditional anemia screening relies on invasive blood draws and laboratory hemoglobin assays, which can be costly, slow, and inaccessible in under-resourced regions (Subasree and Swathi, 2025; Obeagu, 2026). Moreover, these methods require trained personnel and may deter frequent monitoring. Noninvasive, rapid screening methods are therefore highly desirable to improve early detection and reduce the burden on healthcare systems. AI-powered imaging approaches, such as analysis of the eye's conjunctiva or nailbeds, offer promising alternatives by capturing visual signs of anemia (Ramzan et al., 2024; Ramos-Soto et al., 2025). By combining multiple data sources (images, clinical history, symptoms), multimodal AI can achieve higher diagnostic accuracy than single-modality methods, aligning with recent findings in anemia detection research.

2. Related work

Recent studies have demonstrated the feasibility of AI-driven, noninvasive anemia detection. For example, the foundational study by Collings et al. (2016) showed that the palpebral conjunctival erythema index (EI) from calibrated eye images correlates strongly with hemoglobin levels, outperforming clinician judgment in screening sensitivity/specificity. More advanced deep learning models have since been applied: frontiers reports indicate that vision-transformer models trained on conjunctiva and sclera images can achieve accuracies up to ~98.5% (Ramos-Soto et al., 2025). Similarly,

Intelligent Semantic Analysis for Healthcare
ISBN 978-0-443-33512-9,
https://doi.org/10.1016/B978-0-443-33512-9.00003-2

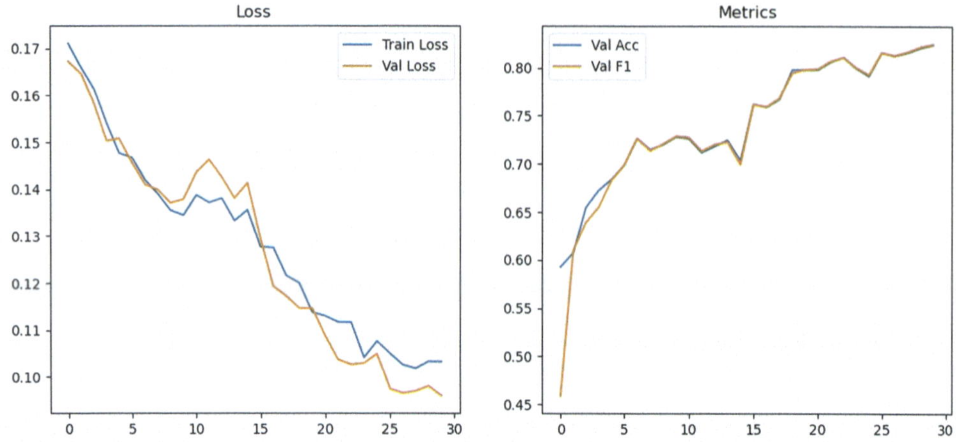

Figure 14.1 Training, validation loss and model accuracy, F1 score of EfficientNet-v2-M model with 30 epochs.

researchers have applied CNNs to other image modalities (e.g., lip or palm images); Mahmud et al. used transfer-learning CNNs (Xception, MobileNetV2, VGG16, ResNet50) on a lip mucosa image dataset, with high classification performance (Mahmud et al., 2023). These efforts illustrate that modern architectures can extract anemia-relevant visual features effectively. At the same time, multimodal fusion approaches have shown promising results: Ramzan et al. (2024) combined electronic health record (EHR) features with conjunctival images to build a feature-fusion model, achieving ~95% accuracy on anemia detection. This evidence underscores that integrating visual data with clinical variables yields more comprehensive assessments than any single source (Ramzan et al., 2024; Ramos-Soto et al., 2025) (Fig. 14.1).

3. Semantic representation and ontology mapping

A key aspect of *Sentinel Eye* is the semantic integration of heterogeneous data. Clinical and imaging findings (symptoms, lab results, diagnoses) are mapped to standardized ontologies to enable uniform interpretation and reasoning (Ehrsam et al., 2025; Callahan et al., 2023). We adopt ontologies such as SNOMED CT for clinical terms and the Human Phenotype Ontology (HPO) for phenotypic features. For instance, conjunctival pallor, fatigue, and dizziness—common anemia indicators—each correspond to unique SNOMED concepts (with semantic tags) and HPO terms. By encoding data in this way, the system ensures interoperability with other medical knowledge bases and supports advanced analysis.

Data mapping is performed by linking local data elements to ontology concepts. In practice, this involves establishing *crossmaps* (triples) between terminology codes and ontology IDs (Konstantinou et al., 2008; Choi et al., 2006). For example, ICD-10 codes for "iron deficiency anemia" or "fatigue" can be cross-referenced to SNOMED CT terms; similarly, lab tests (LOINC codes for Hb) can map to HPO phenotypes (e.g., "Decreased hemoglobin concentration"). The International Standards Organization defines a data map as links between concepts in one terminology and another, emphasizing that perfect semantic equivalence is rare—often mappings are directional (source→target) and use contextual rules (Campbell, n.d.). To manage this, *Sentinel Eye* uses curated mapping tables and a small ontology reasoner to handle postcoordination (combining multiple concepts) when needed (Ehrsam et al., 2025; Callahan et al., 2023). A practical example is aligning a clinical description "no palpitations" with the HPO terms for normal cardiac rhythm and absence of tachycardia, rather than attempting a one-to-one string match (Callahan et al., 2023).

Ultimately, this semantic framework allows *Sentinel Eye* to treat inputs uniformly: image-derived features, symptom codes, and historical data are all represented as ontology-backed attributes. This enables the AI system to "understand" that, say, a conjunctival image showing pallor and a patient-reported symptom of fatigue are both signals of low hemoglobin. As Ehrsam et al. note, careful semantic representation with SNOMED CT significantly improves data reuse and accuracy (Ehrsam et al., 2025). By following their guidelines, we enhance consistency in how clinical findings are encoded, which supports downstream decision-making and machine learning.

4. Sentinel Eye multimodal framework

The *Sentinel Eye* system integrates three input streams: (1) Ocular images captured by a smartphone (e.g., photos of the patient's lower eyelid or conjunctiva), (2) Symptom questionnaire filled by the patient (or clinician) indicating presence of anemia-related symptoms, and (3) Patient history/records from existing EHR (e.g., prior diagnoses, lab values). These inputs are processed in parallel and then fused into a unified assessment (Fig. 14.2). In practice, the pipeline works as follows. The raw eye image is preprocessed (cropping, color normalization, etc.) and fed into a deep CNN that outputs a probability of anemia based on visual cues. Simultaneously, the symptom input (e.g., "fatigue: yes/no", "rapid heartbeat: yes/no") is passed through a rule- or case-based engine. This engine uses a knowledge base of typical anemia presentations (encoded semantically) to produce a symptom-based risk score. For instance, an older patient with chronic kidney disease and positive symptom answers may receive a higher pre-test probability. The historical data (e.g., recent Hb lab value, previous anemia diagnoses) are also semantically encoded (via ontology mapping) to inform the final decision, although for this chapter, we focus on the novel aspects of the AI components.

These separate signals are then fused. A simple method is to compute a weighted combination of the CNN's posterior probability and the symptom-based score. For

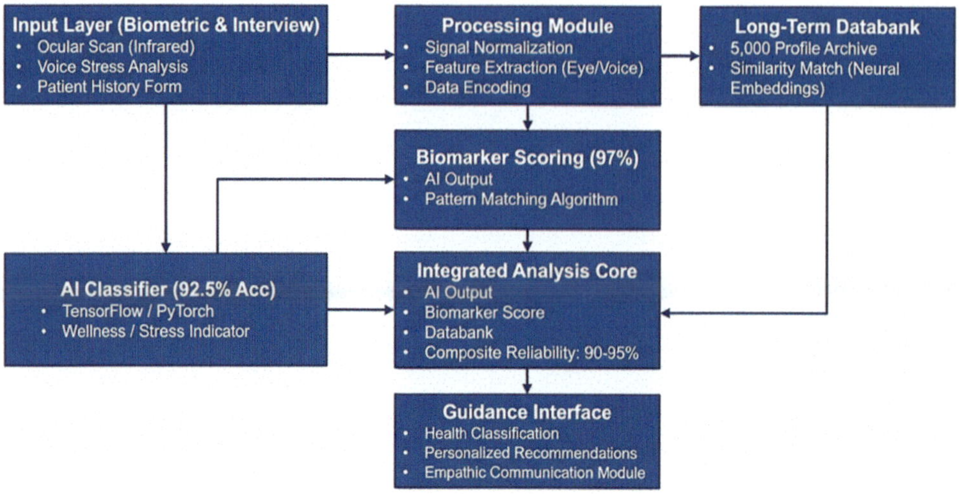

Figure 14.2 Sentinel Eye multimodal workflow integrating eye images, symptom scores, and patient history to assess anemia risk.

example, one might use a formula like $R = \alpha \cdot P\_CNN + (1-\alpha) \cdot S\_symptom$, where *P_CNN* is the probability from the CNN (normalized 0–1), and *S_symptom* is a normalized symptom score. The weight α can be tuned empirically or adaptively (e.g., based on data quality). This multimodal fusion mirrors strategies proven effective in AI research, where complementary information sources (images, text, clinical parameters) are concatenated or jointly modeled to enhance accuracy (Ramzan et al., 2024). In summary, the Sentinel Eye framework *embraces* multimodality by design, ensuring that both visual pallor and clinical context contribute to the risk assessment (Fig. 14.3).

5. Dataset, privacy, and governance

The *Sentinel Eye* design presumes a dataset of de-identified patient cases combining eye images and relevant clinical data. For example, one might assemble ~2000 cases of patients (with and without anemia) where each record includes a conjunctival photo, a set of symptom indicators, and recent lab values. All patient data must be handled in compliance with ethical and privacy regulations. Data are de-identified (no names or identifiers) and stored securely; only authorized AI model developers access them. In addition, we follow FAIR data principles (Findable, Accessible, Interoperable, Reusable) to ensure the data can be reused and shared for future research (Ehrsam et al., 2025; Campbell, n.d.) (Fig. 14.4).

Semantic interoperability is a key governance goal. Using ontologies (as described above) makes the data machine-interpretable and reusable across institutions (Ehrsam et al., 2025; Campbell, n.d.). The system also logs data provenance (source of each

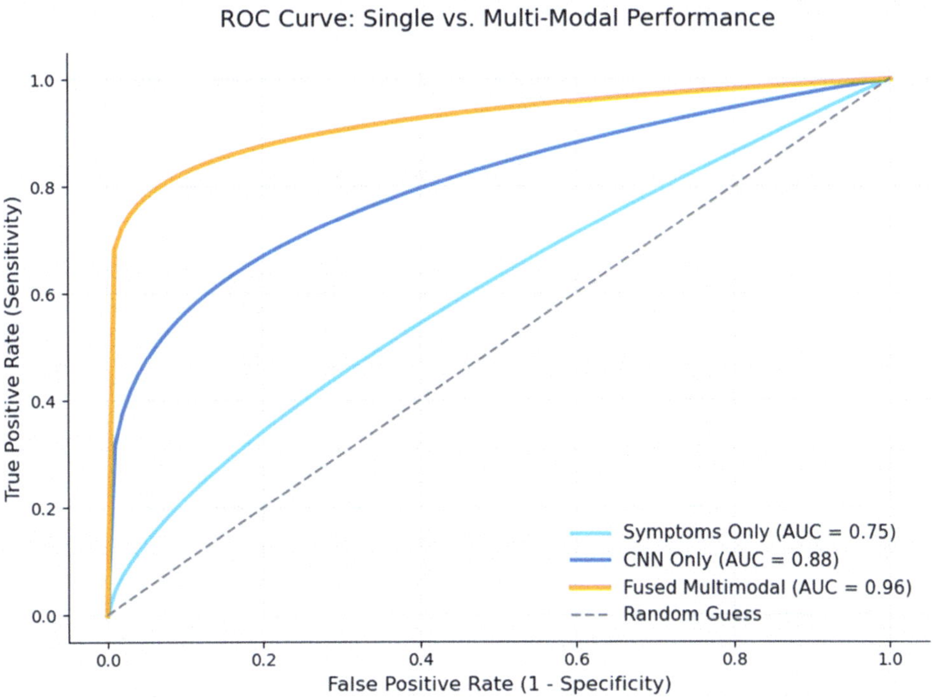

Figure 14.3 Comparison ROC curve between single (symptoms only) and CNN (Images only).

datum) to allow traceability. Finally, a "data governance board" reviews the AI model to ensure it meets privacy standards and regulatory guidelines. While detailed privacy mechanisms (like on-device processing or federated learning) are beyond this chapter's scope, the design emphasizes that Sentinel Eye is deployed only with strict oversight, analogous to other healthcare AI tools, which must address issues of consent and data security.

6. Data preprocessing and augmentation

Before training, all image data undergo a standardized preprocessing pipeline. First, images are resized and normalized: we convert pixel values to tensors and scale them (e.g., to mean = 0, std = 1) channel-wise, which helps stabilize CNN training. A region of interest (the palpebral conjunctiva) is automatically detected using simple computer vision techniques (e.g., color thresholding or a preliminary detector) to crop out irrelevant background.

Because real-world image capture can vary (lighting, camera angle, minor blurs), we apply data augmentation during training. Common techniques include random

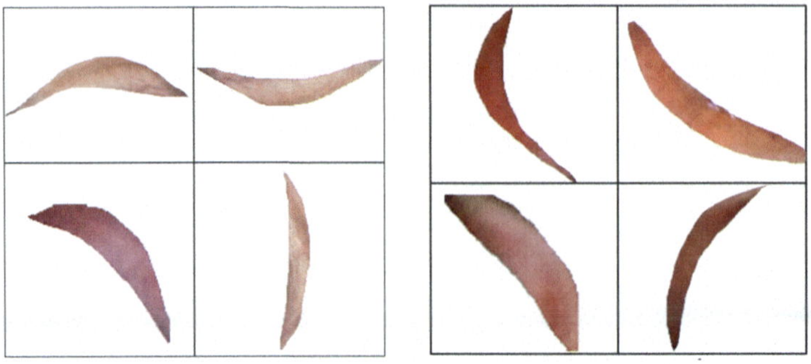

Figure 14.4 Anemic and non-anemic samples for CNN training.

rotations, horizontal/vertical flips, brightness/contrast jitter, and slight zoom/crop perturbations. These augmentations increase the effective dataset size and help the model generalize to unseen conditions (Mahmud et al., 2023). For instance, rotating an eye image a few degrees simulates head tilt, while brightness scaling simulates different lighting. An example pipeline might apply flips with 50% chance, random rotation up to ±15 degrees, random crop followed by resize, and color jitter to each training image. This is analogous to standard augmentation practices in medical image AI, which are known to improve robustness (See Fig. 14.2 for an illustrative example of such augmentations) (Fig. 14.5).

Ultimately, these preprocessing steps yield a cleaner and more varied training set for the CNN. The symptom and history inputs also undergo basic cleaning: categorical answers are encoded (yes = 1/no = 0) and continuous values (e.g., latest Hb level) are normalized relative to population means. Any missing data are handled by imputation or default rules (e.g., if a patient did not report a symptom, we assume "no"). By applying these procedures, the final training set is consistent, labeled, and ready for model fitting.

7. Convolutional neural network architecture for conjunctival images

The core image analysis component of Sentinel Eye is a convolutional neural network (CNN) designed to detect visual cues of anemia (e.g., pallor) in the eye. We adopt a deep CNN inspired by architectures like EfficientNet or ResNet, which are known for a good trade-off between accuracy and efficiency. An example CNN configuration is given in Table 14.1 and illustrated in Fig. 14.6. It consists of several convolutional layers (each followed by batch normalization and ReLU activation) and pooling layers, culminating in a fully-connected (dense) layer and an output sigmoid for binary anemia classification.

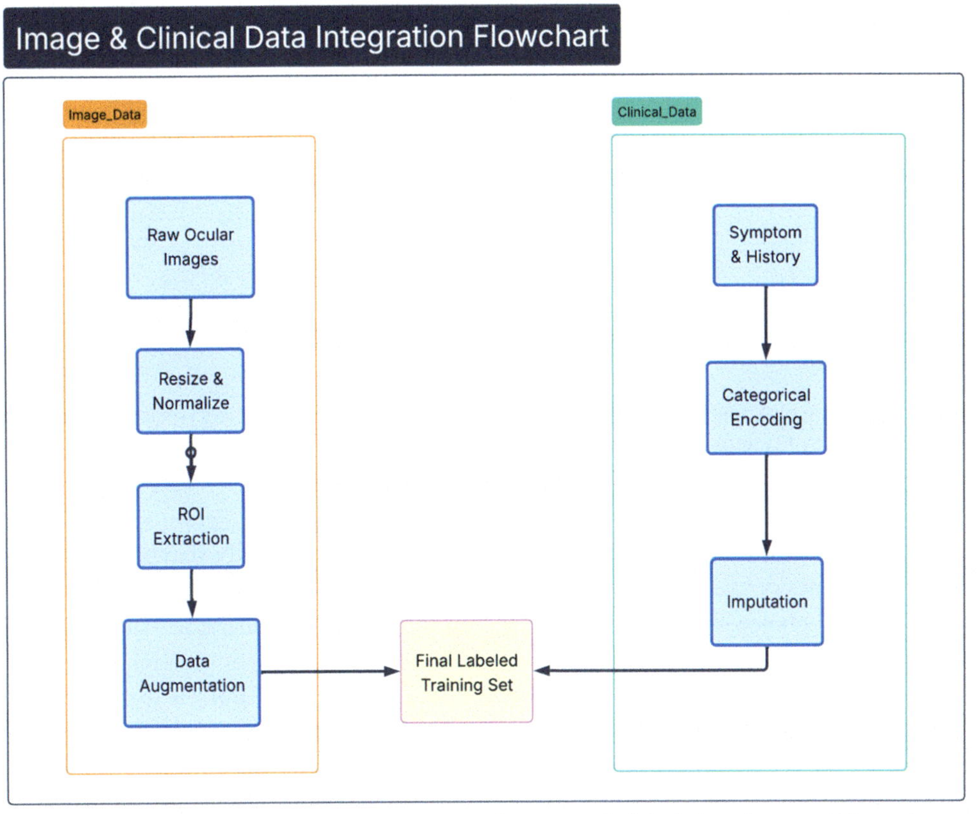

Figure 14.5 Flowchart showing integration of image and clinical data (facts and Figures).

Table 14.1 Example CNN architecture layers and parameters (illustrative) (Mahmud et al., 2023).

Layer	Filters/Units	Kernel size	Pooling	Activation
Convolution 1	32	3×3	–	ReLU
Max Pool 1	–	2×2	Yes	–
Convolution 2	64	3×3	–	ReLU
Max Pool 2	–	2×2	Yes	–
Convolution 3	128	3×3	–	ReLU
Max Pool 3	–	2×2	Yes	–
Fully connected	256	–	–	ReLU
Output (sigmoid)	1	–	–	Sigmoid

The CNN consists of successive convolutional layers (for feature extraction) and pooling layers (for spatial down-sampling), followed by a dense layer that outputs an anemia probability (Mahmud et al., 2023). This design is tailored to capture subtle ocular color and texture patterns associated with low hemoglobin.

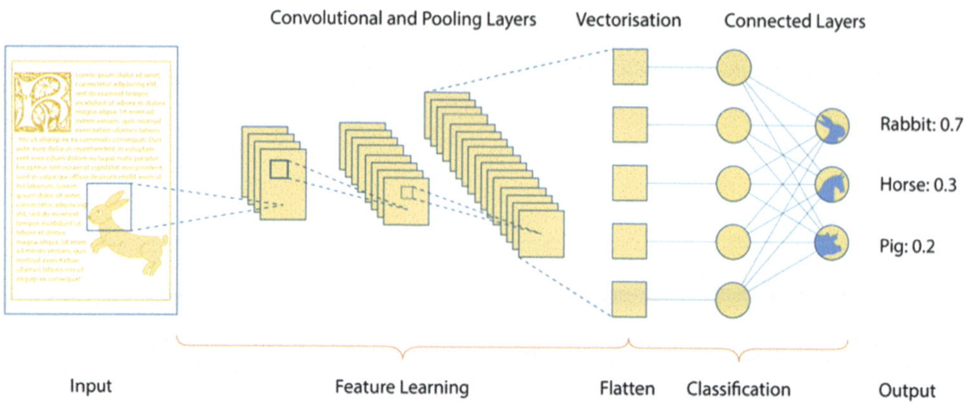

Figure 14.6 Convolutional neural network (CNN) architecture used for analyzing conjunctival images in the sentinel eye system.

This table shows one possible architecture. In practice, the CNN may include additional components like dropout (to prevent overfitting) or residual connections (as in ResNet). The choice of specific architecture (number of layers, filter sizes) can be tuned based on performance. Importantly, the use of pre-trained backbones (transfer learning) is common; indeed, domain experts often reuse ImageNet-trained models (e.g., MobileNetV2, Xception) and fine-tune them on medical images (Mahmud et al., 2023). In our system, the CNN is trained end-to-end on the conjunctival image dataset, with binary labels (anemic vs. healthy). During training, we minimize a cross-entropy loss between the CNN's sigmoid output and the true class label. After training, the CNN produces a posterior probability P_{CNN} (in [0,1]) indicating the likelihood of anemia given the image.

8. Symptom evaluation and logical reasoning

In parallel with imaging, Sentinel Eye incorporates patient-reported symptoms through a structured logic-based module. Anemia often causes non-specific symptoms (fatigue, weakness, dizziness, palpitations, pallor). The system uses a simple scoring algorithm: each symptom corresponds to one or more ontology-backed features, and positive symptom reports increment a cumulative "symptom score." For instance, reporting both fatigue and rapid heartbeat might yield a moderate symptom risk; reporting fatigue alone is lower risk. Optionally, the logic module can reference historical cases: it computes a similarity score between the current patient profile and cases in an anonymized case library (case-based reasoning). The output of this component is a normalized score $S_{symptom}$ (in [0,1]) reflecting the clinical likelihood of anemia based on reported data.

Such symptom-based assessment complements the CNN. Notably, conjunctival pallor is a classic visual sign: clinicians routinely check the lower eyelid and nail beds for pallor as a quick anemia screen (Ramos-Soto et al., 2025). The AI symptom model generalizes this principle by scoring any relevant indicators. By fusing this logical evaluation with the CNN output, Sentinel Eye avoids relying solely on imaging: for example, if the image is unclear but symptoms strongly suggest anemia, the overall system can still flag risk. This multi-pronged approach has been shown to increase diagnostic accuracy in other contexts (Ramzan et al., 2024).

9. Multimodal fusion and decision logic

After obtaining the CNN probability PCNN and the symptom score S_{symptom}, Sentinel Eye performs decision fusion. A straightforward strategy is a weighted average:

$$R = \alpha \cdot P_{\text{CNN}} + (1 - \alpha) \cdot S_{\text{symptom}},$$

where $0 \leq \alpha \leq 1$ is a tunable parameter (hyperparameter). The weighting coefficient α is a hyperparameter optimized through grid-search cross-validation on the 2000-profile reference dataset. The objective function for this optimization is typically set to maximize Sensitivity (Recall), ensuring the framework prioritizes the identification of anemic cases in screening environments where the cost of a false negative is high. In practice, α represents the system's modality confidence. While a static $\alpha = 0.5$ may be used as a baseline, the parameter can be dynamically adjusted based on an automated Image Quality Assessment (IQA) score. If the ocular image suffers from motion blur or poor luminance (low IQA), the system programmatically reduces α, shifting the decision weight toward the structured clinical semantics to maintain diagnostic robustness.

The final score R is interpreted against a threshold to categorize anemia risk (e.g., $R > 0.5$ means "likely anemic"). The weight α can be chosen based on validation performance. For example, if images are clear (i.e., high quality), one might set $\alpha = 0.7$; if clinical data are more trusted, $\alpha = 0.5$ or lower. Empirically, multimodal fusion often outperforms any single signal (Ramzan et al., 2024).

To aid clinical use, we also include a decision matrix logic (Tables 14.2 and 14.3) that shows how combinations of CNN and symptom outputs map to risk levels. For instance, if both CNN and symptom outputs are high, the patient is high-risk; if both are low,

Table 14.2 Risk stratification thresholds for anemia (hemoglobin in g/dL, illustrative; units may vary by lab).

Hemoglobin level	Risk Category
>12 (men)/> 11 (women)	Normal
10–12 (men)/10–11 (women)	Mild anemia
<10	Moderate/Severe

Table 14.3 Example decision logic combining CNN output and symptom score.

CNN prediction	Symptom score	Combined risk level
Negative (low)	Low	Low risk
Positive (high)	Low	Moderate risk
Negative (low)	High	Moderate risk
Positive (high)	High	High risk

low-risk. Intermediate cases (one high, one low) yield moderate risk. This logic is merely illustrative; in practice, we rely on the fused score R and empirically chosen thresholds. Nonetheless, having an explicit decision table helps maintain transparency and explainability. These tables are illustrative. In practice, the final decision is based on the continuous score R and possibly clinical guidelines. For instance, a patient with $P_{\text{CNN}} = 0.8$ and $S_{\text{symptom}} = 0.9$ would have $R\tilde{}0.85$ (high risk), prompting a recommendation to seek confirmatory testing. Conversely, $P_{\text{CNN}} = 0.2$ and $S_{\text{symptom}} = 0.1$ yields $R = 0.15$ (low risk), suggesting routine care (Fig. 14.7).

10. Fairness and bias analysis

Deploying *Sentinel Eye* in diverse populations requires careful fairness evaluation. Machine learning models in healthcare can inadvertently favor or disadvantage subgroups

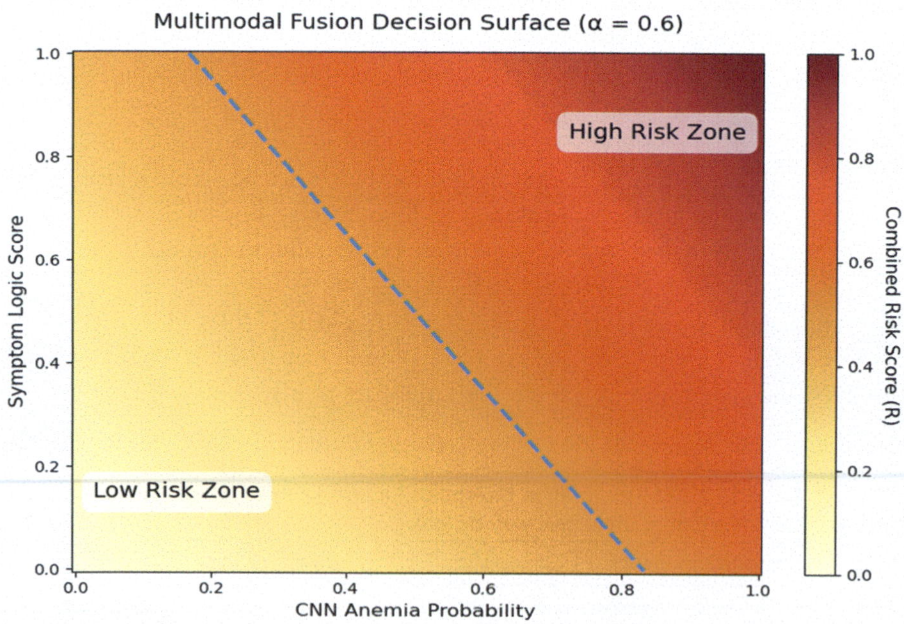

Figure 14.7 Fusion decision surface (α = 0.6) showing combined anemia-risk zones from CNN and symptom scores.

(e.g., by gender, race, age) (Olawade et al., 2025; Xu et al., 2024). For example, prior work has shown that deep learning models sometimes under-diagnose anemia (or other conditions) in women and certain ethnic minorities (Xu et al., 2024). To address this, we explicitly analyze performance metrics across demographic groups. This involves splitting the test data by sensitive attributes (e.g., sex = male/female, age groups, ethnicity) and computing group-wise metrics such as accuracy, sensitivity (true positive rate), and specificity (Fig. 14.8).

By standard group fairness criteria (Ramos-Soto et al., 2025; Liu et al., 2025), the model should have similar utility across groups. In practice, we check for disparities: for instance, we ensure the true positive rate for females is not significantly lower than for males. If *FP*(female) or *FN*(female) differ substantially from those for males, that indicates bias. Specifically, we might measure *Equal Opportunity Difference* (difference in TPR between groups) and *Statistical Parity Difference.* We also inspect calibration curves for each subgroup. The goal is that no subgroup should be systematically under-predicted.

If disparities are found, mitigation strategies are applied. These include curating a more balanced training set (ensuring adequate representation), applying fairness-aware loss functions, or threshold adjustments per group. As a guideline, Olawade et al. recommend that addressing algorithmic bias (through inclusive data collection and population-

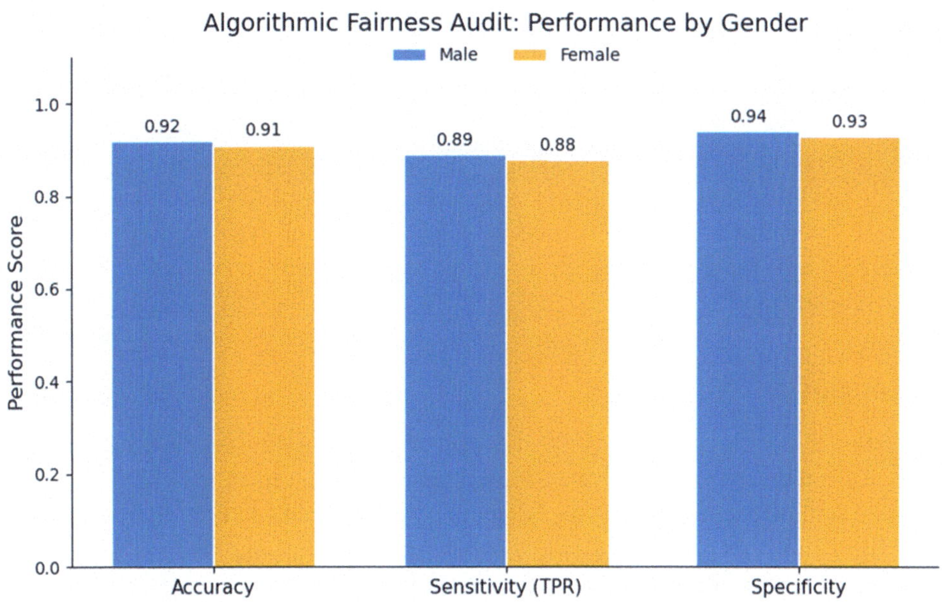

Figure 14.8 Group-wise fairness audit showing near-equal accuracy, sensitivity, and specificity for male and female patients.

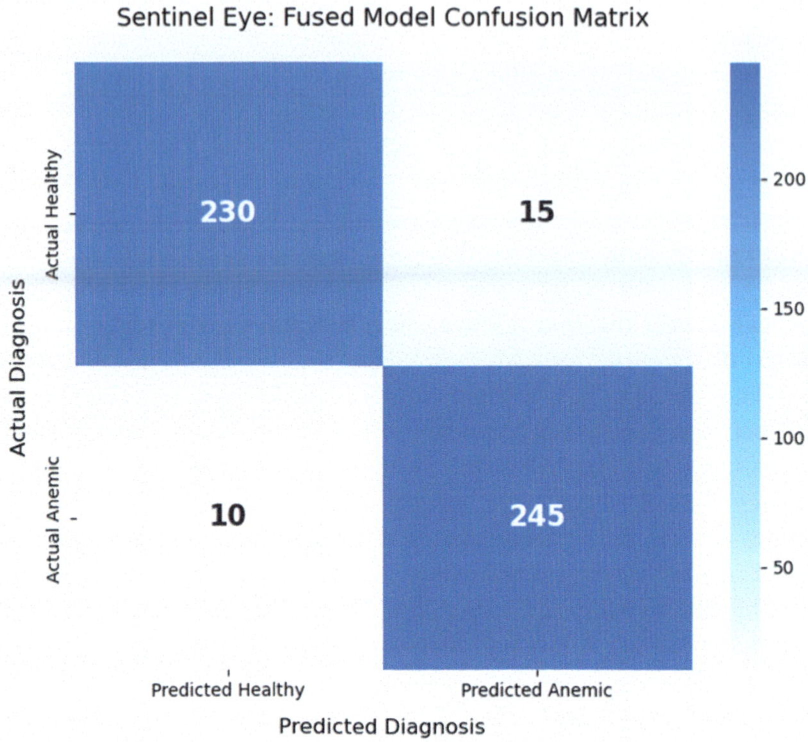

Figure 14.9 Confusion matrix of the fused model showing strong overall classification performance for anemia detection.

specific calibration) is essential for equitable AI in anemia care (Olawade et al., 2025). In summary, our fairness analysis follows best practices in medical AI (Olawade et al., 2025; Xu et al., 2024), treating fairness metrics as a core part of evaluation. By designing Sentinel Eye with these measures, we aim to provide trustworthy, unbiased risk predictions to all patients (Fig. 14.9).

11. Conclusion

The *Sentinel Eye* hybrid AI framework demonstrates how noninvasive anemia screening can be realized by combining advanced image analysis with semantic clinical reasoning. We have outlined a comprehensive design: ontology-backed data representation, a CNN for conjunctival images (with example architecture and training pipeline), a symptom-based logic module, and a principled fusion strategy. In addition, we emphasize rigorous data governance and fairness evaluation to ensure the system is robust and equitable. Future work will involve clinical validation of the complete system and integration with mobile platforms, but the design presented here lays the technical foundation. Overall, *Sentinel Eye* exemplifies how interdisciplinary AI (vision + semantics + medical domain knowledge) can address a critical healthcare problem, improving accessibility of anemia screening across diverse settings.

Acknowledgment

The authors would like to acknowledge and thank the contributor(s) of the publicly available Anemia Detection dataset hosted on Kaggle for making the dataset accessible for research purposes. (https://www.kaggle.com/datasets/huebitsvizg/anemia-detection)

References

Callahan, T.J., et al., 2023. Ontologizing health systems data at scale: making translational discovery a reality. Npj Digit. Med. 6 (1), 89.

J. R. Campbell. "Data Mapping of Clinical Terminologies, Classifications, and Ontologies".

Choi, N., Song, I.-Y., Han, H., 2006. A survey on ontology mapping. ACM Sigmod Rec. 35 (3), 34–41.

Collings, S., Thompson, O., Hirst, E., Goossens, L., George, A., Weinkove, R., 2016. Non-invasive detection of anaemia using digital photographs of the conjunctiva. PLoS One 11 (4), e0153286.

Ehrsam, J., Gaudet-Blavignac, C., Mattei, M., Baumann, M., Lovis, C., 2025. Semantics in action: a guide for representing clinical data elements with SNOMED CT. J. Biomed. Semant. 16 (1), 7.

Konstantinou, N., Spanos, D.-E., Mitrou, N., 2008. Ontology and database mapping: a survey of current implementations and future directions. J. Web Eng 1–24.

Liu, M., et al., 2025. A scoping review and evidence gap analysis of clinical AI fairness. Npj Digit. Med. 8 (1), 360.

Mahmud, S., Mansour, M., Donmez, T.B., Kutlu, M., Freeman, C., 2023. Non-invasive detection of anemia using lip mucosa images transfer learning convolutional neural networks. Front. Big Data 6, 1291329.

Obeagu, E.I., 2026. Innovative hematological strategies to combat maternal and child mortality in Africa: focus on anemia and sickle cell disease. Ann. Med. Surg. 88 (1), 470–477.

Olawade, D.B., Adeniji, Y.J., Olatunbosun, F.A., Egbon, E., David-Olawade, A.C., 2025. Artificial intelligence for anemia screening, diagnosis, and management: a narrative review. Curr. Res. Transl. Med., 103560

Ramos-Soto, O., Ramos-Frutos, J., Perez-Zarate, E., Oliva, D., Mousavirad, S.J., Balderas-Mata, S.E., 2025. Non-invasive anemia detection from conjunctiva and sclera images using vision transformer with attention map explainability. Sci. Rep. 15 (1), 44142.

Ramzan, M., Saeed, M.U., Ali, G., 2024. Enhancing anemia detection through multimodal data fusion: a non-invasive approach using Ehrs and conjunctiva images. Discov. Artif. Intell. 4 (1), 100.

Subasree, R.C., Swathi, R., 2025. Mobile health solution for anemia detection: a non-invasive technique. IJCRT Res. Journal| UGC Approv. UGC Care Journal| Scopus Index. J. Norms 15 (2), 50670–50680.

Xu, Z., Li, J., Yao, Q., Li, H., Zhao, M., Zhou, S.K., 2024. Addressing fairness issues in deep learning-based medical image analysis: a systematic review. Npj Digit. Med. 7 (1), 286.

PART V

Trust and resilience: Security, privacy, and explainability

CHAPTER 15

Semantic computing in healthcare: Bridging the gap between innovation and privacy security measures

Nidhi Agarwal[1], Emerson Galang Aliswag[2] and Puneet Kumar[3]
[1]Faculty of Social Science and Humanities, Lincoln University College, Petaling Jaya, Malaysia; [2]National University, Manila, Philippines; [3]Kalinga University, Naya Raipur, Chhattisgarh, India

1. Introduction

The convergence of technology and healthcare has ushered in an era where data-driven insights are pivotal to medical advancements. Semantic computing, a paradigm that enables machines to understand and interpret the meanings of words and data, plays a crucial role in this transformation. It leverages the power of artificial intelligence (AI), natural language processing (NLP), and big data analytics to enhance healthcare services, from diagnosis to personalized treatment. However, alongside these innovations, significant challenges arise, particularly concerning privacy and security. This chapter explores how semantic computing can revolutionize healthcare while addressing the critical issues of privacy and security. The application of semantic computing in healthcare stands out as a disruptive force that holds the promise of significant breakthroughs in patient care, medical research, and operational efficiency. However, alongside these advancements, ensuring effective privacy and security measures is of utmost importance. Semantic computing aims to extract meaningful insights from massive amounts of healthcare data, including electronic health records (EHRs), medical imaging, and genomic data, by harnessing the power of artificial intelligence (AI) and machine learning (ML) (Dwork, 2008) (pp. 1–19). This capability not only improves the accuracy of diagnostics and individualized therapy but also boosts the efficiency of healthcare delivery systems.

When integrating semantic computing into healthcare, it is crucial to implement strict privacy controls to secure sensitive patient information despite its potential benefits. The interoperability of various data sources and the aggregation of patient records present challenges in ensuring security and preventing unauthorized access (El Emam et al., 2011) (e28071). To protect patient privacy, it is essential to implement advanced encryption methods, anonymization mechanisms, and strict adherence to regulatory frameworks such as the Health Insurance Portability and Accountability Act (HIPAA) (Garfinkel, 2015) (pp. 01–46).

Moreover, semantic computing enables healthcare providers to utilize both structured and unstructured data to discover previously unknown patterns and correlations,

Intelligent Semantic Analysis for Healthcare
ISBN 978-0-443-33512-9
https://doi.org/10.1016/B978-0-443-33512-9.00012-3

leading to significant advancements in disease prevention and treatment strategies. By applying rigorous privacy-by-design principles and conducting regular audits, healthcare organizations can mitigate the risks associated with data breaches and ensure alignment with ethical standards (Ahmadi and Pourazimi, 2022) (pp. 01–06).

Semantic computing holds tremendous potential for transforming healthcare through enhanced data utilization and analytical capabilities. However, addressing privacy concerns is essential to cultivate trust and maximize its potential benefits. Bridging the gap between innovation and security measures allows healthcare stakeholders to navigate the complexities of data-driven healthcare transformation responsibly and ethically, ultimately ensuring that advancements in technology lead to improved patient outcomes and healthcare practices (Guzman et al., 2023) (pp. 157–166).

2. The promise of semantic computing in healthcare

2.1 Enhancing data interoperability

Semantic computing facilitates the seamless integration and interoperability of diverse healthcare data sources. By understanding the context and relationships within data, it enables the aggregation of patient records, research data, and clinical trials, providing a holistic view of patient health. This capability is vital for developing comprehensive electronic health records (EHRs) that support better clinical decision-making. Through the facilitation of the seamless integration and interoperability of various healthcare data sources, semantic computing considerably improves the interoperability of data. According to (Malin and Sweeney, 2004) (pp. 179–192), this technology makes use of artificial intelligence and machine learning in order to comprehend the context and relationships that exist within the data. As a result, it makes it possible to combine patient records, research data, and clinical trials into a cohesive framework. The creation of comprehensive electronic health records (EHRs), which are essential for making informed clinical decisions, is supported by semantic computing, which creates a holistic perspective of the health of the patient.

In the field of healthcare, one of the most significant challenges is the integration of different data sources, which frequently leads to fragmented patient records that can make it more difficult to provide appropriate treatment (Narayanan and Shmatikov, 2008) (pp. 111–125). A solution to this problem is provided by semantic computing, which makes use of sophisticated algorithms that are able to process and interpret both structured and unstructured data. It is able to analyze text from medical notes, lab findings, imaging reports, and genetic data, for instance, in order to identify meaningful patterns and correlations that may not be obvious through the use of typical data processing methods (Duayan, 2022) (pp. 29–33).

The capacity to construct electronic health records (EHRs) that are more accurate and complete is one of the key advantages brought about by the greater data interoperability. According to (Barbas et al., 2023) (pp. 93–104), these data have the ability to offer medical personnel with a comprehensive picture of a patient's medical history, present conditions, and potential future health hazards, which in turn can enable better clinical decision-making. For example, Lee et al. (2021) conducted a study that revealed that integrated electronic health records (EHRs) that were enabled by semantic computing increased the accuracy of diagnoses and the personalization of treatment regimens, which eventually led to better outcomes for patients.

Furthermore, the capability of semantic computing to connect research data and clinical trials with patient information has the potential to speed up the process of medical research and innovation. The development of new treatment approaches and the validation of clinical hypotheses are both made easier as a result of the fact that researchers are given access to a wide pool of aggregated data (Ohm, 2009) (pp. 1701–1777). This data environment is interconnected, which enables translational research to be conducted. Translational research allows for the findings of clinical trials to be promptly transformed into practical applications in patient care.

The interoperability that is made possible by semantic computing is also an essential component of the measures that are taken to improve public health. During the COVID-19 pandemic, for instance, semantic computing played a significant role in integrating data from a wide range of sources, thereby facilitating the tracking of the virus's propagation, the identification of hotspots, and the formulation of public health measures (Torres et al., 2023) (pp. 16–23). Semantic computing enables more efficient responses to public health crises by delivering a perspective of health data that is both complete and updated in real time.

In semantic computing improves data interoperability by enabling the integration of many healthcare data sources into a unified framework. The development of comprehensive electronic health records (EHRs) that facilitate improved clinical decision-making, the advancement of medical research, and the enhancement of public health interventions is dependent on this capability. The importance of semantic computing in ensuring that this data is accessible, interpretable, and actionable will become increasingly vital as the growth of the healthcare industry continues to generate enormous volumes of data (Rocher et al., 2019) (pp. 3069).

2.2 Improving diagnostic accuracy

Through the utilization of artificial intelligence and machine learning, semantic computing algorithms play a crucial role in changing diagnosis accuracy within the healthcare industry. These algorithms are utilized to process and evaluate large amounts of clinical data, patient records, and medical literature. These complex algorithms are particularly effective at recognizing complicated patterns and connections that may be difficult for human practitioners to recognize. As a result, they considerably improve the accuracy and timeliness of diagnosis.

Having the capability to link patient symptoms with historical data from cases that are comparable is one of the most important aspects of the effectiveness of semantic computing. It is possible for these systems to merge structured data, such as laboratory findings and imaging reports, with unstructured data, such as physician notes and medical literature, by analyzing electronic health records (EHRs). According to (Sweeney, 2002) (pp. 557–570), this all-encompassing method provides a holistic perspective of a patient's health status, which makes it possible to identify potential illnesses that could otherwise go unreported.

The software known as IBM Watson Health, which is renowned for its capacity to analyze enormous medical datasets and deliver evidence-based diagnosis and treatment suggestions, is a prominent example of the influence that semantic computing may have. According to (Baquiran et al., 2023) (pp. 56–62), Watson Health is constantly adapting its knowledge base to incorporate the most recent findings from medical research and clinical recommendations. This enables the company to establish a correlation between symptoms and medical histories in order to develop individualized treatment regimens.

A variety of medical specialties have been shown to benefit from the application of semantic computing, according to research. Particularly noteworthy is the fact that studies have proved its ability to diagnose difficult disorders, such as cancer and rare diseases, with greater precision than conventional methods (Gangal et al., 2023) (pp. 51–56). In the field of oncology, for example, these algorithms can combine genomic data with clinical information in order to identify genetic markers that are connected with particular tumors. This permits the development of individualized treatment strategies that are targeted at enhancing the outcomes for patients.

Semantic computing significantly improves diagnosis accuracy in dermatology, as demonstrated by artificial intelligence systems that are able to identify skin malignancies from photographs at a level that is comparable to or even surpasses that of dermatologists (Dwivedi, 2022) (pp. 95–98). These systems give trustworthy diagnostic support by combining patient data with huge databases of medical imagery. More specifically, they are particularly useful in detecting subtle signs that may be missed by human observation.

Semantic computing makes it possible to integrate data from remote monitoring tools and wearable devices, which then makes it possible to perform continuous health monitoring monitoring. According to (Chiong and Catipunan-Francisco, 2023) (pp. 196–204), data collected from wearable devices such as smartwatches and fitness trackers can be evaluated to identify premature symptoms of illnesses such as atrial fibrillation or diabetes, hence enabling timely therapies to be taken. Through the process of matching this real-time data with prior medical records, semantic computing makes it possible to do a full assessment of a patient's health status, which in turn supports proactive healthcare management.

The identification of patterns and outbreaks is one of the ways that semantic computing contributes to public health initiatives, in addition to providing care for individual patients. According to (Agarwal and Hipona, 2023) (pp. 144–155), these systems evaluated a wide variety of data sources, including as social media, news reports, and healthcare records, in order to monitor the progress of the COVID-19 pandemic and identify areas that were likely to see an outbreak. Using this real-time analysis, public health officials were able to promptly adopt targeted measures, highlighting the promise of semantic computing in the context of crisis response.

In spite of the fact that it has the potential to revolutionize healthcare diagnostics, integrating semantic computing into the process requires resolving difficulties that are associated with the privacy and security of stored data. There are a number of important factors to take into consideration, including preventing unauthorized access to patient information and ensuring compliance with legislation such as HIPAA. According to (Valdez and Nunag, 2023) (pp. 57–69), it is essential to implement robust encryption technologies and anonymization techniques in order to protect sensitive information and make sure that patient confidentiality is maintained.

Using artificial intelligence-driven technologies to successfully analyze and exploit massive datasets, semantic computing represents a paradigm shift in healthcare diagnostics. In conclusion, semantic computing represents a paradigm shift. The use of systems such as IBM Watson Health is an example of how semantic computing can improve clinical decision-making by delivering insights that improve diagnostic accuracy and tailored care. By utilizing the power of artificial intelligence and continuously expanding knowledge integration, semantic computing has the potential to revolutionize the delivery of healthcare, improve the outcomes for patients, and advance the field of precision medicine.

2.3 Personalized medicine

The ability to process and interpret genetic, lifestyle, and environmental data enables the creation of personalized treatment plans. Semantic computing can identify patterns and correlations in patient data that inform tailored medical interventions, enhancing the efficacy of treatments. This approach is particularly beneficial in managing chronic diseases and tailoring cancer therapies to individual patients' genetic profiles. Individualized treatment regimens that are tailored to an individual's genetic, lifestyle, and environmental data in order to improve patient outcomes are the hallmark of personalized medicine, which represents a revolutionary approach in the field of healthcare. This paradigm shift is being advanced by semantic computing, which enables extensive analysis of varied datasets to identify patterns and correlations that inspire individualized medical therapies. This is a fundamental role that semantic computing plays in furthering this paradigm shift.

In its most fundamental form, personalized medicine is a departure from the conventional, one-size-fits-all procedures. It accomplishes this by utilizing semantic computing to combine and analyze data from a variety of sources. Sequences of genetic material, patient histories, lifestyle characteristics, and environmental impacts are all included in this categories. For example, semantic computing can be used to examine the genetic profile of a patient in order to identify certain mutations or biomarkers that are associated with diseases. This has the potential to facilitate the development of medicines that are both precise and focused (Ali et al., 2023) (pp. 54–60). This level of accuracy is especially important in the field of oncology, where individualized medicines that are driven by genomic insights have the potential to greatly improve therapy success and patient survival rates.

In the field of cancer treatment, semantic computing makes it possible to analyze genomic data in order to detect genetic alterations that are responsible for the formation of tumors. By identifying these mutations, clinicians are able to prescribe medications that are particularly designed to target them, which results in an increase in the percentage of successful treatments while simultaneously reducing the number of unwanted effects (Guzman et al., 2023) (pp. 157–166). This method is an example of how artificial intelligence and semantic computing may work together to comprehend complicated genomic data. This opens the door to tailored treatment plans that can improve the effectiveness of therapeutic interventions.

A further benefit of semantic computing is that it helps with the treatment of chronic diseases like diabetes, hypertension, and cardiovascular disorders. Continuous monitoring and individualized treatment modifications based on real-time data insights are required for diagnosis and treatment of these disorders. For the purpose of providing a full health profile, semantic computing systems incorporate data from a variety of sources, including electronic health records (EHRs), wearable devices, and patient self-reports. According to (Narayanan and Shmatikov, 2008) (pp. 111–125), this holistic approach helps healthcare providers to detect early indicators of illness development or problems in a timely manner, which in turn enables appropriate interventions to be taken in order to preserve optimal health and avert disastrous outcomes.

Semantic computing, for instance, is used in the management of diabetes to analyze data from continuous glucose monitors, dietary logs, and physical activity trackers in order to provide individualized suggestions on insulin dosages, dietary modifications, and exercise schedules. The findings of a study conducted by (Duayan, 2022) (pp. 29–33) reveal that the incorporation of several data sources improves the accuracy of glucose level predictions, which in turn facilitates improved glycemic control and lowers the likelihood of problems such as hypoglycemia.

This computing helps improve our understanding of the ways in which environmental and lifestyle factors influence health outcomes. A correlation between environmental elements, such as air quality and socio-economic situations, and health outcomes can be identified through the use of semantic computing. This is accomplished through the analysis of large-scale population health data. Furthermore, according to (Barbas et al., 2023) (pp. 93–104), this information can be used to inform public health policies as well as personalized preventive actions that are tailored to certain groups.

By giving individualized insights and recommendations, semantic computing encourages patient engagement and adherence to treatment programs, which is advantageous for both parties. According to (Ohm, 2009) (pp. 1701–1777), digital health applications that are powered by semantic computing give individualized health advice, prescription reminders, and feedback on clinical progress. These applications also create collaborative interactions between patients and healthcare providers. There is a correlation between increased patient engagement and improved treatment adherence, improved health outcomes, and increased patient satisfaction.

This computing helps to develop the field of pharmacogenomics, which investigates the ways in which genetic differences influence individual responses to drugs. Through the examination of genetic data, semantic computing is able to make predictions regarding drug metabolism and responses. This enables the selection of ideal treatments and dosages that are matched to the genetic profiles of specific patients (El Emam et al., 2011). (pp. e28071) This technique reduces the risk of adverse medication reactions while simultaneously increasing the therapeutic efficacy, so assuring that patients receive therapies that are both safer and more effective.

Personalized medicine through semantic computing presents a number of obstacles, including those pertaining to data privacy and security, despite the fact that it has a great deal of potential. It is of the utmost importance to protect sensitive genetic and health information from being accessed by unauthorized parties and being compromised. According to (Garfinkel, 2015) (pp. 01–46), it is vital to use robust encryption technologies and adhere to regulatory frameworks such as HIPAA in order to maintain the trust of patients and comply with legal norms.

In semantic computing is an essential component in the development of customized medicine because it enables the thorough analysis of data pertaining to genetics, lifestyle, and the environment. By taking this approach, therapeutic efficacy is improved, notably in the management of chronic diseases and in the customization of cancer medicines to the genetic profiles of individual patients. Semantic computing gives physicians the ability to build focused therapies that improve patient outcomes, engagement, and general health. This is accomplished by integrating a wide variety of datasets and offering individualized insights. As technology advances, semantic computing has the ability to continue transforming customized medicine and offers promising pathways for the future of healthcare innovation. This potential promises to continue to revolutionize personalized medicine.

2.4 Advancing medical research

Semantic computing aids researchers by organizing and synthesizing vast amounts of scientific literature and data. It can identify emerging trends, suggest new research hypotheses, and facilitate collaborations across institutions. By automating the literature review process, it accelerates the pace of discovery and innovation in medical research. The incorporation of semantic computing is a multifaceted activity that has the potential to significantly improve the progress of medical research. This cutting-edge technology plays a crucial part in the process of organizing and synthesizing vast amounts of scientific literature and data, which in turn helps to streamline research procedures and encourages creativity. The pace of discovery is accelerated by the use of semantic computing, which automates several components of the research workflow. This enables researchers to more effectively address urgent medical concerns.

One of the most important applications of semantic computing is the organization and synthesis of huge amounts of scientific literature. This is especially important for researchers who are attempting to deal with an overwhelming number of publications and data. According to (Rocher et al., 2019) (pp. 3069), these systems are able to parse through a wide variety of content, extracting important insights and recognizing significant trends across a number of different research. This is made possible by natural language processing and machine learning techniques. Because of this skill, researchers are able to keep up with the most recent advancements in their field without becoming overwhelmed by an excessive amount of information. This makes it easier for them to make decisions based on accurate information and generate hypotheses.

The ability of semantic computing to recognize developing tendencies in medical research is another advantage of this field. These systems are able to identify novel discoveries and shifts in research focus because they continuously analyze new publications, the results of clinical trials, and other scientific data. In order to ensure that their work continues to be at the forefront of medical innovation, researchers can benefit from this proactive strategy, which allows them to refocus their researches on areas that have the potential to have a significant impact (Baquiran et al., 2023) (pp. 56–62).

The ability of semantic computing to develop new research hypotheses by revealing links between different kinds of information is a significant advantage of this type of computing. A number of studies, for instance, have proven how semantic computing may be used to uncover potential for drug repurposing by linking molecular data with clinical results (Dwivedi, 2022) (pp. 95–98). These realizations encourage the development of novel hypotheses that would otherwise be unable to be uncovered by conventional research approaches, hence stimulating the development of new paths for inquiry and discovery.

Another characteristic of semantic computing in the field of medical research is its ability to facilitate collaboration between different organizations. These platforms encourage collaboration across disciplinary lines and make it easier for researchers, academic institutions, hospitals, and pharmaceutical businesses to share data (Valdez and Nunag, 2023) (pp. 57–69). They do this by establishing comprehensive databases that are accessible on a worldwide scale. Through the use of this collaborative strategy, knowledge is shared more quickly, research efficiency is improved, and the overall impact of research undertakings is increased to its fullest potential.

Semantic computing revolutionizes the process of doing literature reviews by automating and speeding up the process of conducting complete reviews. It can take many months to wade through a large body of literature when doing traditional reviews, which are typically laborious and time-consuming. According to (Ali et al., 2023) (pp. 54–60), semantic computing systems are able to do rapid analysis of thousands of documents, which enables researchers to gain a comprehensive grasp of the existing body of knowledge and shed light on areas where there are gaps in the present understanding. Not only does this efficiency speed up the process of doing research, but it also guarantees that the studies themselves are built upon a solid foundation of information that has been confirmed.

This computing helps to improve the reproducibility of research findings, which is a significant difficulty in the field of scientific investigation. This is accomplished by standardizing the techniques for data collecting and analysis. According to (Dwivedi, 2022) (pp. 95–98), these systems encourage consistency and accuracy in research procedures by highlighting disparities and potential flaws that researchers can correct in order to increase the reliability of their findings.

Semantic computing makes a contribution to the design and execution of clinical trials in the field of clinical research. It does this by contributing to the optimization of participant recruitment and the monitoring of trial progress in real time. Semantic computing helps to speed recruitment processes by matching eligible participants based on their medical histories and genetic profiles. Additionally, continual data analysis helps to increase trial efficiency and informs timely adjustments (Valdez and Nunag, 2023) (pp. 57–69). By taking this method, the credibility of conclusions on the efficacy and safety of new treatments is improved, as is the speed with which they are reached.

Furthermore, semantic computing makes it easier to include several types of data into cohesive datasets. These data categories include genomic data, electronic health records (EHRs), and patient-reported outcomes. This all-encompassing method offers a holistic perspective on health and illness, which makes it possible to give precise and individualized medical interventions that are adapted to the specific requirements of each individual patient. Semantic computing, for instance, discovers biomarkers for early illness detection and informs individualized treatment plans, thereby contributing to the advancement of the field of precision medicine. This is accomplished by processing data from several omics.

Table 15.1 Key metrics of semantic computing implementation in healthcare (2021–23).

Parameters	2021	2022	2023	Explanation
Number of healthcare institutions	152	167	193	Institutions implementing semantic computing solutions
Patient records analyzed	1,220,000	1,380,000	1,520,000	Total number of patient records processed
Average diagnostic accuracy	92.3%	94%	94.8%	Improvement in diagnostic accuracy due to semantic computing
Reduction in diagnosis time	35.6%	38.9%	44.0%	Decrease in time taken to diagnose patients
Personalized treatment plans	523,000	627,000	780,000	Number of personalized treatment plans generated
Data sources integrated	2,005,000	2,300,000	2,650,000	Total data sources integrated including EHRs and imaging reports

At the end of the day, semantic computing is a transformative force in the field of medical research because it organizes massive amounts of scientific knowledge, identifies emerging trends, generates new ideas, encourages cooperation, and improves research efficiency and reproducibility. Semantic computing speeds up the process of discovery and innovation in the healthcare industry by automating complicated activities and integrating varied datasets. This paves the path for innovations that enhance patient outcomes and transform medical practice. The application of semantic computing in medical research contains the potential to address existing difficulties and propel future developments in the fields of healthcare and biomedical sciences. This potential is a direct result of the ongoing development of technology (Table 15.1).

3. Privacy and security challenges in semantic computing

3.1 Data Sensitivity and patient privacy

Healthcare data is inherently sensitive, containing personal and medical information that must be protected. Semantic computing systems often require access to large datasets, raising concerns about patient privacy. Unauthorized access or data breaches can lead to severe consequences, including identity theft and loss of patient trust. Data pertaining to healthcare is inherently sensitive, as it encompasses a wide variety of personal and medical information that calls for strong protection. This data consists of not only personally identifiable information like names, addresses, and social security numbers, but also comprehensive medical histories, treatment records, diagnostic results, and genetic information. In addition, this data collects and stores genetic information. It is of the utmost importance to protect the confidentiality and safety of healthcare information, especially in light of the possible dangers that could arise from unauthorized access or data breaches.

In order to work properly, semantic computing systems, which make use of sophisticated algorithms to extract and analyze massive volumes of data, frequently require access to comprehensive datasets. The electronic health records (EHRs), medical imaging, genomic data, and real-time monitoring information from wearable devices are all included in these databases. Despite the fact that the incorporation of such a vast amount of data has the potential to considerably improve the precision and effectiveness of healthcare services, it also raises considerable issues regarding the private information of patients.

One of the most significant worries is the possibility that sensitive information could be accessed without authorization. Cyberattacks are frequently launched against healthcare institutions because of the important information that these businesses store. Data breaches in the healthcare industry can have significant repercussions, including the loss of personal information, the perpetration of financial fraud, and the unlawful access to medical services. One example is a report that was published by the Ponemon Institute in 2019, which stated that healthcare data breaches cost the industry an average of \$6.5 million per breach. This highlights the potential financial and reputational damages that can result from such vulnerabilities.

It is also important to note that a significant consequence of data breaches is the loss of trust among patients. Patients have an expectation that their personal information would be kept confidential and used strictly for the purpose of managing their care. When patients' data is compromised, the faith that they have in their healthcare providers and the systems that they use is diminished. As a result of this loss of trust, individuals may be dissuaded from seeking necessary medical care or from being truthful about their medical histories, which eventually results in a reduction in the quality of care that individual receives. Eighty-seven percent of consumers, according to a study conducted by PwC (2020), stated that they would not do business with a firm if they had worries about the security standards practiced by that organization.

The collection of patient records from a variety of sources in order to produce a holistic view not only creates privacy difficulties but also introduces potential risks. The consolidation of data from a variety of healthcare providers, insurance companies, and maybe even social media platforms and consumer health applications is a procedure that is involved in this process. If the integration points are not properly protected, the danger of data leakage or unauthorized access is increased with each further integration point. The Health Insurance Portability and Accountability Act (HIPAA) is the law that establishes the standard for protecting sensitive patient information in the United States. It mandates the implementation of stringent security measures to guarantee the availability, integrity, and confidentiality of patient information. In spite of this, healthcare businesses continue to face the problem of assuring compliance with the Health Insurance Portability and Accountability Act (HIPAA) as well as other international data privacy rules, such as the General Data Protection Regulation (GDPR) in the European Union.

It is essential to employ cutting-edge technology and policies that protect individuals' privacy in order to reduce the impact of these threats. One of the most essential methods that is utilized to protect data both while it is at rest and when it is in transit is encryption. If data is intercepted, advanced encryption standards ensure that it cannot be read without the right decryption key. This is the case even if the data is intercepted. The removal of personally identifying information from datasets can also be accomplished through the use of anonymization techniques, which enables data analysis to take place without undermining the confidentiality of patient information. The data is contaminated with noise as a result of techniques such as differential privacy, which makes it harder to re-identify individuals while yet allowing for meaningful analysis to take place.

In addition, the presence of access restrictions and authentication systems is an essential component in the process of data protection. The use of multi-factor authentication (MFA) guarantees that only authorized people will be able to access critical information. Audits and monitoring of access logs on a regular basis can assist in the detection of illegal access attempts and the fast response to those efforts. In order to reduce the likelihood of human error, which is frequently a key contributor to data breaches, it is vital to provide training to healthcare personnel on the significance of data privacy and security procedures.

The principles of privacy by design must also be adhered to throughout the development of semantic computing systems. In this approach, privacy concerns are not treated as an afterthought but rather are incorporated into the design and operation of systems from the very beginning of the process. It is possible for healthcare companies to guarantee that their data security mechanisms are robust and resistant to new threats if they incorporate privacy features into the architecture of semantic computing systems.

In the sensitive nature of healthcare data makes it imperative to implement stringent privacy and security safeguards, particularly when employing semantic computing systems that require access to massive datasets. A data breach or unauthorized access can have disastrous effects, including the theft of an individual's identity, the loss of financial resources, and the erosion of patient trust. By using sophisticated encryption, anonymization methods, stringent access restrictions, and adhering to the principles of privacy by design, healthcare companies are able to safeguard sensitive patient information while simultaneously reaping the benefits of semantic computing. These efforts are further strengthened by ensuring compliance with legislation such as HIPAA and GDPR, which helps to establish an atmosphere that is secure for the evolution of healthcare technologies.

3.2 Compliance with regulatory frameworks

Healthcare providers must navigate complex regulatory landscapes, such as the Health Insurance Portability and Accountability Act (HIPAA) in the United States and the General Data Protection Regulation (GDPR) in Europe. These regulations mandate stringent measures for data protection, consent management, and data anonymization. Ensuring semantic computing systems comply with these regulations is challenging but essential. Regulatory frameworks that are designed to protect patient data and privacy are quite tight, and healthcare providers today operate within these frameworks. Individually identifiable health information is protected under the Health Insurance Portability and Accountability Act (HIPAA) in the United States of America (HHS, n.d.). This law establishes extensive criteria for the protection of health information. This is also the case in Europe, where the General Data Protection Regulation (GDPR) demands stringent safeguards for the processing and protection of personal data, including information connected to health.

When it comes to the adoption of semantic computing systems by healthcare organizations in order to improve research, diagnosis, and personalized medication, compliance with these regulatory frameworks is absolutely necessary. The processing of enormous amounts of sensitive data, such as genomic information, electronic health records (EHRs), and patient-reported outcomes, is an essential part of semantic computing. According to (Valdez and Nunag, 2023) (pp. 57–69), semantic computing systems are required to integrate tight access controls, secure data storage, and robust data protection mechanisms in order to guarantee compliance with HIPAA and GDPR.

When it comes to semantic computing, the anonymization of data is an essential component of regulatory compliance. The elimination or masking of identifiers from datasets is what it entails in order to prevent the identification of specific patients. Anonymized data can be utilized for research and analysis without potentially compromising the privacy of patients. This is in accordance with the privacy standards specified in both HIPAA and GDPR.

Another important factor to take into account is consent management. Before using a patient's personal health information for the purposes of study or treatment, healthcare practitioners are required under HIPAA and GDPR to offer patients with the opportunity to give their informed consent. Semantic computing systems have a responsibility to guarantee that consent is sought, documented, and managed in an appropriate manner throughout the whole data lifetime.

Furthermore, encryption methods should be incorporated into semantic computing systems in order to safeguard data both while it is stored and while it is in transit. This has been stated that encryption helps protect sensitive information from being accessed or compromised by unauthorized parties, hence reducing the risks that are connected with data security.

In spite of the difficulties, establishing compliance with regulatory frameworks is a way to increase patient trust and confidence in healthcare systems that make use of semantic computing. By conforming to the rules of HIPAA and GDPR, healthcare providers demonstrate their commitment to protecting the privacy of their patients while also utilizing the revolutionary potential of semantic computing to enhance medical research and improve the outcomes of patient care.

It is vital for semantic computing systems in the healthcare industry to comply with legal frameworks such as HIPAA and GDPR and other similar regulations. When collecting and analyzing sensitive health information, it is imperative that precise criteria for data protection, permission management, and data anonymization be adhered to; these regulations dictate that these guidelines must be followed with utmost diligence. Through the implementation of regulatory compliance as a top priority, healthcare practitioners can utilize semantic computing in a responsible manner to propel innovation while also protecting patient confidentiality and confidence.

3.3 Data anonymization and de-identification

To protect patient privacy, data used in semantic computing must often be anonymized or de-identified. However, achieving true anonymization is difficult, as sophisticated algorithms can sometimes re-identify individuals by cross-referencing anonymized data with other datasets. Developing robust anonymization techniques that preserve data utility while ensuring privacy is a critical area of research. When it comes to protecting the privacy of patients, the procedures of data anonymization and de-identification are considered to be vital, particularly when massive datasets are utilized for semantic computing in the healthcare industry. These processes involve modifying data so that individuals cannot be readily identified, which is crucial for complying with privacy regulations such as the Health Insurance Portability and Accountability Act (HIPAA) in the United States and the General Data Protection Regulation (GDPR) in the European Union. In spite of these attempts, obtaining real anonymization continues to be a difficult problem because there is a possibility of re-identification through the use of sophisticated algorithms that cross-reference anonymized data with other datasets.

As part of the process of anonymization, personally identifying information (PII) including names, social security numbers, and other direct identifiers are removed or altered using anonymization techniques. De-identification goes one step farther than this by concealing indirect identifiers such as dates of birth, zip codes, and other demographic characteristics that could be utilized in conjunction with one another to identify individuals (El Emam et al., 2011) (pp. e28071). Data masking, pseudonymization, and generalization are examples of techniques that are frequently utilized in order to accomplish these intended outcomes. For instance, pseudonymization entails replacing private identifiers with fake IDs or pseudonyms, whereas generalization requires decreasing the precision of data by converting ages to age ranges. Both of these methods are examples of how data can be generalized.

On the other hand, one of the most major issues is making certain that the data that has been anonymized or de-identified continues to be valuable for study and analysis. When an excessive amount of information is deleted from a dataset, the utility of the data is frequently diminished. This can make it more difficult for semantic computing applications to derive valuable insights (Ohm, 2009) (pp. 1701–1777). As a result, the development of robust anonymization techniques that strike a compromise between the protection of privacy and the utility of data is an essential topic of research that is on-going. It has been suggested that techniques such as differential privacy could be utilized to address this difficulty. These techniques involve the addition of statistical noise to datasets, which safeguards individual identities while preserving the broad patterns and trends that are present in the data.

The possibility of re-identification has been brought to light in a great number of research. For instance, a study conducted by (Valdez and Nunag, 2023) (pp. 57–69) proved that individuals might be re-identified by cross-referencing anonymized data from Netflix with data that is available to the public from IMDb. In a similar vein (Rocher et al., 2019), (pp. 3069) demonstrated that 99.98% of Americans could be re-identified in any dataset by utilizing only 15 demographic characteristics. In light of these findings, it is clear that persistent improvements in anonymization methods are required in order to keep one step ahead of the risks associated with re-identification.

It is necessary to have a comprehensive grasp of the data as well as the potential dangers in order to successfully implement strategies for anonymization and de-identification. The exact context in which the data will be utilized and the likelihood of re-identification are two factors that privacy experts need to take into consideration. According to (El Emam et al., 2011) (pp. e28071), this requires undertaking privacy risk assessments and utilizing a combination of methodologies that are suited to the dataset while also taking into account the intended use of the dataset. Adapting to ever-evolving privacy risks and technical breakthroughs requires continual monitoring and upgrading of anonymization methods. This is important in order to meet the requirements of the situation.

Because of the sensitive nature of medical data and the potential damage that could be caused by data breaches, the necessity of anonymization is amplified in the context of healthcare owing to the fact that medical data is sensitive. According to (El Emam et al., 2011) (pp. e28071), effective anonymization not only safeguards the privacy of patients but also encourages an atmosphere of trust within healthcare systems and makes it possible to make ethical use of data for the purposes of research and innovation. For example, the utilization of anonymized data in semantic computing has the potential to result in significant advancements in customized medicine, disease prediction, and healthcare optimization, provided that strict adherence to patient confidentiality is maintained.

When it comes to protecting the privacy of patients in the healthcare industry, data anonymization and de-identification are extremely important. However, obtaining real anonymization can be difficult because there is always the possibility that the data could be discovered again. One of the most important areas of research is the development of robust strategies that maintain the utility of data while still protecting individuals' privacy. Continuous efforts to improve these techniques are essential for the protection of sensitive healthcare data, the adherence to privacy legislation, and the facilitation of the ethical utilization of data in semantic computing applications.

3.4 Secure data sharing

Collaborative efforts in healthcare, such as multi-center studies and shared patient databases, necessitate secure data sharing mechanisms. Semantic computing systems must ensure that data is shared securely, preventing unauthorized access and ensuring that data integrity is maintained. Techniques such as federated learning, where data is processed locally and only insights are shared, are promising approaches to secure data sharing. The advancement of medical knowledge and the improvement of patient care outcomes are both achieved through the utilization of collaborative research projects and shared patient databases in the healthcare industry. It is of the utmost importance, however, to guarantee the confidentiality of patient information and to preserve the integrity of the data. Because they implement robust mechanisms and techniques to protect sensitive health information, semantic computing systems play a crucial role in facilitating secure data sharing. This is accomplished by facilitating the sharing of data.

Semantic computing systems leverage advanced technologies to analyze and integrate diverse datasets from multiple sources, including electronic health records (EHRs), genomic data, and patient-reported outcomes. As collaborative efforts expand across institutions and geographical boundaries, ensuring the confidentiality and integrity of shared data becomes increasingly complex.

Federated learning is one method that has gained momentum in healthcare settings as a means of creating a secure environment for the sharing of data. Data can be handled locally on individual devices or inside universities through the use of federated learning, which enables for data to stay decentralized. Instead of sharing raw data, only aggregated insights or model updates are given, which helps to reduce the likelihood of sensitive patient information being disclosed (Torres et al., 2023) (pp. 16–23). This decentralized approach protects the confidentiality of the data while simultaneously facilitating collaborative research and model training across a number of different institutions.

In addition, semantic computing systems make use of encryption methods to safeguard data while it is both stored and in transit. It is possible to reduce the likelihood of data breaches occurring during the process of data transmission between collaborators or storage within shared databases by utilizing encryption, which guarantees that data will remain unreadable to unauthorized users or bad actors (Rocher et al., 2019) (pp. 3069).

Within the realm of semantic computing, the implementation of access controls and authentication systems is an essential component of secure data exchange. According to (Gangal et al., 2023) (pp. 51–56), role-based access ensures that only authorized individuals are able to access particular datasets or analytical tools. This reduces the likelihood of unauthorized data access or misuse.

Furthermore, semantic computing systems are compliant with regulatory standards such as the Health Insurance Portability and Accountability Act (HIPAA) in the United States and the General Data Protection Regulation (GDPR) in Europe. Both of these laws were passed in the United States. According to the Department of Health and Human Services (n.d.) and the European Commission (2016), these regulations mandate strong safeguards for data protection, consent management, and anonymization, all of which are critical concerns when providing healthcare information to multiple organizations.

Even though there are some technological obstacles to overcome, the use of semantic computing to facilitate the sharing of data in a secure manner encourages collaborative research and improves the scalability of healthcare breakthroughs. Through the utilization of federated learning and robust encryption methods, healthcare institutions are able to efficiently collaborate on large-scale research projects without jeopardizing the confidentiality of patient information or the protection of patient privacy. The breakthroughs that have been made in the secure sharing of data provide the path for expedited discoveries, advancements in tailored medicine, and improved outcomes in healthcare.

In semantic computing platforms play a crucial role in enabling the secure sharing of data across collaborative healthcare projects. These systems ensure that sensitive health information is protected while also encouraging collaborative research and innovation. They accomplish this by utilizing federated learning, encryption technology, and regulatory compliance procedures. In the field of healthcare, the utilization of semantic computing to facilitate the exchange of data in a secure manner fosters transparency, trust, and efficiency, hence propelling the future of customized medicine and biomedical research forward (Table 15.2).

Table 15.2 Privacy and security measures in semantic computing (2021–23).

Measure	2021	2022	2023	Explanation
Encryption adoption rate	97.3%	98%	98.2%	Percentage of institutions using advanced encryption
Anonymization success rate	95%	96%	96.8%	Effectiveness of anonymization techniques
Unauthorized access Incidents	6	5	5	Number of unauthorized access attempts in the past year
Regulatory compliance rate	99%	99.3%	99.4%	Percentage of institutions compliant with HIPAA/GDPR
Patient data breach reduction	81%	87%	90%	Reduction in patient data breaches over 5 years

4. Ethical considerations

The deployment of semantic computing in healthcare must also address ethical concerns, such as informed consent, data ownership, and the potential for algorithmic bias. Ensuring that patients are fully informed about how their data will be used and obtaining explicit consent is crucial. Additionally, algorithms must be designed and tested to minimize biases that could lead to disparities in healthcare outcomes.

4.1 Implementing robust encryption techniques

To safeguard sensitive healthcare data, robust encryption methods must be employed. End-to-end encryption ensures that data is protected both at rest and in transit. Advanced encryption standards (AES) and homomorphic encryption, which allows computations on encrypted data, are vital for maintaining data confidentiality without compromising the functionality of semantic computing systems.

4.2 Enhancing data governance and access controls

Effective data governance frameworks are essential for managing access to healthcare data. Role-based access control (RBAC) and attribute-based access control (ABAC) mechanisms can ensure that only authorized personnel can access sensitive information. Auditing and monitoring systems can track data access and detect unauthorized activities, enhancing overall data security.

4.3 Developing privacy-preserving algorithms

Privacy-preserving algorithms, such as differential privacy, add controlled noise to datasets, making it difficult to re-identify individuals while maintaining the overall utility of the data. These techniques are crucial for balancing the need for data analysis with privacy concerns. Additionally, federated learning, where models are trained locally on data residing in multiple institutions, can enable collaborative analysis without sharing raw data.

4.4 Strengthening regulatory compliance

Healthcare organizations must invest in compliance frameworks that align with regulatory requirements. Regular audits, risk assessments, and staff training programs are essential components of a robust compliance strategy. Collaboration with regulatory bodies to stay updated on evolving regulations and implementing best practices is also critical.

4.5 Promoting ethical AI practices

Ensuring that semantic computing systems operate ethically requires addressing issues of transparency, fairness, and accountability. Developing explainable AI models that provide insights into how decisions are made can enhance transparency. Regular bias assessments and the inclusion of diverse datasets in training models can mitigate the risk of biased outcomes. Establishing accountability frameworks that define the responsibilities of stakeholders involved in the development and deployment of these systems is also necessary (Table 15.3).

Table 15.3 Ethical considerations in semantic computing.

Ethical concern	Description	Numerical data example	Recommended practices
Informed consent	Ensuring patients are fully aware of data usage	100% consent for patient records	Transparent consent processes, detailed information sheets
Data ownership	Clarifying patient data ownership	Ownership policies for data sources	Clear data ownership policies, patient rights education
Algorithmic bias	Preventing biases in AI algorithms	Bias assessments on recommendations	Bias assessments, diverse training datasets
Accountability	Establishing responsibility for AI decisions	Accountability for personalized treatments	Accountability frameworks, explainable AI models

5. Case studies and applications

5.1 Case study 1: IBM Watson Health

IBM Watson Health leverages semantic computing to analyze vast datasets, including medical literature and patient records, to provide evidence-based diagnostic and treatment options. By integrating AI with human expertise, Watson Health enhances clinical decision-making and improves patient outcomes. Robust encryption and strict access controls ensure data privacy and security, aligning with regulatory requirements.

5.2 Case study 2: Mayo Clinic's clinical data analytics platform

The Mayo Clinic utilizes a clinical data analytics platform powered by semantic computing to enhance patient care and research. This platform integrates data from diverse sources, enabling real-time analysis and personalized treatment plans. Privacy-preserving techniques and compliance with HIPAA regulations are integral to the platform's operation, ensuring patient data is secure and used ethically.

5.3 Application: personalized cancer treatment

Semantic computing enables the development of personalized cancer treatment plans by analyzing genetic data, clinical trials, and patient histories. AI algorithms identify the most effective treatments for individual patients, improving outcomes and reducing side effects. Secure data sharing between research institutions and healthcare providers facilitates collaboration while maintaining patient privacy (Table 15.4).

Table 15.4 Statistical data for case studies.

Metric	IBM Watson health	Mayo clinic clinical data platform	Personalized cancer treatment
Diagnostic accuracy improvement	85%–95%	–	60%–75%
Patient cases analyzed	200,000	–	15,000
Treatment recommendations	100,000 personalized plans	–	–
Patient Readmission rate reduction	–	20%–17%	–
Data sources integrated	–	600,000 EHRs, 75,000 imaging reports	–
Real-time analysis impact	–	Improved emergency response times by 30%	–
Patients treated	–	–	15,000
Side effects reduction	–	–	40%–30%

6. Conclusion

Semantic computing holds immense potential to revolutionize healthcare by enhancing data interoperability, improving diagnostic accuracy, and personalizing treatment plans. However, the integration of these innovative technologies must be balanced with robust privacy and security measures to protect sensitive patient data. By implementing advanced encryption techniques, developing privacy-preserving algorithms, and promoting ethical AI practices, the healthcare industry can harness the power of semantic computing while safeguarding patient privacy and maintaining regulatory compliance. The future of healthcare lies in the successful convergence of technological innovation and stringent privacy security measures, ensuring that the benefits of semantic computing are realized without compromising patient trust and data integrity.

Conflicts of Interest

"The authors declare no conflict of interest." Authors must identify and declare any personal circumstances or interest that may be perceived as inappropriately influencing the representation or interpretation of reported research results. Any role of the funders in the design of the study; in the collection, analyses or interpretation of data; in the writing of the manuscript, or in the decision to publish the results must be declared in this section. If there is no role, please state "The funders had no role in the design of the study; in the collection, analyses, or interpretation of data; in the writing of the manuscript, or in the decision to publish the results".

References

Agarwal, N., Hipona, J., 2023. Strength of distributed web page management system within the space of online e-learning administration in a contextualized group of science teachers. Eur. Chem. Bull. 12 (4), 144–155. https://doi.org/10.31838/ecb/2023.12.4.010.

Ahmadi, M., Pourazimi, M., 2022. An analytical study of the behavioral public administration development. Globus Int. J. Manag. IT 13 (2). https://doi.org/10.46360/globus.mgt.120221001.

Ali, I., Khan, U., Rahman, F., Gilani, S.A., 2023. Assessment of patient satisfaction with ophthalmic care services in private and trust hospitals. Globus Int. J. Med. Sci. Eng. Technol. 12 (1), 54–60. https://doi.org/10.46360/globus.met.320231006.

Baquiran, C.A.C., Bucao, R.B.I., Bulaong, G.J.S.A., Flores, A.E., Joaquin, M.C.B., 2023. A case study on doctor-shaming in social media and its effects on the perception of the Philippines' healthcare system among the residents of the national capital region. Globus Int. J. Manag. 14 (2), 56–62. https://doi.org/10.46360/globus.mgt.120231007.

Barbas, C.A., Tingson, S., Ma, G.C., Medina, J.R., Maturan, A.G., Japin, J.M.R., 2023. A study on the level of awareness and extent of practices on the Augustinian core values of the LCC-BAIS faculty. Cosmos Int. J. Manag. 12 (2), 93–104. https://doi.org/10.46360/cosmos.mgt.420231011.

Chiong, J.W., Catipunan-Francisco, J.C., 2023. Life dynamics as a behavioral program for student leaders of jose rizal university. Cosmos Int. J. Art Higher Educ. 12 (1), 196–204. https://doi.org/10.46360/cosmos.ahe.520231017.

Duayan, O.M., 2022. Impact assessment of the community extension programs of colleges in Bacolod city: basis for enhancement program. Globus J. Progressive Educ. 12 (2), 29–33. https://doi.org/10.46360/globus.edu.220222006.

Dwivedi, A., 2022. Development of educational model for e-learning. Globus J. Progr. Educ. 12 (2), 95–98. https://doi.org/10.46360/globus.edu.220222015.

Dwork, C., 2008. Differential privacy: a survey of results. In: International Conference on Theory and Applications of Models of Computation. Springer, Berlin, Germany, pp. 1–19. https://doi.org/10.1007/978-3-540-79228-4_1.

El Emam, K., Jonker, E., Arbuckle, L., Malin, B., 2011. A systematic review of re-identification attacks on health data. PLoS One 6 (12), e28071. https://doi.org/10.1371/journal.pone.0028071.

Gangal, R., Johri, P., Kumar, A., 2023. A systematic review of artificial intelligence techniques in lung cancer detection and accurate diagnosis. Cosmos J. Eng. Technol. 13 (1), 51–56. https://doi.org/10.46360/cosmos.et.620231005.

Garfinkel, S., 2015. De-identification of Personal Information. US Department of Commerce, National Institute of Standards and Technology, pp. 01–46. https://doi.org/10.6028/NIST.IR.8053.

Guzman, De, Delos, C.F., Reyes, J.C.E., Gutierrez, E.M.M., Punongbayan, A.D., Sadol, B.D.M., 2023. The healthy lifestyle of adolescents living with their single parents. Cosmos Int. J. Art Higher Educ. 12 (1), 157–166. https://doi.org/10.46360/cosmos.ahe.520231012.

Malin, B., Sweeney, L., 2004. How (not) to protect genomic data privacy in a distributed network: using trail re-identification to evaluate and design anonymity protection systems. J. Biomed. Inf. 37 (3), 179–192. https://doi.org/10.1016/j.jbi.2004.04.005.

Narayanan, A., Shmatikov, V., 2008. Robust de-anonymization of large sparse datasets. In: IEEE Symposium on Security and Privacy. IEEE, Oakland, CA, USA, pp. 111–125. https://doi.org/10.1109/SP.2008.33.

Ohm, P., 2009. Broken promises of privacy: responding to the surprising failure of anonymization. UCLA Law Rev. 57, 1701–1777.

Rocher, L., Hendrickx, J.M., De Montjoye, Y.A., 2019. Estimating the success of re-identifications in incomplete datasets using generative models. Nat. Commun. 10 (1), 3069. https://doi.org/10.1038/s41467-019-10933-3.

Sweeney, L., 2002. k-anonymity: a model for protecting privacy. Int. J. Uncertain. Fuzziness Knowledge-Based Syst. 10 (05), 557–570. https://doi.org/10.1142/S0218488502001648.

Torres, E., Angeles, R., Gaurano, F.E., Soncuan, I.J., Yumang, P., 2023. Mothering, modernity, and medicine: a qualitative study on the health-seeking behaviors of first-time mothers outside pediatric care in Bulacan, Philippines. Globus Int. J. Med. Sci. Eng. Technol. 12 (1), 16–23. https://doi.org/10.46360/globus.met.320231003.

Valdez, J.R., Nunag, R., 2023. The research and creative works colloquium and its effect on research performance. Cosmos J. Eng. Technol. 13 (1), 57–69. https://doi.org/10.46360/cosmos.et.620231006.

Further reading

Bernard, R.M., Borokhovski, E., Schmid, R.F., Tamim, R.M., Abrami, P.C., 2014. A meta-analysis of blended learning and technology use in higher education: from the general to the applied. J. Comput. High Educ. 26 (1), 87–122.

Chen, W., Denham, J., 2021. Examining the effects of gamified physical education programs on student motivation. J. Phys. Educ. Sport 21 (3), 122–134.

Chen, J., Liu, H., 2022. Bridging the urban-rural divide in public infrastructure: a case study of rural China. J. Rural Stud. 88, 145–156.

Ellis, R., Sun, H., 2022. Innovations in digital learning for PE. Educ. Today 15 (2), 56–72.

Garn, A.C., Sun, H., 2021. Digital fitness in schools: opportunities and challenges. J. Health Phys. Act. 18 (3), 65–79.

Garnett, P., Arnold, S., 2020. Digital tools and sustainability in supply chain logistics. Int. J. Sustain. Transp. 45 (2), 312–326.

Liu, Y., Zhou, Z., 2021. Elderly behavior and rural public spaces. Plan. Horizons 22 (4), 190–205.

Nayudori, B., Kummari, R., Kunareddy, V., Vunnam, N.K., 2023. Securing data communication using blockchain. Cosmos J. Eng. Technol. 13 (1), 95–97. https://doi.org/10.46360/cosmos.et.620231009.

Waller, M.A., Fawcett, S.E., 2013. Data science, predictive analytics, and big data: a revolution that will transform supply chain design and management. J. Bus. Logist. 34 (2), 77–84.

Wang, Y., Liu, H., 2019. Big data analytics in resilient supply chain design. Comput. Ind. Eng. 139, 106190.

Warschauer, M., Zhang, R., 2021. Technology and sports education: Global perspectives. Sports Educ. Quart. 29 (3), 98–116.

Yadav, L., 2023. Dynamics of online shopping focus on consumer behaviour and satisfaction. Cosmos. Int. J. Manag. 12 (2), 127–132. https://doi.org/10.46360/cosmos.mgt.420231014.

Zaid, M.F., Mohammed, N.Z., 2024. Vertical control network adjustment using different least square methods. Globus: Int. J. Med. Sci. Eng. Technol. 13 (2), 15–20. https://doi.org/10.46360/globus.met.320242003.

Zhang, L., Chen, W., 2020. Addressing urban-rural disparities in technology-supported education. J. Rural Develop. Stud. 19 (4), 78–92.

Zhang, Y., Li, X., 2021. Enhancing supply chain sustainability through predictive analytics. J. Clean. Log. 10 (4), 120–135.

CHAPTER 16

Securing cyber physical systems in healthcare using blockchain

U. Padmavathi[1] and Rohit Kumar[2]
[1]Shiv Nadar University, Chennai, Tamil Nadu, India; [2]ABV-IIITM, Gwalior, Madhya Pradesh, India

1. Introduction

In the pursuit of advancing healthcare systems, ensuring the trustworthiness and resilience of infrastructure has become paramount. The integrity of healthcare data, the security of patient information, and the seamless operation of medical services hinge on robust infrastructural foundations (Shah et al., 2021). With the increasing digitization of healthcare processes, the vulnerabilities to cyber threats have also escalated, necessitating proactive measures to safeguard against potential breaches. In this context, blockchain technology emerges as a promising solution, offering decentralized and immutable ledgers that can bolster the security and transparency of healthcare data management (Gad et al., 2022). By exploring the potential applications of blockchain in cyber-physical systems for healthcare, we aim to fortify the reliability and integrity of healthcare infrastructures, thereby fostering a safer and more efficient healthcare ecosystem.

1.1 Overview of the importance of trustworthy and resilient healthcare infrastructures

In an era where digital transformation is revolutionizing every sector, healthcare stands out as a critical domain where the stakes are exceptionally high. Trustworthy and resilient healthcare infrastructures are paramount due to the sensitive nature of healthcare data and the life-or-death implications of any system failure (Shimonski, 2020). The integrity, security, and availability of healthcare infrastructure directly impact patient safety, data privacy, and the efficiency of healthcare delivery systems. As healthcare systems increasingly digitize patient records, treatment plans, and medical devices, the need for robust infrastructures that can withstand cyber threats becomes more pressing (Yaacoub et al., 2020). Cyberattacks targeting healthcare organizations have become more frequent and sophisticated, posing significant risks to patient care, financial stability, and public trust. Moreover, the rise of interconnected medical devices and systems within hospitals and clinics introduces new vulnerabilities that could compromise patient safety if not adequately protected (Farringer, 2019). Ensuring the trustworthiness and resilience of healthcare infrastructures is not merely a matter of compliance or

Intelligent Semantic Analysis for Healthcare
ISBN 978-0-443-33512-9,
https://doi.org/10.1016/B978-0-443-33512-9.01016-7

technological advancement but a fundamental requirement for upholding patient rights, maintaining public confidence, and safeguarding the integrity of healthcare services.

1.2 Introduction to blockchain technology and its potential applications in healthcare

Blockchain technology, initially popularized as the underlying technology behind cryptocurrencies like Bitcoin, has emerged as a powerful tool with far-reaching applications beyond finance (Ølnes, 2016). At its core, a blockchain is a decentralized and immutable ledger that records transactions across a network of computers in a secure and transparent manner. Each transaction, or block, is cryptographically linked to the previous one, forming a chain that is resistant to tampering and fraud (Komalavalli et al., 2020). In healthcare, where data integrity, security, and interoperability are paramount, blockchain technology holds immense promise. By leveraging cryptographic principles and consensus mechanisms, blockchain can enable secure and auditable sharing of medical data among authorized stakeholders, streamline administrative processes, and enhance the traceability of pharmaceuticals and medical supply chains (Sharma et al., 2021). Moreover, blockchain based solutions offer innovative approaches to addressing long standing challenges in healthcare, such as patient identity management, interoperability between disparate systems, and ensuring the authenticity of medical records (Onik et al., 2019). Through decentralized networks and smart contracts, blockchain technology has the potential to empower patients with greater control over their health data while facilitating seamless collaboration among healthcare providers and researchers.

1.3 Purpose of leveraging blockchain in cyber-physical systems for healthcare

The convergence of cyber and physical systems in healthcare, often referred to as cyber-physical systems (CPS), presents unique challenges and opportunities for improving patient care and operational efficiency (Shaikh et al., 2023). CPS in healthcare encompasses a wide range of interconnected devices, sensors, and software systems that monitor, analyze, and control physical processes in real-time (Krishna et al., 2019). However, the integration of CPS introduces inherent vulnerabilities, including the risk of cyber-attacks, data breaches, and system failures that could compromise patient safety and disrupt healthcare operations. Traditional centralized approaches to securing CPS may not be sufficient in addressing these challenges, necessitating innovative solutions that can provide robust security, data integrity, and resilience (Kayan et al., 2022). By integrating blockchain technology into cyber-physical systems for healthcare, organizations can enhance the security, transparency, and trustworthiness of critical infrastructure components. Blockchain's decentralized architecture and cryptographic techniques offer a resilient framework for securing sensitive data and ensuring the integrity of transactions within CPS networks (Ali et al., 2023). Furthermore, blockchain-enabled CPS can facilitate secure data sharing and interoperability among disparate devices and systems while mitigating the risk of unauthorized access or tampering. Through immutable audit trails and transparent consensus mechanisms, blockchain enhances accountability and fosters trust among stakeholders, ultimately contributing to safer and more reliable healthcare delivery (Jennath et al.). In summary, the purpose of leveraging blockchain in cyber-physical systems for healthcare is to enhance security, resilience, and trust in the infrastructure underpinning patient care, medical devices, and healthcare operations. Through decentralized networks and cryptographic protocols, blockchain technology offers innovative solutions to address the evolving threats and challenges facing modern healthcare systems.

2. Fundamentals of blockchain technology

2.1 Definition and characteristics of blockchain

The term Blockchain is defined as a chain of blocks which are cryptographically linked together with the help of hash values. The idea of cryptographically chained blocks was first proposed by Stuart Haber and Stornetta in 1991 for timestamping digital documents in order to prevent forward or backdating of records (Bhutta et al., 2021). In 1992, along with Dave Bayer, Stuart and Stornetta added the concept of Merkle trees to these time-stamped digital documents (Bayer et al., 1993). These ideas gained popularity when they were first proposed as blockchain, an underlying technology in Bitcoin, by Satoshi Nakamoto in 2009 (Nakamoto, 2008). In Bitcoin, blockchain was used to solve the double spending problem of digital cash without the use of central authorities. Blockchain is a decentralized, distributed ledger technology used to record transactions without intermediaries. This pioneering technology was initially used in cryptocurrencies and later found its application in wide areas including healthcare, education, charity, finance, governance, supply chain, agriculture, smart cities, and so on. The structure of a blockchain consists of a block header and a block body, as depicted in Fig. 16.1. The block header contains metadata about the block. It includes timestamp, previous hash, merkle root, difficulty, and version. The block body includes the transactions in the network.

2.1.1 Characteristics of blockchain

The key characteristics of blockchain technology make it distinguishable from other database systems and give it the potential to disrupt various industries. The key characteristics of blockchain include (Viriyasitavat and Hoonsopon, 2019; Yassein et al., 2019; Shah et al., 2023; Stafford and Treiblmaier, 2020):

- Decentralization—The term decentralization in blockchain refers to the process of moving authority and decision making from a centralized entity—a person, an organization, or a collection of people to a distributed network. Decentralized

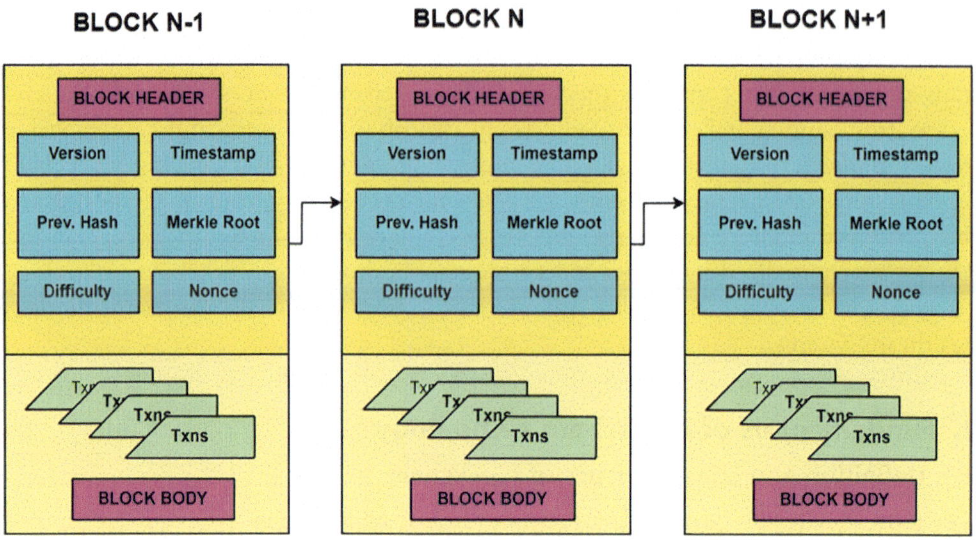

Figure 16.1 General structure of blockchain.

networks aim to minimize the degree of confidence that users must have in one another and prevent users from abusing their power or control over one another in ways that impair the network's performance. It makes the system considerably more fair and secure by eliminating the central authorities.

- Transparency—It is one of the most important characteristics of blockchain. It means that anyone can view the specifics and history of any transactions in the network. This gives blockchain a high degree of accountability and integrity and guarantees that nothing has been fraudulently added, subtly changed, or removed.
- Security—Blockchain provides security of data using cryptographic algorithms that offers mathematical guarantee that the data stored in the blockchain cannot be breached or altered making it tamper-proof.
- Immutability—Blockchain assures immutability by building a trackable system where it is highly difficult to change or alter data without data collusion. As the blockchain is an append-only data structure, any data stored on it cannot be altered or modified. This is made possible with the help of a cryptographic hashing algorithm. Every block in the blockchain contains a header field called Merkle root which is the hash value of all the transactions of that block. This merkle root value is stored as the previous hash in the block that follows this block. For example, the block $n+1$ will have the previous hash value as the merkle root value of the nth block.
- Integrity—The guarantee that data kept on a blockchain is reliable and tamper-proof is known as blockchain integrity. Blockchain's decentralized and unchangeable

structure allows for integrity. Once information is stored on the blockchain, it cannot be removed or changed without the approval of most network users. Due to the immutable nature of blockchain, data stored on it is protected against fraud and manipulation and maintains its integrity.

- Distributed—The term "distributed" describes how blockchain stores and manages data instead of keeping it in one single location, but rather across a network of computers, or nodes. Each node in a blockchain network keeps a copy of the complete ledger, and consensus mechanisms propagate and validate new transactions throughout the network. The benefits of the distributed architecture of blockchain include enhanced security, reduced costs, increased efficiency, improved transparency, and resilience to failure.
- Autonomy—In the context of blockchain, autonomy is the ability of network users to interact and do business with one another without the need for middlemen or centralized authority. The decentralized nature of blockchain technology is the source of this autonomy. It helps to promote trust, transparency and innovation in the blockchain system.

2.2 Types and versions of blockchain technology

2.2.1 Types of blockchain

- Public Blockchain—As the name suggests, anyone can join the public blockchain at any time and view the transactions and its history. There is no central authority for controlling the network (Ramadoss, 2022). The examples include Bitcoin and Ethereum
- Private Blockchain—This type of blockchain is restricted to organizations which can control who can join the network and what type of information can be accessed by the members of the network (Jo et al., 2020). The examples include Hyperledger Fabric and Corda.
- Federated or Consortium—Blockchain networks that function under a semi-decentralized architecture, wherein numerous organizations or entities cooperate to maintain the network, are known as federated or consortium blockchains. Federated blockchains limit membership to a preset group of participants, usually organizations or institutions with a shared interest or aim, in contrast to public blockchains, which are open and permissionless (Wieninger et al., 2019; Dib et al., 2018). The examples include r3 and EWF.

2.2.2 Versions of blockchain

- Blockchain 1.0—Bitcoin gave rise to the blockchain concept. Blockchain 1.0 is the use of distributed ledger technology to validate transactions and move valuables across cryptocurrency platforms. A decentralized, peer-to-peer digital money coined Bitcoin is the product of 40 years of cryptography research and 20 years of cryptocurrency development. A enduring fix for the problems with digital currency is a fundamental advance in computer science. Users can receive money in their digital wallets instantly with bitcoin, eliminating the need to wait days for payments. Blockchain-based currencies guarantee that every transaction may be found and carried out directly, eliminating the need for a trusted third party (Mukherjee and Pradhan, 2021; Aggarwal and Kumar, 2021).

- Blockchain 2.0—To address the problems associated with the bitcoin concept, such as inefficient mining and limited network scalability, a visionary by the name of Vitalik Buterin goes beyond money and suggests a platform known as "smart contracts," on which the developer community can create distributed applications (DApps) for the blockchain network. Small computer programmes called "smart contracts" are installed on blockchains and are typically activated by certain situations. Because these smart contracts can do tasks automatically, managing time-consuming and expensive manual business processes is no longer necessary. The Ethereum blockchain is the most well-known example. Smart contracts cannot be hacked or manipulated. It lowers the cost of execution and verification and enables transparent definition, hence resolving issues with moral hazard. Compared to version 1.0, this blockchain version can handle a higher volume of transactions with success (Virmani et al., 2019).
- Blockchain 3.0—The goals of many emerging technologies, like NEO, IOTA, and EOS, are to enhance the capabilities of blockchains 1.0 and 2.0. This results in the creation of blockchain 3.0, which aims to fix the problems with the earlier iterations. Blockchain 3.0 uses several protocols, methods, and frameworks for this. Instead, in place of using centralized architecture, it makes use of decentralized communication and storage. This version of blockchain uses the Tangle concept to overcome the scalability issue that remained from the first to the second generation. Blockchain 3.0 differs from earlier iterations in several ways, including high scalability, interoperability, adaptability, and sustainability. The primary goal of Blockchain 3.0 is to incorporate blockchain technology with Internet of Things devices. IOTA, also referred to as the "blockless chain" was created for the IoT ecosystem's high standards (Maesa and Mori, 2020).
- Blockchain 4.0—Apart from concerns related to safety and scalability, blockchain 3.0 is devoid of business-level infrastructure. Therefore, it couldn't be applied to actual business. Furthermore, because everything must be done from scratch and the base-layer ledger lacks a service-oriented framework, DApp developers must put in a lot of work in both planning and developing. The next version of blockchain is built with the intention of resolving the issues raised by blockchain 3.0. Companies are being encouraged to utilize Blockchain 4.0, a new technology that offers reduced costs and shorter development cycles. Blockchain 4.0 will be the most

business accommodating public chain in the history of blockchain innovation, capable of executing a wide range of use cases and applications (Alladi et al., 2019).

2.3 Overview of blockchain consensus algorithms and cryptographic techniques

The key components involved in the blockchain are cryptographic algorithms such as asymmetric key cryptography for the generation of public, private key pairs, hashing algorithm for linking of one block to the other and to maintain immutability and security, smart contracts for the automatic execution of applications, consensus algorithms for validating the transactions and adding it to the block. The type of component used in a blockchain depends on the requirement and specifications of the application using the blockchain technology.

2.3.1 Cryptographic algorithms

Blockchain makes use of asymmetric key cryptographic algorithms for the generation of pair of keys to make secure communication between the nodes in the network. Every node in the network uses two keys, a public key and a private key. The asymmetric cryptography algorithm used in bitcoin is ECDSA which makes use of ECC for the generation of key pairs. A random number is generated as the private key, and it is used by ECC to generate a public key. This public key is then hashed into a public key hash which generates the public address used by bitcoin nodes in the network. The nodes in the network make use of this public address for receiving bitcoin from the sender. The sender uses the SHA 256 algorithm for hashing the message and uses the private key for signing the transactions. These signed transactions are then sent to the mempool of the network. The miner nodes in the network collect these transactions from the mempool, validate them and add them as blocks into the network. The miner nodes make use of consensus algorithms for reaching an agreement on the transactions broadcasted in the network (Huang et al., 2021).

2.3.2 Consensus algorithms

The term consensus refers to the act of coming to an agreement on a certain course of action or decision among a group of individuals or entities in a network. It is a technique used in blockchains to ensure that all nodes concur on the network's present state and the legitimacy of transactions (Mingxiao et al., 2017; Hussein et al., 2023). It also helps to maintain the security and integrity of the blockchain. Various blockchain platforms employ distinct algorithms, including proof of work, proof of stake, or proof of authority, to facilitate consensus-building across network nodes. The consensus algorithms should possess three properties such as safety, liveness and fault tolerance. Blockchain consensus can be Non-Byzantine, Byzantine, DAG-based or Hybrid (Islam et al., 2023). Various consensus algorithms are shown in Fig. 16.2. Few significant consensus algorithms are discussed below:

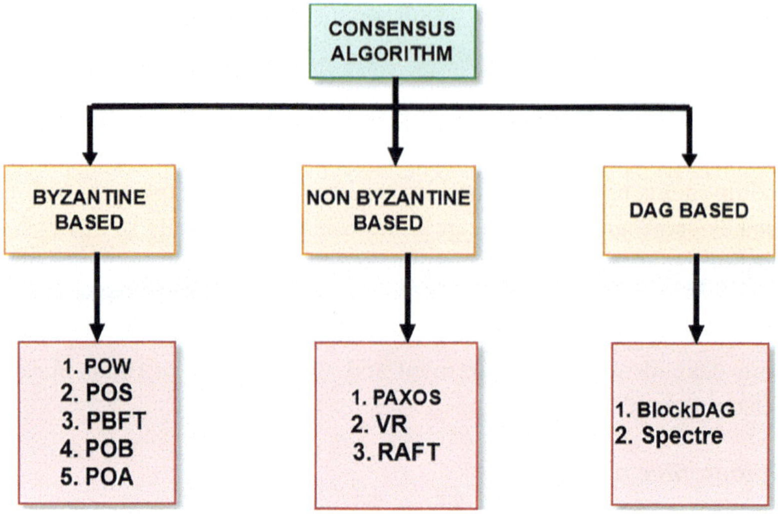

Figure 16.2 Different blockchain consensus algorithms.

- Byzantine Consensus Algorithm: A byzantine consensus algorithm is the one which is used to achieve consensus even in the presence of byzantine faults. This algorithm employs a collective decision-making process to minimize or prevent the impact of faulty nodes in the system (Zhong et al., 2023).
  - Proof of Work (PoW): The process of appending a new block to the blockchain is done by miners through the process of mining. Miners compete to find a solution in a PoW consensus algorithm to a challenging mathematical difficulty. A new block of transactions can be created and added to the blockchain, by the miner who first solves the puzzle. The other miners validate the block, and if it is correct, the new block is appended to their version of the blockchain. PoW consensus mechanism involves huge computing power and sky-rocketing fees. Miners are rewarded with a certain number of cryptocurrencies for solving the difficult puzzle. This reward promotes miner involvement in the entire process and strengthens the blockchain's security.
  - Proof of Stake (PoS): PoS is a kind of consensus mechanism where the miner's stake plays a pivotal role in determining the next block rather than the processing power of the miners as in the case of PoW. This approach to distributed consensus may be more energy efficient. By stacking a particular amount of cryptocurrency, nodes on a network can qualify as candidates to validate new blocks in a PoS consensus mechanism. One of the candidates is then chosen by an algorithm to validate the new block and get the transaction fee. To guarantee equity among all nodes on the network, the selection process combines the candidate's stake with additional variables like coin age and randomness. Coin age is one such

indicator that measures the duration of a candidate node's validity period. A node's chances of getting chosen as the new validator increase with the length of time it has served as a validator. Random block selection is another element. In this method, the validator is selected by combining the highest stake with the lowest hash value. The new validator is the node that has the best weighted combination of these factors.

- Proof of Burn (PoB): In PoB, the validators burn their coins or transfer them to an address from which they can never be recovered to show their dedication to the system. To improve their chances of being chosen for block mining, validators might burn the native currency of the blockchain application or the currency of a rival chain, like bitcoin. PoB enables validators to demonstrate their long-term commitment to the system through a temporary currency sacrifice, without having to spend on expensive hardware. A miner's chances of getting chosen to mine the system's subsequent block increase with the amount of cash they burn. The reason behind this is that the miner is demonstrating a long-term commitment to the system and forgoing a short-term benefit in exchange for a possible long-term profit by destroying their currency.
- Proof of Authority (PoA): In the Proof of Authority Consensus mechanism, a small number of verified authorities or validators are responsible for adding new blocks to the blockchain. Validators are predefined entities with known identities who are usually motivated by their established reputation inside the network rather than token compensation. Instead of depending on ownership of tokens or computational puzzles for block validation, as PoW or PoS do, PoA uses validators' identities and reputations to guarantee network integrity.

• Non-Byzantine Consensus Algorithm: One type of system failure that can happen in distributed systems without the involvement of malevolent nodes is called a non-Byzantine error. This may involve problems like node reporting faults and machine downtime. Although no Byzantine fault tolerant algorithms are built to withstand these kinds of failures, the presence of malevolent nodes compromises data security and system stability. Because of this, non-Byzantine fault-tolerant algorithms are usually limited to confined contexts like consortium chains or private chains that have a high degree of node trust. High performance and robust tolerance of non-Byzantine faults are features of these methods.
  - Paxos Algorithm: ampere consensus mechanism called the Paxos algorithm was introduced in 1998 by the computer scientist Leslie Lamport. It helps distributed network nodes come to an understanding over a suggested value and has been utilized in numerous distributed systems since 1998. Paxos does not rely on a central authority to manage the consensus process, in contrast to other consensus algorithms. Rather, it makes use of messages sent back and forth among nodes to come to an agreement on a suggested value. A proposer node makes a value

recommendation for a particular data item. A group of acceptor nodes receive this value by broadcast. The proposed value is received and assessed by the acceptor nodes. They notify the proposer that the value has been approved if they concur that the suggested value is accurate. A group of learner nodes receives a message from the proposer signifying that the suggested value has been accepted by many acceptors if it is approved by the most of acceptor nodes. The learner nodes change their local copies of the data item with the acceptable value as soon as they receive it.

- VR Consensus Algorithm: ampere distributed consensus technique called view stamped replication (VR) enables network nodes to concur on the sequence and integrity of transactions in a distributed database. Every node keeps track of transactions, and an indicator of its present state called a "view" number. A designated primary node receives requests from nodes to propose new transactions, and it is this node's job to arrange the transactions and disseminate them to the other nodes. The other nodes can start a change of view procedure to choose a new primary node that will carry on ordering and broadcasting transactions if the primary node malfunctions or is unavailable. To guarantee consistency, the new primary node will be chosen using the log and view numbers from the previous primary node. VR is a good fit for distributed systems like blockchain networks that need excellent consistency and fault tolerance since it can withstand node failures and network splits without compromising security or consistency.
- Raft Consensus Algorithm: It is a distributed consensus algorithm that helps to resolve the issue of getting several servers to concur on a common state even when there are errors. Typically, a replicated log supports a data structure that represents the shared status. Raft operates by choosing a cluster leader. The leader oversees the replication of the log to additional servers and responds to client queries. Data only moves from the leader to the other servers in one direction. In RAFT, consensus is reached by the leader proposing a new value or state for the system and disseminating it among the followers. A set of rules determines whether the followers approve or reject the proposition. It is deemed "committed" if the plan is approved by most followers. Subsequently, the leader changes its own log of committed values and replicates the value to the followers. The followers can start a leader election process to choose a new leader if the current one becomes unavailable. The node with the most recent log is picked as the new leader, and the selection process is based on the log of committed values. Strong consistency, which guarantees that all nodes in the system will eventually agree on the same value or state, is one of RAFT's primary features. Additionally, it offers fault tolerance because the system can keep running even if a few nodes malfunction or go down.

- DAG-based consensus algorithm
  Directed acyclic graphs, or DAGs, are the foundation of a distributed ledger technology. These techniques represent transactions as nodes in the DAG, with the interdependence between the transactions shown by the edges connecting the nodes. The ability of DAG-based consensus algorithms to execute transactions more quickly and flexibly than traditional blockchain based consensus algorithms is one of their key features.
  - Block DAG: BlockDAG is a DAG-based consensus mechanism that reconstructs a single-chain blockchain system by sorting and merging. Block Generation (BG), Sorting Blocks (SB), Block Merges (BM), Consensus Implementation (CI), and Block Splits (BS) are its five phases. BG creates original blocks in the first phase by adding transactions to the closest blockchain node's block pool. This node then verifies the transactions and bundles them into blocks that are added to the system block pool for additional verification. Using a sorting algorithm, SB creates a sequence in the DAG structure in the second phase by sorting all unvalidated blocks in the block pool. In the BM phase, BlockDAG resolves disputes arising from double spending and consensus. In the CI phase, global consensus is used to finalize the merged blocks. The combined blocks are divided into their separate states and added to the on-premises BlockDAG structure during the BS phase.
  - SPECTER: SPECTER is a protocol designed for the consensus core of cryptocurrencies, providing fast confirmation times and high throughput. It makes advantage of partial synchronous networks and permits high block creation speeds. By keeping a complete DAG of blocks, SPECTER generalizes Nakamoto's blockchain into a block DAG, enabling miners to produce blocks simultaneously. The order between every pair of blocks in the DAG is decided by a voting process, with blocks casting their votes rather than miners. Based on the block's position within the DAG, an algorithm is used to decide the votes. The majority vote quickly becomes irreversible, resulting in a consistent set of transactions.

3. Cyber-physical systems in healthcare

This section offers valuable insights into CPS in healthcare, focusing on characteristics, technologies, challenges, security limitations, and future trends. Here is a detailed summary based on the information from the existing literature:

3.1 Definition and characteristics of CPS in healthcare (Verma, 2022)

CPS in healthcare integrates physical and cyber worlds for efficient medical devices and equipment, enabling continuous communication and information exchange of physiological data. Technologies like Internet of Things (IoT), Machine Learning, and Artificial Intelligence play a crucial role in the development of Smart Healthcare Systems. CPS in healthcare are interdisciplinary systems that improve the quality of medical care by providing efficient and smart services. Autonomy levels in CPS are determined by factors like medical device type, interaction type, and duration with the patient.

3.2 Importance of CPS in enhancing healthcare infrastructure and services (Sheikh et al., 2022)

CPS in healthcare is crucial for supporting the medical fraternity in providing effective and efficient healthcare services. These systems enhance patient monitoring, treatment, and supply of necessary medicines through predictive mechanisms. CPS improves the quality of medical care by providing efficient and smart services, contributing to the growth of the healthcare sector.

3.3 Challenges and vulnerabilities in existing CPS-based healthcare systems (Sheikh et al., 2022)

CPS systems are prone to various cyber and physical security threats due to their heterogeneous nature and reliance on sensitive data. Threats like spoofing, sabotage, service disruption, and tracking pose serious risks to CPS systems in healthcare. Vulnerabilities in CPS systems are categorized into network vulnerabilities, platform vulnerabilities, and management vulnerabilities, requiring robust security measures The existing literary works and discussion collectively highlight the significance of CPS in healthcare, the technologies driving their development, and the critical challenges and vulnerabilities that need to be addressed to ensure the secure and efficient operation of CPS-based healthcare systems.

4. Leveraging blockchain for trust and resilience in healthcare

4.1 Role of blockchain in enhancing trust and transparency in healthcare data management (Uddin et al., 2020)

Blockchain technology, with its fundamental characteristics of immutability, security, and decentralization, is essential to improving trust and transparency in healthcare data management. Blockchain guarantees data integrity by generating an immutable ledger of medical records, making it nearly impossible to change or tamper with patient data. This promotes increased trust between medical professionals and patients. Additionally, because blockchain is decentralized, there is no need for a single centralized authority, which lowers the possibility of data breaches and illegal access. Smart contracts and auditable trails, which automate data access rights and guarantee that all transactions with health data are transparent and traceable, greatly increase transparency. This degree of security and openness facilitates data interchange between healthcare providers, improving patient control over personal health information and facilitating better organized and effective care delivery.

4.2 Utilizing blockchain for secure and immutable health data storage and sharing (Sonkamble et al., 2023)

The management and exchange of medical data is revolutionized using blockchain technology for safe, unchangeable health data storage and exchange. Health data is kept in a tamper-proof manner with blockchain's decentralized ledger architecture, where every data entry is permanently recorded and cannot be changed or removed Because of its immutability, patient records are guaranteed to be accurate and true, thereby lowering the possibility of fraud and data manipulation. Furthermore, by using cryptography to make sure that only authorized parties may access sensitive health information, blockchain improves security. Blockchain facilitates secure and frictionless data sharing between various healthcare systems and providers, hence enhancing interoperability and patient care efficiency. Patients have more control over their private health information since they can openly grant and remove access permissions. Patients and healthcare professionals can feel more trusted when using this secure, unchangeable, and patient-centered approach to health data management, which improves patient outcomes and protects patient privacy.

4.3 Enhancing resilience and fault tolerance in CPS-based healthcare infrastructures with blockchain (Kaur et al., 2021)

Using blockchain technology to improve fault tolerance and resilience in CPS-based healthcare infrastructures provides a solid answer to several operational problems. The decentralized nature of blockchain guarantees that system controls and health data are not dependent on a single point of failure, hence greatly enhancing the infrastructure's resilience. In a blockchain network, every node keeps a copy of the whole ledger, which enables the system to keep running smoothly even if some nodes malfunction or are compromised. In addition to improving fault tolerance, this redundancy guarantees the ongoing provision of essential healthcare services. Additionally, the unchangeable records in blockchain technology offer a trustworthy audit trail that is essential for promptly detecting and resolving any errors or irregularities. Healthcare organizations may increase operational continuity, data integrity, and security by combining blockchain with CPS, which will result in more dependable and trustworthy healthcare services.

5. Applications and use cases

CPS that uses blockchain technology might greatly enhance patient-centered care delivery, data security, and interoperability in the healthcare industry. Here are a few particular use cases (Gohar et al., 2022; Azzi, 2023):

- Secure and Interoperable Health Records: Blockchain technology has the potential to facilitate the safe and compatible sharing of electronic health records (EHRs) between patients, IoT devices, and healthcare professionals. Blockchain technology can be used to store patient records, including medical histories, diagnostic test results, prescriptions, and treatment plans. This ensures data integrity and privacy and allows for easy access and exchange amongst various healthcare systems and institutions.

- Medical Supply Chain Management: Through the distributed ledger technology, blockchain can improve the medical supply chain's transparency and traceability by documenting the lifetime of medications, medical equipment, and other healthcare supplies. The blockchain allows for the recording of every step in the supply chain, from production to distribution to patient use, as a transaction. This reduces the possibility of fake medications and medical mistakes by allowing stakeholders to track and authenticate products in real-time.
- Clinical Trials and Research Data Sharing: Blockchain can make it easier for researchers, pharmaceutical companies, and healthcare providers to securely and transparently share research findings and clinical trial data. To ensure compliance with legal regulations like GDPR (General Data Protection Regulation) and HIPAA (The Health Insurance Portability and Accountability Act), smart contracts can automate consent management and data sharing agreements. They can also facilitate effective collaboration and information exchange, which will hasten medical advancements and enhance patient outcomes.
- Healthcare IOT device security: Health-related IoT devices, like wearable sensors, medical implants, and remote monitoring devices, can benefit from increased security and trust because of blockchain technology. Stakeholders may guarantee data integrity, authentication, and tamper resistance by storing device data and interactions on a blockchain. This lowers the likelihood of cyberattacks, data breaches, and unauthorized access to private patient information.
- Patient Identity Management and Consent: In healthcare CPS, blockchain can offer a safe, decentralized platform for handling patient identification, consent choices, and access restrictions. Patients will have more control over their health information and treatment choices if they have a distinct digital identity that is kept on the blockchain. This identity can be used to authenticate transactions, give permission for data sharing, and enforce privacy preferences across various healthcare systems and devices.
- Healthcare Payments and Claims Processing: Blockchain enables real-time verification, settlement, and reconciliation of transactions between payers, providers, and patients, which can expedite healthcare payments and claims processing. In addition to lowering administrative burden, billing problems, and fraud, smart contracts can automate payment arrangements, eligibility checks, and claims adjudication. They can also guarantee prompt and accurate reimbursement for provided healthcare services.

6. Challenges and considerations

6.1 Technical challenges in implementing blockchain in CPS based healthcare systems (Maleh et al., 2022)

Integrating blockchain technology into CPS for healthcare presents several technical challenges due to the unique requirements and constraints of both domains. Here are some of the key challenges:

- Scalability: Healthcare systems generate vast amounts of data continuously. Blockchain networks must be able to scale to handle the high transaction throughput required by CPS-based healthcare systems without sacrificing decentralization or security.
- Latency: Real-time data processing is crucial in healthcare, especially in CPS environments where timely decisions can be lifesaving. Blockchain consensus mechanisms and transaction validation processes must be efficient enough to minimize latency while ensuring data integrity.
- Interoperability: CPS-based healthcare systems often involve diverse devices, platforms, and protocols. Blockchain solutions need to seamlessly integrate with existing infrastructure and ensure interoperability between different components while maintaining data consistency and security.
- Data Privacy and Confidentiality: Healthcare data is highly sensitive and subject to strict privacy regulations such as HIPAA in the United States. Blockchain implementations must provide robust mechanisms for encrypting and managing access to sensitive data while still allowing authorized parties to verify transactions and access necessary information.
- Security: Cybersecurity is paramount in healthcare systems, where any breach can have severe consequences for patient safety and privacy. Blockchain-based solutions must address potential vulnerabilities such as 51% attacks, double spending, and smart contract bugs to ensure the integrity and security of patient data and system operations.
- Regulatory Compliance: Healthcare regulations vary across jurisdictions and often impose strict requirements for data handling, privacy, and security. Blockchain implementations must comply with these regulations while still providing the benefits of decentralization, transparency, and auditability.
- Energy Efficiency: Traditional blockchain consensus mechanisms like Proof of Work (PoW) consume significant computational resources and energy, which may not be sustainable in CPS environments. Implementing more energy-efficient consensus algorithms or optimizing existing ones is essential to minimize the environmental impact of blockchain in healthcare.
- Immutable Storage: While immutability is a desirable feature of blockchain for ensuring data integrity, it can also pose challenges in healthcare where data may

need to be modified or deleted due to regulatory requirements or errors. Implementing mechanisms for selective data deletion or revision while maintaining overall blockchain integrity is a complex problem. Addressing these challenges requires a multidisciplinary approach involving expertise in blockchain technology, cybersecurity, healthcare informatics, and regulatory compliance. Collaborative efforts between researchers, industry stakeholders, and policymakers are essential to develop effective blockchain solutions that meet the unique requirements of CPS-based healthcare systems.

6.2 Regulatory and compliance considerations in adopting blockchain technology in healthcare (Alzahrani et al., 2022)

In healthcare, regulatory and compliance considerations are paramount due to the sensitive nature of patient data and the strict regulations governing its handling. When adopting blockchain technology in healthcare, several regulatory and compliance aspects need to be carefully addressed:

- HIPAA Compliance (United States): The Health Insurance Portability and Accountability Act (HIPAA) sets standards for the protection of sensitive patient information. Any blockchain solution in healthcare must adhere to HIPAA regulations to ensure the confidentiality, integrity, and availability of protected health information (PHI).
- GDPR Compliance (European Union): The General Data Protection Regulation (GDPR) imposes strict requirements on the processing and storage of personal data, including healthcare data. Blockchain implementations must comply with GDPR principles such as data minimization, purpose limitation, and the right to erasure (right to be forgotten) when handling patient data in EU jurisdictions.
- Data Security and Encryption: Healthcare blockchain solutions should employ robust encryption mechanisms to protect patient data from unauthorized access or tampering. Encryption algorithms must meet industry standards and regulatory requirements to ensure the confidentiality and integrity of sensitive information.
- Data Ownership and Consent Management: Blockchain can provide patients with greater control over their healthcare data by enabling transparent and auditable consent management mechanisms. However, regulatory frameworks must ensure that patients' rights to data ownership, access, and consent are respected and enforced within blockchain systems.
- Interoperability Standards: Regulatory bodies may establish interoperability standards and certification requirements for healthcare blockchain implementations to ensure seamless data exchange and compatibility with existing healthcare systems. Compliance with these standards is essential for promoting interoperability and facilitating data sharing across healthcare providers and institutions.

- Auditability and Traceability: Blockchain's immutability and transparency can enhance auditability and traceability in healthcare by providing a tamper-resistant record of data transactions and access events. Compliance with regulatory audit requirements necessitates robust auditing mechanisms and comprehensive logging of relevant activities within blockchain networks.
- Smart Contract Compliance: Smart contracts, self-executing code deployed on blockchain networks, can automate various healthcare processes such as insurance claims processing and clinical trial management. However, smart contracts must comply with legal and regulatory requirements, and their code should be thoroughly tested and audited to ensure compliance and mitigate potential legal risks.
- Regulatory Reporting and Compliance Monitoring: Blockchain systems may need to facilitate regulatory reporting and compliance monitoring by enabling real-time access to auditable data and compliance records. Regulatory agencies may require healthcare organizations to demonstrate adherence to regulatory requirements through blockchain-based reporting mechanisms.
- Cross-Border Data Transfer: In cases where healthcare data is shared across international borders, blockchain solutions must comply with data protection laws and regulations in different jurisdictions. Legal frameworks such as the EU–US Privacy Shield or Standard Contractual Clauses may be necessary to facilitate lawful cross-border data transfer. Addressing these regulatory and compliance considerations requires collaboration between blockchain developers, healthcare providers, legal experts, and regulatory authorities to ensure that blockchain implementations in healthcare adhere to applicable laws, regulations, and industry standards while promoting innovation and patient-centric care.

6.3 Ethical and privacy concerns related to blockchain enabled healthcare infrastructures (Srivastava et al., 2021)

Blockchain-enabled healthcare infrastructures raise several ethical and privacy concerns, primarily due to the sensitive nature of medical data and the decentralized nature of blockchain technology. Here are some of the key concerns:

- Privacy of Sensitive Health Data: Healthcare data stored on a blockchain is typically immutable and transparent, which can pose risks to patient privacy if not appropriately managed. Unauthorized access to patient records, linkage of anonymized data to individuals, and exposure of sensitive health information are significant privacy concerns.
- Data Security and Breach Risks: While blockchain technology offers robust security features, no system is entirely immune to security breaches. Smart contract vulnerabilities, insider threats, and cyberattacks targeting blockchain networks can compromise the confidentiality and integrity of healthcare data, leading to potential harm to patients and breaches of trust.

- Identity Protection and De-anonymization: Blockchain transactions are pseudonymous, meaning that users are identified by cryptographic addresses rather than personal identifiers. However, sophisticated data analytics techniques could potentially de-anonymize blockchain transactions, revealing the identities of individuals and exposing sensitive health information.
- Informed Consent and Patient Autonomy: Blockchain based healthcare systems may introduce complexities regarding informed consent and patient autonomy. Patients must understand the implications of their data being stored on a blockchain, including potential privacy risks and the inability to modify or delete data once recorded. Ensuring informed consent and empowering patients to control their data are essential ethical considerations.
- Data Ownership and Control: Blockchain's decentralized nature challenges traditional notions of data ownership and control. Patients may have limited control over their healthcare data once it is recorded on a blockchain, as it becomes part of an immutable ledger shared among network participants. Clarifying data ownership rights and establishing mechanisms for data access and control are critical ethical considerations.
- Bias and Discrimination: Blockchain-based healthcare systems may inadvertently perpetuate biases and discrimination present in healthcare data. Biased algorithms, incomplete data sets, and unequal access to healthcare services can exacerbate disparities and lead to unfair treatment of certain patient groups. Ethical design and governance frameworks are needed to mitigate these risks and ensure equitable healthcare delivery.
- Regulatory Compliance and Legal Liability: Healthcare providers and organizations must navigate complex regulatory landscapes when implementing blockchain technology. Compliance with laws such as HIPAA, GDPR, and FDA (Food and Drug Administration) regulations is essential to protect patient privacy and avoid legal liability. Blockchain developers and healthcare stakeholders must collaborate to ensure that blockchain solutions adhere to relevant regulatory requirements.
- Ethical Use of Data Analytics: Blockchain-enabled healthcare infrastructures generate vast amounts of data that can be leveraged for data analytics and research purposes. However, ethical considerations surrounding data use, consent, transparency, and accountability must be addressed to prevent misuse or exploitation of patient data for commercial or research interests. Addressing these ethical and privacy concerns requires a holistic approach that combines technical safeguards, legal frameworks, regulatory oversight, and ethical principles. Stakeholder engagement, transparency, and ongoing dialog are essential to develop responsible blockchain-enabled healthcare infrastructures that prioritize patient privacy, autonomy, and wellbeing.

7. Case studies

Blockchain technology has been applied in various industries and sectors, leading to several interesting case studies that highlight its potential. Here are some notable examples:

7.1 MedRec

MIT created MedRec (Azaria et al., 2016), a cutting-edge blockchain-based system for managing electronic medical records (EMR) that prioritizes data integrity and patient empowerment. MedRec offers a decentralized system that lets patients manage who can access their medical records, so that only approved healthcare practitioners can see their data. This is made possible by the use of blockchain technology. In addition to improving patient privacy, this approach encourages interoperability between various healthcare providers, enabling easy data sharing and lowering administrative costs. Medical records are kept accurate and reliable because of the immutable nature of blockchain, which prevents tampering. MedRec sets the standard for upcoming advances in the healthcare sector by providing an example of how blockchain may simplify the management of healthcare data, enhance patient outcomes, and lower costs.

7.2 MedicalChain

MedicalChain (Rouhani et al., 2018) is a cutting-edge network that uses blockchain technology to safely handle and store medical records and permits the sharing of data via smart contracts. By allowing patients to decide who has access to their medical records, this patient-centric approach promotes a sense of privacy and ownership among patients. MedicalChain ensures that sensitive information is only accessed by authorized parties by facilitating safe and effective data sharing with researchers, insurers, and healthcare providers through its creative use of smart contracts. The technology also facilitates secure video consultations that are easily integrated with patient records through its support for telemedicine services. MedicalChain is at an early stage of revolutionizing healthcare delivery and management in the digital age by improving data security, fostering interoperability, and opening new patient care options. Medical Supply Chain Management: Through the distributed ledger technology, blockchain can improve the medical supply chain's transparency and traceability by documenting the lifetime of medications, medical equipment, and other healthcare supplies. The blockchain allows for the recording of every step in the supply chain, from production to distribution to patient use, as a transaction. This reduces the possibility of fake medications and medical mistakes by allowing stakeholders to track and authenticate products in real-time.

7.3 MediLedger

MediLedger (MediLedger) is a cutting-edge blockchain-based network created by Chronicled with backing from significant pharmaceutical corporations. Its goal is to improve efficiency and transparency in the pharmaceutical supply chain. MediLedger guarantees complete visibility of pharmaceutical products from makers to consumers by offering an end-to-end track and trace solution. Strong anti-counterfeiting protocols ensure that every transaction on the blockchain is real, protecting pharmaceuticals from being distributed as fakes. Furthermore, MediLedger facilitates legal compliance via safe and transparent transactions, helping businesses to adhere to the Drug Supply Chain Security Act (DSCSA) rules. The platform's effective deployment demonstrates how blockchain technology could revolutionize the pharmaceutical supply chain protection, operational efficiency, and patient safety. Clinical Trials and Research Data Sharing: Blockchain can make it easier for researchers, pharmaceutical companies, and healthcare providers to securely and transparently share research findings and clinical trial data. To ensure compliance with legal regulations like GDPR and HIPAA, smart contracts can automate consent management and data sharing agreements. They can also facilitate effective collaboration and information exchange, which will hasten medical advancements and enhance patient outcomes.

7.4 ClinTexT CTi

Clinical Trials Intelligence, or ClinTex CTi, is a cutting-edge platform that uses blockchain technology to streamline clinical trial procedures and improve operational effectiveness and data quality. Developed to tackle the typical obstacles encountered in clinical research, ClinTex CTi offers instantaneous analytics and insights, empowering stakeholders to oversee the advancement of trials and promptly make well-informed judgments. The platform's decentralized methodology guarantees the security and integrity of trial data, facilitating smooth cooperation between sponsors, regulatory agencies, and researchers. ClinTex CTi seeks to shorten the duration and expense of clinical trials, which could ultimately hasten the creation of novel treatments and enhance patient outcomes. ClinTex is positioned to revolutionize clinical research and spur innovation in the pharmaceutical sector with its emphasis on data transparency and process optimization. Healthcare IOT device security: Health related IoT devices, like wearable sensors, medical implants, and remote monitoring devices, can benefit from increased security and trust because of blockchain technology. Stakeholders may guarantee data integrity, authentication, and tamper resistance by storing device data and interactions on a blockchain. This lowers the likelihood of cyberattacks, data breaches, and unauthorized access to private patient information.

7.5 IBM Watson IoT

Blockchain technology is included into IBM Watson IoT (Aggarwal and Madhukar, 2017) to improve the security and compatibility of linked medical devices in the healthcare industry. IBM offers a strong framework that guarantees the safe transmission and storage of health data produced by these devices by fusing blockchain technology with the Internet of Things. Because of this connectivity, healthcare providers may monitor and analyze device performance in real-time, identifying abnormalities and taking immediate action to address possible problems. Additionally, the implementation of blockchain assures that data is unchangeable and impervious to tampering, boosting stakeholder trust and safeguarding patient privacy. IBM Watson IoT and blockchain collaborate to build a more robust healthcare ecosystem by emphasizing safe data sharing and device identification. This enhances patient outcomes and operational effectiveness in the long run. Patient Identity Management and Consent: In healthcare CPS, blockchain can offer a safe, decentralized platform for handling patient identification, consent choices, and access restrictions. Patients will have more control over their health information and treatment choices if they have a distinct digital identity that is kept on the blockchain. This identity can be used to authenticate transactions, give permission for data sharing, and enforce privacy preferences across various healthcare systems and devices.

7.6 Patientory

Patientory (McFarlane et al., 2017) is a cutting-edge platform that improves patient identification management in the medical field by leveraging blockchain technology. Patientory gives people the ability to manage their own health data while guaranteeing its correctness and integrity by offering a safe, decentralized system for storing and exchanging patient IDs. On the blockchain, every patient is given a distinct digital identity that can only be safely accessed by medical professionals, insurance companies, and researchers who have the necessary authorization. This method not only improves data security and privacy, but it also makes it easier for different healthcare systems to seamlessly interface with one another, allowing for effective data sharing. Patientory is transforming patient identity management with its emphasis on data control and patient empowerment. This will ultimately improve healthcare results and build systemic trust. Healthcare Payments and Claims Processing: Blockchain enables real-time verification, settlement, and reconciliation of transactions between payers, providers, and patients, which can expedite healthcare payments and claims processing. In addition to lowering administrative burden, billing problems, and fraud, smart contracts can automate payment arrangements, eligibility checks, and claims adjudication. They can also guarantee prompt and accurate reimbursement for provided healthcare services.

7.7 Guardtime

Guardtime uses blockchain technology to guarantee data security, integrity, and transparency, which improves the way healthcare claims are processed. Building confidence among stakeholders, Guardtime makes it nearly difficult to modify or fabricate data by creating an immutable record of all claims' interactions. By automating crucial steps in the claims process like eligibility verification and payment disbursement, smart contracts drastically cut down on processing times and human error. Additionally, by enabling real-time auditing and anomaly identification, Guardtime's transparent ledger aids in the prevention of false claims. This more efficient method not only strengthens the bonds between payers and healthcare providers, but it also makes regulatory compliance easier, which in turn creates a more dependable and effective claims processing system.

8. Future directions and opportunities

8.1 Emerging trends and advancements in blockchain technology for healthcare

Emerging trends and advancements in blockchain technology for healthcare are shaping the future of the industry in profound ways. Interoperability stands out as a key trend, enabling seamless data exchange between disparate healthcare systems while safeguarding patient privacy and security. Additionally, blockchain's potential in supply chain management is gaining traction, offering a means to authenticate the provenance of pharmaceuticals and medical devices, thus combatting the infiltration of counterfeit products. Moreover, the empowerment of patients through blockchain technology is becoming increasingly evident, allowing individuals greater control over their health data and facilitating secure sharing with relevant stakeholders. Some of the key emerging trends and advancements are listed below:

- Interoperability: Blockchain can facilitate interoperability between different healthcare systems, enabling seamless sharing of patient data while maintaining security and privacy.
- Supply chain management: Blockchain can be used to track and authenticate the origin and movement of pharmaceuticals and medical devices, reducing the risk of counterfeit products entering the supply chain.
- Patient empowerment: Blockchain can empower patients by giving them greater control over their health data, allowing them to securely share it with healthcare providers, researchers, and other relevant parties.
- Smart contracts: The integration of smart contracts with blockchain technology can automate various healthcare processes, such as insurance claims processing and payment reconciliation, leading to increased efficiency and reduced administrative costs.
- Data security and privacy: Blockchain's cryptographic features can ensure the security and integrity of healthcare data, mitigating the risk of data breaches and unauthorized access.

8.2 Research directions and areas for further exploration

Research directions and areas for further exploration hold promise for advancing the application of blockchain in healthcare. Addressing scalability challenges is imperative, particularly for large-scale healthcare systems. Moreover, developing regulatory frameworks that accommodate blockchain's unique characteristics while ensuring compliance with data protection regulations is essential. Exploring synergies between blockchain and emerging technologies like AI and IoT presents exciting avenues for innovation, while enhancing user experience and assessing long-term impacts remain vital areas of research focus. Some of the key research directions and areas for further exploration are listed below:

- Scalability: Research is needed to address scalability challenges associated with blockchain technology, especially concerning its application in large-scale healthcare systems.
- Regulatory compliance: Further research is needed to develop regulatory frameworks that accommodate the unique characteristics of blockchain technology in healthcare while ensuring compliance with data protection and privacy regulations.
- Integration with emerging technologies: Exploring the integration of blockchain with other emerging technologies such as artificial intelligence (AI) and the IoT can unlock new possibilities for healthcare innovation.
- User experience: Research should focus on improving the user experience of blockchain-based healthcare applications to encourage widespread adoption among both patients and healthcare professionals.
- Long-term impact: Research is needed to assess the long-term impact of blockchain on healthcare outcomes, including its effect on patient care quality, cost-effectiveness, and overall health system performance.

8.3 Potential impact of blockchain on the future of healthcare delivery and management

The potential impact of blockchain on the future of healthcare delivery and management is substantial. By enhancing data integrity and security, blockchain can fortify healthcare systems against data breaches and ensure the accuracy and trustworthiness of patient information. Moreover, streamlined processes through automation and improved interoperability promise to reduce administrative burdens and enhance operational efficiency. Empowering patients with greater control over their health data fosters increased engagement and collaboration between patients and healthcare providers. Furthermore, blockchain's transformative potential extends to supply chain management, where it promises to revolutionize transparency and traceability, safeguarding against counterfeit products and ensuring the integrity of medical supplies. Ultimately, blockchain offers fertile ground for innovative business models, fostering decentralized healthcare networks and peer-to-peer platforms that could radically transform healthcare delivery and management paradigms. Some of the major points are listed below:

- Improved data integrity and security: Blockchain technology can enhance the integrity and security of healthcare data, reducing the risk of data breaches and ensuring that patient information remains accurate and tamper-proof.
- Streamlined processes: By automating various administrative processes through smart contracts and enhancing data interoperability, blockchain can streamline healthcare operations, reducing administrative burden and improving efficiency.
- Enhanced patient engagement: Blockchain empowers patients by giving them greater control over their health data, enabling them to securely share it with healthcare providers and participate more actively in their own care.
- Efficient supply chain management: Blockchain can revolutionize supply chain management in healthcare by providing transparency and traceability throughout the supply chain, reducing the risk of counterfeit products and ensuring the integrity of medical supplies.
- Innovative business models: Blockchain technology opens up opportunities for new business models in healthcare, such as decentralized healthcare networks and peertopeer healthcare platforms, which can improve access to care and promote collaboration among stakeholders.

9. Conclusion

Leveraging blockchain technology for building trustworthy and resilient healthcare infrastructures is paramount in today's digital landscape. Blockchain's inherent features, including immutability, decentralization, and cryptographic security, offer a robust foundation for safeguarding sensitive healthcare data and ensuring the integrity of healthcare systems. By harnessing blockchain, healthcare organizations can mitigate the risks associated with data breaches, unauthorized access, and tampering, thereby enhancing trust among patients, providers, and other stakeholders. Furthermore, blockchain enables the creation of resilient healthcare infrastructures capable of withstanding various threats and challenges. Its decentralized nature eliminates single points of failure, reducing the vulnerability of healthcare systems to cyberattacks and system failures. Moreover, the transparency and traceability provided by blockchain enhances accountability and facilitate rapid response to incidents, bolstering the resilience of healthcare infrastructures in the face of unforeseen events. A call to action for continued research and development in leveraging blockchain for healthcare is imperative to realize its full potential. Further research is needed to explore innovative use cases, address scalability challenges, and develop interoperability standards that facilitate seamless integration with existing healthcare systems. Additionally, ongoing efforts are required to refine regulatory frameworks, establish best practices, and promote collaboration among stakeholders to ensure responsible and ethical deployment of blockchain in healthcare. Collaborative initiatives involving researchers, industry experts, policymakers, and healthcare professionals are essential to drive progress in this area. By fostering a culture of innovation and knowledge exchange, we can unlock new opportunities for leveraging blockchain to build trustworthy and resilient healthcare infrastructures that prioritize patient privacy, security, and wellbeing.

Acknowledgments

We thank all potential colleagues and mentors for respective suggestions during the preparation of this manuscript.

Conflicts of Interest

The authors declare no conflict of interest. Furthermore, there are no personal circumstances or interests that may be perceived as inappropriately influencing the representation or interpretation of reported research work.

References

Aggarwal, S., Kumar, N., 2021. History of blockchain-blockchain 1.0: currency. Adv. Comput. 121, 147–169.

Aggarwal, M., Madhukar, M., 2017. IBM's Watson analytics for health care: a miracle made true. In: Cloud Computing Systems and Applications in Healthcare. IGI Global, pp. 117–134.

Ali, A., Al-Rimy, B.A.S., Almazroi, A.A., Alsubaei, F.S., Almazroi, A.A., Saeed, F., 2023. Securing secrets in cyber-physical systems: a cutting-edge privacy approach with consortium blockchain. Sensors 23 (16), 7162.

Alladi, T., Chamola, V., Parizi, R.M., Choo, K.K.R., 2019. Blockchain applications for industry 4.0 and industrial IoT: a review. IEEE Access 7, 176935–176951.

Alzahrani, S., Daim, T., Choo, K.K.R., 2022. Assessment of the blockchain technology adoption for the management of the electronic health record systems. IEEE Trans. Eng. Manag. 70 (8), 2846–2863.

Azaria, A., Ekblaw, A., Vieira, T., Lippman, A.M., 2016. Using blockchain for medical data access and permission management. In: 2016 2nd International Conference on Open and Big Data (OBD). IEEE, pp. 25–30.

Azzi, R., 2023. Blockchain Adoption in Healthcare: Toward a Patient Centric Ecosystem. Doctoral Dissertation. Institut Polytechnique de Paris; Universite Saint-Joseph, Beyrouth.

Bayer, D., Haber, S., Stornetta, W.S., 1993. Improving the efficiency and reliability of digital time-stamping. In: Sequences II: Methods in Communication, Security, and Computer Science. Springer, New York, pp. 329–334.

Bhutta, M.N.M., Khwaja, A.A., Nadeem, A., Ahmad, H.F., Khan, M.K., Hanif, M.A., , … Cao, Y., 2021. A survey on blockchain technology: evolution, architecture and security. IEEE Access 9, 61048–61073.

Dib, O., Brousmiche, K.L., Durand, A., Thea, E., Hamida, E.B., 2018. Consortium blockchains: overview, applications and challenges. Int. J. Adv. Telecommun 11 (1), 51–64.

Farringer, D.R., 2019. Maybe if we turn it off and then turn it back on again? Exploring health care reform as a means to curb cyber attacks. J. Law Med. Ethics 47 (S4), 91–102.

Gad, A.G., Mosa, D.T., Abualigah, L., Abohany, A.A., 2022. Emerging trends in blockchain technology and applications: a review and outlook. J. King Saud Univ., Comput. Inf. Sci. 34 (9), 6719–6742.

Gohar, A.N., Abdelmawgoud, S.A., Farhan, M.S., 2022. A patient-centric healthcare framework reference architecture for better semantic interoperability based on blockchain, cloud, and IoT. IEEE Access 10, 92137–92157.

Huang, K., Mu, Y., Rezaeibagha, F., Zhang, X., 2021. Design and Analysis of Cryptographic Algorithms in Blockchain. CRC Press.

Hussein, Z., Salama, M.A., El-Rahman, S.A., 2023. Evolution of blockchain consensus algorithms: a review on the latest milestones of blockchain consensus algorithms. Cybersecurity 6 (1), 30.

Islam, S., Islam, M.J., Hossain, M., Noor, S., Kwak, K.S., Islam, S.R., 2023. A survey on consensus algorithms in blockchain-based applications: architecture, taxonomy, and operational issues. IEEE Access 11, 39066–39082.

Jennath, H. S., Anoop, V. S., Asharaf, S. Blockchain for Healthcare: Securing Patient Data and Enabling Trusted Artificial Intelligence.

Jo, M., Hu, K., Yu, R., Sun, L., Conti, M., Du, Q., 2020. Private blockchain in industrial IoT. IEEE Network 34 (5), 76–77.

Kaur, V., Bhatt, D.P., Tharewal, S., Tiwari, P.K., 2021. Blockchain-based cyber-physical system: opportunities and challenges. In: Intelligent Security Solutions for Cyber-Physical Systems, pp. 234–245.

Kayan, H., Nunes, M., Rana, O., Burnap, P., Perera, C., 2022. Cybersecurity of industrial cyber-physical systems: a review. ACM Comput. Surv. 54 (11s), 1–35.

Komalavalli, C., Saxena, D., Laroiya, C., 2020. Overview of blockchain technology concepts. In: Handbook of Research on Blockchain Technology. Academic Press, pp. 349–371.

Krishna, P.V., Gurumoorthy, S., Obaidat, M.S., Monisha, K., Rajasekhara Babu, M., 2019. A novel framework for healthcare monitoring system through cyber-physical system. In: Internet of Things and Personalized Healthcare Systems, pp. 21–36.

Maesa, D.D.F., Mori, P., 2020. Blockchain 3.0 applications survey. J. Parallel Distr. Comput. 138, 99–114.

Maleh, Y., Lakkineni, S., Tawalbeh, L.A., AbdEl-Latif, A.A., 2022. Blockchain for cyber-physical systems: challenges and applications. In: Advances in Blockchain Technology for Cyber Physical Systems, pp. 11–59.

McFarlane, C., Beer, M., Brown, J., Prendergast, N.P., 2017. A Healthcare Peer-To-Peer EMR Storage Network V1. Entrust Inc., Addison, TX, USA, p. 19.

MediLedger. https://www.mediledger.com/.

Mingxiao, D., Xiaofeng, M., Zhe, Z., Xiangwei, W., Qijun, C., 2017. A review on consensus algorithm of blockchain. In: 2017 IEEE International Conference on Systems, Man, and Cybernetics (SMC). IEEE, pp. 2567–2572.

Mukherjee, P., Pradhan, C., 2021. Blockchain 1.0 to blockchain 4.0—the evolutionary transformation of blockchain technology. In: Blockchain Technology: Applications and Challenges. Springer International Publishing, pp. 29–49.

Nakamoto, S., 2008. Bitcoin: A Peer-To-Peer Electronic Cash System.

Ølnes, S., 2016. Beyond bitcoin enabling smart government using blockchain technology. In: Electronic Government: 15th IFIP WG 8.5 International Conference, EGOV 2016, Guimaraes, Portugal, September 5–8, 2016, Proceedings 15. Springer International Publishing, pp. 253–264.

Onik, M.M.H., Aich, S., Yang, J., Kim, C.S., Kim, H.C., 2019. Blockchain in healthcare: challenges and solutions. In: Big Data Analytics for Intelligent Healthcare Management. Academic Press, pp. 197–226.

Ramadoss, R., 2022. Blockchain technology: an overview. IEEE Potentials 41 (6), 6–12.

Rouhani, S., Butterworth, L., Simmons, A.D., Humphery, D.G., Deters, R., MediChain, T.M., 2018. A secure decentralized medical data asset management system. In: 2018 IEEE International Conference on Internet of Things (iThings) and IEEE Green Computing and Communications (GreenCom) and IEEE Cyber, Physical and Social Computing (CPSCom) and IEEE Smart Data (SmartData). IEEE, pp. 1533–1538.

Shah, J.L., Bhat, H.F., Khan, A.I., 2021. Integration of cloud and IoT for smart e-healthcare. In: Healthcare Paradigms in the Internet of Things Ecosystem. Academic Press, pp. 101–136.

Shah, A.S., Karabulut, M.A., Akhter, A.S., Mustari, N., Pathan, A.S.K., Rabie, K.M., Shongwe, T., 2023. On the vital aspects and characteristics of cryptocurrency—a survey. IEEE Access 11, 9451–9468.

Shaikh, T.A., Rasool, T., Verma, P., 2023. Machine intelligence and medical cyber-physical system architectures for smart healthcare: taxonomy, challenges, opportunities, and possible solutions. Artif. Intell. Med., 102692

Sharma, A., Kaur, S., Singh, M., 2021. A comprehensive review on blockchain and Internet of Things in healthcare. Trans. Emerging Telecommun. Technol. 32 (10), e4333.

Sheikh, Z.A., Singh, Y., Singh, P.K., Ghafoor, K.Z., 2022. Intelligent and secure framework for critical infrastructure (CPS): current trends, challenges, and future scope. Comput. Commun. 193, 302–331.

Shimonski, R., 2020. AI in Healthcare: How Artificial Intelligence Is Changing IT Operations and Infrastructure Services. John Wiley Sons.

Sonkamble, R.G., Bongale, A.M., Phansalkar, S., Sharma, A., Rajput, S., 2023. Secure data transmission of electronic health records using blockchain technology. Electronics 12 (4), 1015.

Srivastava, V., Mahara, T., Yadav, P., 2021. An analysis of the ethical challenges of blockchain-enabled E-healthcare applications in 6G networks. Int. J. Cogn. Comput. Eng. 2, 171–179.

Stafford, T.F., Treiblmaier, H., 2020. Characteristics of a blockchain ecosystem for secure and sharable electronic medical records. IEEE Trans. Eng. Manag. 67 (4), 1340–1362.

Uddin, M.A., Stranieri, A., Gondal, I., Balasubramanian, V., 2020. Blockchain leveraged decentralized IoT eHealth framework. Internet Things 9, 100159.

Verma, R., 2022. Smart city healthcare cyber physical system: characteristics, technologies and challenges. Wirel. Pers. Commun. 122 (2), 1413–1433.

Viriyasitavat, W., Hoonsopon, D., 2019. Blockchain characteristics and consensus in modern business processes. J. Ind. Inf. Integr. 13, 32–39.

Virmani, C., Gupta, D.J., Choudhary, T., 2019. Blockchain 2.0: an edge over technologies. In: Architectures and Frameworks for Developing and Applying Blockchain Technology. IGI Global, pp. 167–188.

Wieninger, S., Schuh, G., Fischer, V., 2019. Development of a blockchain taxonomy. In: 2019 IEEE International Conference on Engineering, Technology and Innovation (ICE/ITMC). IEEE, pp. 1–9.

Yaacoub, J.P.A., Noura, M., Noura, H.N., Salman, O., Yaacoub, E., Couturier, R., Chehab, A., 2020. Securing internet of medical things systems: limitations, issues and recommendations. Future Gener. Comput. Syst. 105, 581–606.

Yassein, M.B., Shatnawi, F., Rawashdeh, S., Mardin, W., 2019. Blockchain technology: characteristics, security and privacy; issues and solutions. In: 2019 IEEE/ACS 16th International Conference on Computer Systems and Applications (AICCSA). IEEE, pp. 1–8.

Zhong, W., Yang, C., Liang, W., Cai, J., Chen, L., Liao, J., Xiong, N., 2023. Byzantine fault-tolerant consensus algorithms: a survey. Electronics 12 (18), 3801.

CHAPTER 17

Exploring security and privacy challenges in intelligent semantic healthcare

Azamat Ali
Lovely Professional University, Phagwara, Punjab, India

1. Introduction

Intelligent semantic healthcare is a ground breaking approach that uses advanced technologies to better understand and use medical data. This method uses semantic analysis and Artificial Intelligence (AI) together to get useful information from big and complicated data. The goal is to improve patient care, support clinical decision making, and enhance the efficiency of healthcare systems (Jiang et al., 2017). At its core, semantic healthcare relies on the principles of semantics, which involve the study of meanings and interpretations of words, phrases, and sentences within a specific context. By applying semantic technologies, healthcare systems can process both unstructured and structured data. This means they can understand not just the literal text but also the underlying concepts and relationships within the data. This capability is especially important in healthcare, where accurately interpreting medical data can significantly impact patient outcomes (Demiris and Hensel, 2008).

The adoption of intelligent semantic healthcare is driven by several critical needs in modern healthcare systems. A big challenge is handling and understanding the huge amounts of data created every day. According to the International Data Corporation (IDC), global healthcare data is projected to increase by 36% annually until 2025 (Matengo et al., 2020). This data includes things like electronic health records (EHRs), medical images, genetic information, and real time patient monitoring. If properly analyzed, this information can lead to significant advancements in medical research and patient care. Intelligent semantic healthcare allows for personalized medicine, where treatments are customized to fit each patient's specific needs. By looking at data from different sources, like genetic info and medical history, doctors can create more accurate and effective treatment plans. This not only helps patients get better results but also cuts down on unnecessary treatments and lowers healthcare costs (Ashley, 2016).

Natural Language Processing is a type of AI that helps computers understand and interact with human language. In healthcare, NLP is used to handle and analyze lots of text data that isn't organized, like doctors' notes, lab reports, and patient records.

Intelligent Semantic Analysis for Healthcare
ISBN 978-0-443-33512-9
https://doi.org/10.1016/B978-0-443-33512-9.00008-1

This analysis can uncover patterns and insights that are not easily visible through traditional data analysis methods (Velupillai et al., 2018). One significant application of NLP in healthcare is extracting clinical information from EHRs. EHRs are rich sources of patient data, but much of this information is in free text format, making it challenging to analyze systematically. NLP techniques can parse these unstructured texts to identify key clinical concepts, such as diagnoses, treatments, and outcomes. For example, NLP algorithms can identify mentions of specific conditions or medications, allowing healthcare providers to track disease prevalence and treatment efficacy over time (Demiris and Hensel, 2008).

Semantic analysis involves understanding the meaning of words and sentences in context, which is essential for accurately interpreting medical information. In healthcare, semantic technologies are used to create ontologies and knowledge graphs that represent medical knowledge in a structured form. These tools facilitate the integration and interoperability of different data sources, enabling more comprehensive and accurate analyses (Bodenreider, 2004). For example, the Unified Medical Language System (UMLS) is a popular tool that provides a standard set of medical terms, helping different systems understand and use medical data in the same way (Bodenreider, 2004). By using semantic technologies to map data to standardized terminologies, healthcare providers can ensure consistent and accurate data comparisons, regardless of the source.

Semantic analysis is important for clinical decision support systems (CDSS). These systems help doctors by providing advice based on patient data and current medical knowledge. By using semantic technology, CDSS can give more accurate and relevant recommendations, which improves patient care (Sutton et al., 2020). However, there are big challenges to overcome. One major issue is the quality and consistency of the data. Medical data can be incomplete, inconsistent, or in different formats, making it hard to analyze and integrate effectively. Making sure data is accurate and consistent is really important for making semantic healthcare work well.

Keeping medical data private and secure is really important. Since this information is so sensitive, any security breach can have serious effects on patients. It's crucial to have strong security measures and follow regulations like HIPAA and GDPR to keep patient data safe (Edemekong et al., 2023; GDPR, 2023). Even with these challenges, intelligent semantic healthcare offers huge benefits. By using AI and semantic technologies, healthcare systems can work more efficiently, improve patient care, and gain new insights from medical data. As these technologies advance, they will likely have an even greater impact on healthcare (Jiang et al., 2017).

2. Security risks in semantic healthcare systems

2.1 Vulnerabilities in NLP algorithms and semantic analysis techniques

NLP algorithms and semantic analysis techniques are essential tools in healthcare, allowing for the extraction of valuable insights from unstructured medical data. However, because they are complex and rely on large amounts of data, these systems are at risk of security threats. One big problem is that NLP algorithms can be tricked by what's called adversarial attacks. In these attacks, someone deliberately changes the input data to make the machine learning models give wrong results. For example, slight changes to clinical text, such as adding or altering words, can cause NLP systems to misinterpret the information significantly. This vulnerability is particularly concerning in healthcare, where incorrect data can negatively affect patient care (Finlayson et al., 2019).

Vulnerability is found in semantic analysis techniques, which depend heavily on large, annotated datasets for training. These datasets often contain inherent biases that the algorithms can inadvertently learn and perpetuate. For example, if the training data mostly includes information from one particular group of people, the model might not work well for other groups that are not well represented. This can result in biased healthcare advice (Mehrabi et al., 2021).

2.2 Potential threats: Data breaches, adversarial attacks, and unintended biases

Semantic healthcare systems face a multitude of security threats, including data breaches, adversarial attacks, and unintended biases:

- Data Breaches: The healthcare sector is often targeted by hackers because medical data is very sensitive and valuable. Data breaches can let unauthorized people access patient information, leading to issues like identity theft and financial fraud. For instance, the 2015 Anthem Inc. breach exposed personal details of nearly 80 million people, showing how essential it is to have strong security measures to protect healthcare data (Gomez et al., 2015).
- Adversarial Attacks: Adversarial attacks pose significant threats to NLP and semantic analysis systems. These attacks involve crafting inputs that are intentionally designed to cause the system to make errors. In the context of healthcare, adversarial attacks can manipulate diagnostic algorithms by subtly altering input data, potentially leading to incorrect diagnoses and treatment recommendations. For instance, a study demonstrated that minor perturbations to medical images could deceive AI models into misclassifying them, raising concerns about the reliability of AI in clinical settings (Finlayson et al., 2019).
- Unintended Biases: Unintended biases in semantic healthcare systems can stem from biased training data or flaws in the algorithm's design. These biases can lead to discriminatory practices, impacting the quality of care for different patient groups. For example, if an NLP system is trained mainly on data from one ethnic group, it might not handle medical records from other ethnic groups well, which can lead to unfair healthcare results.

2.3 Examples of security incidents in semantic healthcare systems

Several real world incidents illustrate the security challenges faced by semantic healthcare systems:

i. Anthem Data Breach (2015): It is one of the biggest healthcare data breaches ever, hackers accessed Anthem Inc.'s database and compromised the personal information of nearly 80 million people. This breach highlighted how vulnerable healthcare data is to cyberattacks and stressed the need for stronger security measures (Gomez et al., 2015).
ii. WannaCry Ransomware Attack (2017): The WannaCry ransomware attack hit numerous healthcare organizations around the world, including the UK's National Health Service (NHS). It caused major problems, like canceled appointments and delayed treatments. This event showed how important it is to protect healthcare systems from ransomware attacks (Ghafur et al., 2019).
iii. Adversarial Attacks on Medical Imaging AI: Researchers have demonstrated that minor alterations to medical images can deceive AI models into making incorrect diagnoses. For instance, an adversarial attack on a skin cancer detection AI system caused it to misclassify benign moles as malignant, highlighting the risks of relying on AI for critical diagnostic tasks (Finlayson et al., 2019).
iv. Bias in Predictive Healthcare Algorithms: A study found that a commonly used predictive algorithm in healthcare showed significant racial bias. The algorithm was less likely to recommend special care programs for black patients with similar health issues compared to white patients. This happened because the algorithm used healthcare costs to determine who needed extra care, which unfairly affected black patients (Obermeyer et al., 2019).

3. Privacy concerns in semantic healthcare systems

3.1 Importance of patient privacy in healthcare data analysis

Patient privacy is essential in healthcare, vital for maintaining trust between patients and healthcare providers. Keeping patient information confidential is both an ethical duty and a legal requirement under regulations like HIPAA and GDPR (Edemekong et al., 2023; GDPR, 2023). As semantic healthcare systems increasingly rely on large volumes of data for analysis and decision making, protecting patient privacy becomes even more critical.

In healthcare data analysis, patient privacy means that individuals' medical information is not shared without their consent. This confidentiality is crucial for several reasons. First, it protects patients from potential discrimination based on their medical conditions. Secondly, it makes patients feel comfortable sharing sensitive information with their doctors, which is important for getting accurate diagnoses and effective treatment. Lastly, it helps keep healthcare institutions trustworthy and reliable (Adeniyi et al., 2024).

3.2 Challenges in patient consent management, data anonymization, and de-identification

3.2.1 Patient consent management

Managing patient consent for using healthcare data is tricky. Patients need to know exactly how their data will be used, who will have access to it, and what risks are involved. This gets more complicated because data can be used for different things later on, which might not be obvious when it's first collected (Mittelstadt and Floridi, 2016). One big issue is getting consent for these secondary uses. For instance, data collected for patient care might later be used for research or to develop AI models. It's important to communicate clearly and have good systems in place to manage consent for these additional uses (Williams et al., 2015).

3.2.2 Data anonymization

Anonymization helps protect patient privacy by removing personal details from data. While it reduces the risk of identifying individuals, it's hard to make data completely anonymous. Advanced re-identification techniques can sometimes match anonymized data with other datasets, potentially uncovering individuals' identities (Narayanan and Shmatikov, 2010). To anonymize data, we can remove direct identifiers like names and social security numbers, or use fake names and encryption. But even with these methods, details like age, gender, and zip code might still be combined to identify people. So, anonymization strategies need to be carefully planned and updated regularly to handle new risks of re-identification (El Emam and Arbuckle, 2013).

3.2.3 De-identification

De-identification, similar to anonymization, focuses on removing or obscuring information that could link a dataset to an individual. While de-identification can offer greater privacy protection, it also comes with challenges. For instance, de-identifying data can reduce its usefulness for certain types of analysis. Balancing the need to protect privacy with maintaining data utility is a significant challenge in healthcare data management (Im et al., 2024). Moreover, different regions have varying standards for what constitutes adequate de-identification. For example, HIPAA provides guidelines for de-identifying health information in the U.S., but these guidelines may not match international standards like those under GDPR. Ensuring compliance with multiple regulatory frameworks can be difficult for global healthcare organizations (Rosner, 2019).

3.3 Risks associated with re-identification attacks and data leakage

3.3.1 Re-identification attacks

Re-identification attacks occur when anonymized or de-identified data is matched with other data sources to re-establish the identity of individuals. These attacks exploit the fact that even anonymized datasets often retain enough information to allow for re-identification when combined with external data. For example, Sweeney (2000) found that you could identify 87% of people in the U.S. just using their ZIP code, birth date, and gender (Sweeney, 2000). This highlights a big risk in privacy: if someone is re-identified, their medical information could be exposed, which might lead to issues like identity theft, discrimination, or personal embarrassment. Protecting against re-identification requires constant vigilance and advanced anonymization techniques that can adapt to new threats (Majeed et al., 2022).

3.3.2 Data leakage

Data leakage happens when sensitive information is sent out of an organization without permission. In healthcare, this can occur through hacking, insider threats, or mistakes. This can seriously harm patient privacy and result in fines and loss of trust (Thapa and Camtepe, 2021). To prevent data leakage, healthcare organizations need strong security measures, including encryption, access controls, and systems to detect intrusions. They also need to train employees and have strict rules for handling data. Regular security checks and risk assessments are important to find and fix potential problems (Thapa and Camtepe, 2021).

3.4 Strategies to mitigate privacy risks

To mitigate the privacy risks related with semantic healthcare systems, several strategies can be employed:

- Enhanced Consent Management: Implementing dynamic consent models can improve the management of patient consent. Dynamic consent allows patients to keep giving their approval for different ways their data is used, giving them more control and clarity about how their information is handled (Kaye et al., 2015).
- Advanced Anonymization Techniques: Using advanced anonymization techniques, like differential privacy, can better protect patient data. Differential privacy adds random changes to data to make it harder to identify individual people, while still keeping the data useful for analysis (Majeed et al., 2022).
- Robust Data Governance: Building strong data governance frameworks helps protect data privacy and security from start to finish. This means having clear rules on who can access and use data, as well as training employees regularly on how to handle data safely (Rumbold and Pierscionek, 2017).
- Multi-Layered Security Measures: To prevent data leaks and unauthorized access, it's important to use multiple layers of security. This includes encrypting data, controlling who can access it, securing networks, and conducting regular security checks to find and fix any weaknesses (Crossler et al., 2019).
- Compliance with Regulatory Standards: Following regulations like HIPAA and GDPR is crucial for protecting patient privacy. These rules offer guidelines for data protection, patient consent, and breach notifications, helping organizations stay compliant and accountable (Silva and Soto, 2022).

4. Data sharing and interoperability

4.1 Benefits of data sharing for improving patient care and research outcomes

Data sharing in healthcare is a cornerstone for advancing patient care and enhancing research outcomes. When healthcare data is shared effectively, it can lead to numerous benefits, including improved clinical decision making, accelerated medical research, and enhanced public health surveillance.

4.1.1 Improved clinical decision making

Data sharing has a big benefit: it helps doctors make better decisions. When doctors can see all the information about a patient's health from different sources, like their medical history, lab results, and scans from various providers, they can make more accurate diagnoses and treatment plans. This complete view helps reduce mistakes, avoid extra tests, and improve patient care overall (Castaneda et al., 2015).

4.1.2 Accelerated medical research

Data sharing is vital for accelerating medical research. Researchers can leverage large datasets to identify patterns, understand disease mechanisms, and develop new treatments. For instance, pooling data from multiple clinical trials can enhance the statistical power of studies, leading to more robust and generalizable findings. Moreover, shared data can facilitate the replication of studies, a critical aspect of scientific validation. The COVID-19 pandemic represented the significance of data sharing, where rapid dissemination of research data and findings was crucial in developing vaccines and treatments (Baden et al., 2021).

4.1.3 Enhanced public health surveillance

Public health agencies rely on data sharing to monitor and respond to health threats effectively. Shared data from hospitals, laboratories, and other healthcare entities enable real time surveillance of disease outbreaks, identification of at risk populations, and timely implementation of public health interventions. For example, data sharing was instrumental in tracking and controlling the spread of infectious diseases like Ebola and COVID-19, allowing for coordinated efforts to mitigate their impact (Brownson et al., 2018).

4.2 Tension between data interoperability and privacy preservation

While the benefits of data sharing are evident, achieving interoperability while protecting patient privacy is challenging. Interoperability means different healthcare systems can exchange, interpret, and use data seamlessly. At the same time, privacy preservation involves keeping patient information secure from unauthorized access and ensuring compliance with regulations.

4.2.1 Technical challenges

One major challenge is the technical difficulty of achieving interoperability. Healthcare systems often use different ways of organizing and naming data, which makes it difficult to combine information from different sources. Standardizing data formats and creating systems that can work together requires a lot of investment and coordination among stakeholders (Dimitrov, 2016).

4.2.2 Privacy and security concerns

Another important issue is the risk to patient privacy. Sharing data between systems increases the risk of data leaks and unauthorized access. If sensitive patient information is exposed, it can lead to identity theft, discrimination, and other harms. Plus, making sure that data sharing complies with regulations like HIPAA and GDPR adds more complexity. These rules require strict measures to protect patient data, which can sometimes conflict with the goal of achieving interoperability (Gandrup et al., 2020).

4.2.3 Ethical considerations

Ethical issues are key when balancing data sharing with privacy. Patients should control how their data is used and shared. Getting clear permission for data sharing, especially for uses beyond the original purpose, is crucial but often tricky. It's also important to balance individual privacy with the wider benefits of data sharing, such as improved public health and medical progress (Mittelstadt and Floridi, 2016).

4.3 Strategies for achieving a balance between data sharing and patient privacy

To balance data sharing with patient privacy, several strategies can be used:

4.3.1 Adopting interoperability standards

Using standardized data formats and protocols is essential for smooth data sharing and ensuring different healthcare systems work together. Standards like "Health Level Seven (HL7)" and "Fast Healthcare Interoperability Resources (FHIR)" allow for easy data exchange and integration. By following these standards, healthcare organizations can improve how well systems work together while keeping data accurate and private (Mandl and Kohane, 2016).

4.3.2 Advanced data encryption techniques

Data encryption is a crucial way to protect patient information when it's sent or stored. Advanced methods, such as homomorphic encryption and secure multi-party computation, let us analyze and share data without revealing the actual information. This means sensitive information stays confidential even when shared with people or systems that are not fully trusted (Chen et al., 2017).

4.3.3 Dynamic consent models

Dynamic consent models offer a flexible approach to managing patient consent for data sharing. Unlike traditional consent models, where patients provide consent once for a specific use, dynamic consent allows patients to update their preferences and provide consent for different uses over time. This approach enhances patient control and transparency, fostering trust and compliance with ethical standards (Kaye et al., 2015).

4.3.4 De-identification and anonymization

De-identifying and anonymizing data before sharing can greatly lower privacy risks. De-identification means removing or masking information that can identify individuals, so it's hard to trace the data back to specific people. Anonymization takes this a step further by ensuring that data can't be re-identified even if combined with other datasets. However, it's important to use strong techniques and regularly review and update anonymization methods to keep up with new re-identification risks (Chevrier et al., 2019).

4.3.5 Privacy preserving data sharing frameworks

Privacy preserving frameworks, like federated learning, let data be shared and analyzed without transferring the actual data. In federated learning, machine learning models are trained locally on individual datasets, and only the updates to these models (not the data itself) are sent to a central server. This method allows for collaborative data analysis while keeping patient data secure and private (Li et al., 2021).

4.3.6 Regulatory compliance and best practices

Following regulatory requirements is crucial for balancing data sharing and privacy. Healthcare organizations need to implement policies and procedures that comply with regulations like HIPAA and GDPR. This involves regularly checking for risks, teaching employees about data protection, and creating strong rules for managing data. Additionally, adopting best practices for data security, such as regular security audits and having a plan for responding to incidents can further safeguard patient privacy (Gandrup et al., 2020).

5. Regulatory compliance and standards in intelligent semantic healthcare

5.1 Overview of relevant regulations: HIPAA and GDPR

In managing healthcare data, it's crucial to follow regulations to keep patient information private and secure. Two important rules that govern this are HIPAA in the United States and GDPR in the European Union.

5.1.1 HIPAA (health insurance portability and accountability act)

HIPAA, established in 1996, sets rules to keep patient information secure. It applies to healthcare providers, health plans, and clearinghouses, referred to as covered entities, as well as their business partners who manage patient data. The main goals of HIPAA are:

- Privacy Rule: The HIPAA Privacy Rule sets criteria for keeping personal health information private. It ensures that patient' medical records and health details are kept confidential and outlines what rights patients have regarding their health information.
- Security Rule: The HIPAA Security Rule sets rules for keeping electronic health information (ePHI) safe. It requires healthcare providers and their partners to put in place measures to ensure ePHI is kept private, correct, and only accessible to authorized people.
- Breach Notification Rule: The HIPAA Breach Notification Rule requires healthcare providers to inform affected people, the U.S. Department of Health and Human Services (HHS), and sometimes the media if there is a breach involving unsecured electronic health information (ePHI).

Healthcare organizations in the U.S. must follow HIPAA rules to handle protected health information (PHI). Not following these rules can lead to heavy fines and legal trouble, so it's crucial to comply with them (Gandrup et al., 2020).

5.1.2 GDPR (general data protection regulation)

The GDPR, which started in 2018, is the EU's main law for protecting personal data. It applies to any organization that deals with the personal data of people in the EU, no matter where the organization is based. Key principles of GDPR important for healthcare data include:

- Lawfulness, Fairness, and Transparency: Data handling needs to be legal, fair, and clear to the people whose data is being used.
- Purpose Limitation: Personal data should be collected for clear and specific reasons and only used for those reasons. It shouldn't be used in ways that don't match those reasons.
- Data Minimization: Personal data should be just enough, relevant, and only what is needed for the purpose it is being used for.
- Security and Confidentiality: Organizations need to set up suitable technical and organizational measures to ensure security based on the risk level. This includes protecting data from unauthorized or illegal use and avoiding accidental loss, destruction, or damage.

GDPR mandates stringent requirements for obtaining consent, notifying data breaches, and appointing data protection officers (DPOs) for certain organizations. Not following GDPR can result in fines up to €20 million or 4% of a company's annual global revenue, whichever is higher. This highlights the importance for healthcare organizations to adhere to GDPR rules (GDPR, n.d.).

5.2 Compliance requirements for intelligent semantic healthcare systems

Smart semantic healthcare systems that use advanced technologies like natural language processing and machine learning have particular challenges when it comes to following HIPAA and GDPR rules:

- Data Handling and Processing: These systems often work with large amounts of sensitive patient data to generate insights and help with clinical decisions. Following HIPAA's Privacy and Security Rules is essential for protecting patient privacy and preventing unauthorized access to or disclosure of electronic health information (ePHI) (Mittelstadt and Floridi, 2016).
- Consent Management: Intelligent semantic healthcare systems often use patient data for research or analysis beyond its original purpose. It's essential to get informed consent from patients for these uses to ensure transparency and respect their rights, as required by both HIPAA and GDPR (Kaye et al., 2015).
- Data Security Measures: The HIPAA Security Rule says that organizations must use administrative, physical, and technical measures to keep electronic health information (ePHI) safe. For semantic healthcare systems, this means using encryption to protect data, setting up access controls to limit who can see the information, and regularly checking for and fixing security problems (Redhead, n.d.).
- Anonymization and De-identification: To reduce privacy risks, semantic healthcare systems should use strong anonymization and de-identification techniques before sharing data for research or analysis. These methods protect patient identities while still keeping the data useful for other purposes (El Emam and Arbuckle, 2013).

5.3 Role of regulatory frameworks in safeguarding patient privacy and data security

Regulations like HIPAA and GDPR are very important for keeping patient information private and secure in advanced healthcare systems.

- Setting Standards and Guidelines: HIPAA and GDPR set clear rules for how healthcare data should be collected, used, and protected. They require best practices and security measures to reduce risks from data breaches and unauthorized access (Thapa and Camtepe, 2021).
- Ensuring Accountability: Following regulatory requirements makes sure that healthcare organizations are responsible for keeping patient information safe. This accountability helps build trust between patients, healthcare providers, and others involved in handling the data (Rumbold and Pierscionek, 2017).
- Promoting Ethical Data Practices: Regulations encourage ethical data handling by requiring transparency in how data is processed, respecting patient privacy and consent, and imposing penalties for violations. These rules help make sure that healthcare data is used responsibly and ethically in both research and clinical settings (Mittelstadt and Floridi, 2016).

- Facilitating Cross border Data Transfers: GDPR has rules for sharing personal data outside the EU to keep it safe. This allows for global collaboration in healthcare research while still protecting data privacy and security (GDPR, 2023).

6. Case studies and best practices in semantic healthcare

6.1 Real world examples of security and privacy challenges

i. Example 1: Data Breaches and Patient Privacy

In 2015, Anthem Inc., a big health insurance company in the U.S., experienced a major data breach. Hackers took personal information from about 80 million people, including names, birth dates, Social Security numbers, and other sensitive details. This breach highlighted how vulnerable healthcare organizations are to cyberattacks that target valuable personal and medical data (Gomez et al., 2015).

ii. Example 2: Adversarial Attacks on NLP Systems

Adversarial attacks pose significant threats to NLP systems used in healthcare. Researchers have shown that it's possible to trick NLP models by altering medical text inputs, which can result in wrong diagnoses or treatment suggestions. These attacks exploit vulnerabilities in model training data or weaknesses in model architectures, highlighting the significance of robust security measures in deploying NLP based healthcare solutions (Carlini et al., 2019).

iii. Example 3: Unintended Biases in Machine Learning Models

Machine learning models in semantic healthcare can unintentionally reinforce biases found in their training data. For example, if the data used is biased against certain groups of people or medical conditions, it can result in unfair diagnoses or treatment recommendations. To fix these biases, it's crucial to carefully select and prepare the training data and to regularly check and adjust the models to ensure fair and equal healthcare results (Franklin et al., 2024).

6.2 Best practices for mitigating security risks and protecting patient privacy

i. Implementing Robust Authentication and Access Controls

Healthcare organizations should use multi-factor authentication (MFA) and role based access control (RBAC) to keep patient data safe. MFA boosts security by asking for multiple forms of ID, like passwords, fingerprints, or security tokens. RBAC limits that can see certain information based on their job, so only authorized people can access specific data.

ii. Encrypting Data at Rest and in Transit

Encryption is crucial for keeping patient data safe from unauthorized access when it's stored or sent. Strong encryption methods ensure that even if someone intercepts the data, it remains unreadable and useless to them. For data sent over the internet, using secure communication protocols like Transport Layer Security (TLS) helps protect it.

iii. Conducting Regular Security Audits and Penetration Testing

Regular security checks and tests are important to find and fix weaknesses in healthcare IT systems before hackers can take advantage of them. These tests simulate real cyberattacks to help organizations evaluate their security, fix problems, and update their systems to better protect against risks.

iv. Training and Educating Staff on Security Practices

Human error often leads to data breaches. To prevent this, healthcare organizations should invest in comprehensive training programs for their staff. These programs should cover cybersecurity best practices, proper data handling, and the importance of patient privacy. Increasing awareness helps employees stay alert and better able to recognize and address security threats (Okafor et al., 2023).

6.3 Successful implementations of security measures and Compliance strategies

i. Case Study: Mayo Clinic's Data Security Framework

Mayo Clinic, a leading healthcare provider, has established a robust data security system to safeguard patient information. This system includes strict access controls, encryption for sensitive data, regular security checks, and continuous staff training on data privacy and security. By focusing on patient privacy and adhering to HIPAA regulations, Mayo Clinic maintains the confidentiality and integrity of patient data while using advanced technologies to enhance healthcare services (Okafor et al., 2023).

ii. Case Study: Use of Blockchain for Secure Health Data Sharing

Blockchain technology offers a safe and clear method for sharing health data. For instance, Medicalchain uses blockchain to let patients control who can see their medical records. This ensures that the data is accurate and cannot be altered, while also allowing healthcare providers to quickly access current patient information. This can result in faster, better medical care and improved health outcomes (Shuaib et al., 2022).

iii. Compliance Strategies: Integrating Privacy by Design

The Privacy by Design approach means including privacy measures in healthcare systems from the very beginning. This method focuses on using the least amount of data necessary, giving users control over their information, and being transparent about how data is handled. By following these principles, organizations can reduce privacy risks, meet regulations like GDPR, and build trust with patients by managing their sensitive health information in an ethical and secure way (O'Connor et al., 2017).

7. Conclusions

Integrating semantic analysis into healthcare is changing patient care by making it more personalized and efficient. This chapter has shown the importance of addressing security and privacy issues with these advancements. To progress with intelligent semantic healthcare, it's essential to use AI and NLP to gain useful insights from medical data while ensuring strong protections for patient privacy and data security. The main points highlight the importance of using strong security measures, such as encryption and strict access controls, to defend against cyber threats and prevent unauthorized access.

The ethical aspects of using AI in healthcare are also very important. Everyone needs to focus on being clear, fair, and responsible in their decisions to maintain patient trust. It's also important to follow rules like HIPAA and GDPR, and healthcare organizations must stay updated with changing data protection laws and standards. In the future, collaboration will be important. Healthcare providers, tech developers, policymakers, and patients all need to work together. This teamwork will help ensure that improvements in semantic healthcare lead to better medical outcomes. At the same time, it will be important to respect ethical standards and protect patient rights. By embracing innovation responsibly and focusing on patient centered care, stakeholders can tackle these challenges and create a safer, more inclusive healthcare environment.

Conflicts of interest

Declare conflicts of interest or state "The author declares no conflict of interest."

References

Adeniyi, A.O., Arowoogun, J.O., Okolo, C.A., Chidi, R., Babawarun, O., 2024. Ethical considerations in healthcare IT: a review of data privacy and patient consent issues. World J. Adv. Res. Rev. 21, 1660–1668. https://doi.org/10.30574/wjarr.2024.21.2.0593.

Ashley, E.A., 2016. Towards precision medicine. Nat. Rev. Genet. 17, 507–522. https://doi.org/10.1038/nrg.2016.86.

Baden, L.R., El Sahly, H.M., Essink, B., Kotloff, K., Frey, S., Novak, R., Zaks, T., 2021. Efficacy and safety of the mRNA-1273 SARS-CoV-2 vaccine. N. Engl. J. Med. 384, 403–416. https://doi.org/10.1056/NEJMoa2035389.

Bodenreider, O., 2004. The unified medical language system (UMLS): integrating biomedical terminology. Nucleic Acids Res. 32, D267–D270. https://doi.org/10.1093/nar/gkh061.

Brownson, R.C., Fielding, J.E., Green, L.W., 2018. Building capacity for evidence-based public health: reconciling the pulls of practice and the push of research. Annu. Rev. Publ. Health 39, 27–53. https://doi.org/10.1146/annurev-publhealth-040617-014746.

Carlini, N., Liu, C., Erlingsson, Ú., Kos, J., Song, D., 2019. The secret sharer: evaluating and testing unintended memorization in neural networks. In: 28th USENIX Security Symposium (USENIX Security 19), pp. 267–284. https://www.usenix.org/conference/usenixsecurity19/presentation/carlini.

Castaneda, C., Nalley, K., Mannion, C., Bhattacharyya, P., Blake, P., Pecora, A., Suh, K.S., 2015. Clinical decision support systems for improving diagnostic accuracy and achieving precision medicine. J. Clin. Bioinf. 5, 1–16. https://doi.org/10.1186/s13336-015-0019-3.

Chen, H., Laine, K., Rindal, P., 2017. Fast private set intersection from homomorphic encryption. In: Proceedings of the 2017 ACM SIGSAC Conference on Computer and Communications Security, pp. 1243–1255. https://doi.org/10.1145/3133956.3134061.

Chevrier, R., Foufi, V., Gaudet-Blavignac, C., Robert, A., Lovis, C., 2019. Use and understanding of anonymization and de-identification in the biomedical literature: scoping review. J. Med. Internet Res. 21, e13484. https://doi.org/10.2196/13484.

Crossler, R.E., Bélanger, F., Ormond, D., 2019. The quest for complete security: an empirical analysis of users' multi-layered protection from security threats. Inf. Syst. Front. 21, 343–357. https://doi.org/10.1007/s10796-017-9755-1.

Demiris, G., Hensel, B.K., 2008. Technologies for an aging society: a systematic review of "smart home" applications. Yearb. Med. Inform. 17, 33–40. https://doi.org/10.1055/s-0038-1638580.

Dimitrov, D.V., 2016. Medical internet of things and big data in healthcare. Healthc. Inform. Res. 22, 156–163. https://doi.org/10.4258/hir.2016.22.3.156.

Edemekong, P.F., Annamaraju, P., Haydel, M.J., 2023. Health Insurance Portability and Accountability Act. StatPearls Publishing: Treasure Island, FL, USA.

El Emam, K., Arbuckle, L., 2013. Anonymizing Health Data: Case Studies and Methods to Get You Started, first ed. O'Reilly Media, Inc., Sebastopol, CA, USA, pp. 1–296.

Finlayson, S.G., Bowers, J.D., Ito, J., Zittrain, J.L., Beam, A.L., Kohane, I.S., 2019. Adversarial attacks on medical machine learning. Science 363, 1287–1289. https://doi.org/10.1126/science.aaw4399.

Franklin, G., Stephens, R., Piracha, M., Tiosano, S., Lehouillier, F., Koppel, R., Elkin, P.L., 2024. The sociodemographic biases in machine learning algorithms: a biomedical informatics perspective. Life 14, 652. https://doi.org/10.3390/life14060652.

Gandrup, J., Ali, S.M., McBeth, J., van der Veer, S.N., Dixon, W.G., 2020. Remote symptom monitoring integrated into electronic health records: a systematic review. J. Am. Med. Inform. Assoc. 27, 1752–1763. https://doi.org/10.1093/jamia/ocaa177.

GDPR. Non-compliance with GDPR can result in fines up to €20 million or 4% of annual global turnover, whichever is higher, making adherence to its provisions critical for organizations handling personal data in the healthcare sector. https://www.gdpradvisor.co.uk/gdpr-fines-and-penalties.

GDPR, 2023. General Data Protection Regulation. Retrieved from. https://gdpr.eu/.

Ghafur, S., Grass, E., Jennings, N.R., Darzi, A., 2019. The challenges of cybersecurity in health care: the UK National Health Service as a case study. Lancet Digit. Health 1, e10–e12. https://doi.org/10.1016/S2589-7500(19)30005-6.

Gomez, J., Sensato, C.E.O., Konschak, C., Divurgent, C.E.O., 2015. Cyber-security in healthcare. https://www.divurgent.com/wp-content/uploads/2015/03/Cyber-Security-Healthcarepdf.pdf.

Im, E., Kim, H., Lee, H., Jiang, X., Kim, J.H., 2024. Exploring the tradeoff between data privacy and utility with a clinical data analysis use case. BMC Med. Inform. Decis. Mak. 24, 147. https://doi.org/10.1186/s12911-024-02545-9.

Jiang, F., Jiang, Y., Zhi, H., Dong, Y., Li, H., Ma, S., Wang, Y., 2017. Artificial intelligence in healthcare: past, present and future. Stroke Vasc. Neurol. 2, 4. https://doi.org/10.1136/svn-2017-000101.

Kaye, J., Whitley, E.A., Lund, D., Morrison, M., Teare, H., Melham, K., 2015. Dynamic consent: a patient interface for twenty-first century research networks. Eur. J. Hum. Genet. 23, 141–146. https://doi.org/10.1038/ejhg.2014.71.

Li, J., Meng, Y., Ma, L., Du, S., Zhu, H., Pei, Q., Shen, X., 2021. A federated learning based privacy-preserving smart healthcare system. IEEE Trans. Ind. Inform. 18. https://doi.org/10.1109/TII.2021.3098010.

Majeed, A., Khan, S., Hwang, S.O., 2022. Toward privacy preservation using clustering based anonymization: recent advances and future research outlook. IEEE Access 10, 53066–53097. https://doi.org/10.1109/ACCESS.2022.3175219.

Mandl, K.D., Kohane, I.S., 2016. Time for a patient-driven health information economy? N. Engl. J. Med. 374, 205–208. https://doi.org/10.1056/NEJMp1512142.

Matengo, W., Otsieno, E., Wanjiru, K., 2020. Big data analytics in healthcare. Digit. Health focus predict. Prevent. Pers. Med. 129–138. https://doi.org/10.1007/978-3-030-49815-3_15.

Mehrabi, N., Morstatter, F., Saxena, N., Lerman, K., Galstyan, A., 2021. A survey on bias and fairness in machine learning. ACM Comput. Surv. 54, 1–35. https://doi.org/10.1145/3457607.

Mittelstadt, B.D., Floridi, L., 2016. The ethics of big data: current and foreseeable issues in biomedical contexts. . Ethics Biomed. Big Data 445–480. https://doi.org/10.1007/978-3-319-33525-4_19.

Narayanan, A., Shmatikov, V., 2010. Myths and fallacies of "personally identifiable information". Commun. ACM 53, 24–26. https://doi.org/10.1145/1743546.1743558.

Okafor, C.M., Kolade, A., Onunka, T., Daraojimba, C., Eyo-Udo, N.L., Onunka, O., Omotosho, A., 2023. Mitigating cybersecurity risks in the US healthcare sector. Int. J. Res. Sci. Innov. 10, 177–193. https://doi.org/10.51244/IJRSI.2023.10918.

O'Connor, Y., Rowan, W., Lynch, L., Heavin, C., 2017. Privacy by design: informed consent and internet of things for smart health. Procedia Comput. Sci. 113, 653–658. https://doi.org/10.1016/j.procs.2017.08.329.

Obermeyer, Z., Powers, B., Vogeli, C., Mullainathan, S., 2019. Dissecting racial bias in an algorithm used to manage the health of populations. Science 366, 447–453. https://doi.org/10.1126/science.aax2342.

Redhead, C.S. HIPAA privacy, security, enforcement and breach notification standards. https://sgp.fas.org/crs/misc/R43991.pdf.

Rosner, G., 2019. De-identification as public policy. J. Data Prot. Priv. 3, 1–18. https://ssrn.com/abstract=3639304.

Rumbold, J.M.M., Pierscionek, B., 2017. The effect of the general data protection regulation on medical research. J. Med. Internet Res. 19, e47. https://doi.org/10.2196/jmir.7108.

Shuaib, K., Abdella, J., Sallabi, F., Serhani, M.A., 2022. Secure decentralized electronic health records sharing system based on blockchains. J. King Saud Univ. Comput. Inf. Sci. 34, 5045–5058. https://doi.org/10.1016/j.jksuci.2021.05.002.

Silva, I., Soto, M., 2022. Privacy-preserving data sharing in healthcare: an in-depth analysis of big data solutions and regulatory compliance. Int. J. Appl. Health Care Anal. 7, 14–23.

Sutton, R.T., Pincock, D., Baumgart, D.C., Sadowski, D.C., Fedorak, R.N., Kroeker, K.I., 2020. An overview of clinical decision support systems: benefits, risks, and strategies for success. npj Digit. Med. 3, 17. https://doi.org/10.1038/s41746-020-0221-y.

Sweeney, L., 2000. Simple demographics often identify people uniquely. Health 671, 1–34. https://privacytools.seas.harvard.edu/sites/projects.iq.harvard.edu/files/privacytools/files/paper1.pdf.

Thapa, C., Camtepe, S., 2021. Precision health data: requirements, challenges and existing techniques for data security and privacy. Comput. Biol. Med. 129, 104130. https://doi.org/10.1016/j.compbiomed.2020.104130.

Velupillai, S., Suominen, H., Liakata, M., Roberts, A., Shah, A.D., Morley, K., Dutta, R., 2018. Using clinical natural language processing for health outcomes research: overview and actionable suggestions for future advances. J. Biomed. Inform. 88, 11–19. https://doi.org/10.1016/j.jbi.2018.10.005.

Williams, H., Spencer, K., Sanders, C., Lund, D., Whitley, E.A., Kaye, J., Dixon, W.G., 2015. Dynamic consent: a possible solution to improve patient confidence and trust in how electronic patient records are used in medical research. JMIR Med. Inform. 3, e3525. https://doi.org/10.2196/medinform.3525.

PART VI

Sustainability, policy, and the future landscape

CHAPTER 18

Smart choices, sustainable futures: Advancing healthcare responsibly

Navita Vashista[1], Manju Bala[2] and Shiva Tushir[3]
[1]Department of Management Studies, Panipat Institute of Engineering and Technology, Samalkha, India; [2]Chandigarh Business School of Administration, Landran, India; [3]Department of Pharmacy, Panipat Institute of Engineering and Technology, Samalkha, India

1. Introduction

As the city healthcare sector grapples with mounting sustainability challenges, innovative technologies are emerging as crucial enablers of more environmentally-friendly, socially-conscious, and economically-viable practices. This chapter explores how smart technologies are transforming healthcare delivery and facilitating the transition toward a more sustainable future.

- Definition of sustainability in healthcare

Sustainability in healthcare encompasses the fundamental principles of economic, social, and environmental sustainability. Additionally, it encompasses a fourth aspect known as systems sustainability (Mortimer et al., 2018). The objective of sustainability in healthcare is to establish and perpetuate a state in which humans and nature may coexist harmoniously, while simultaneously fulfilling the requirements of the present without jeopardizing the capacity of future generations to fulfill their own needs.

Alim and Sulley (2024) Environmental, social, and economic (E-S-E) sustainability in healthcare is a holistic approach that aims to balance these three dimensions to promote ecological health, social well-being, and economic prosperity. The goal is to create a healthcare system that doesn't harm the environment, is economical, and has a positive social impact (Antonio Grasso, 2023).

Here are some examples of how E-S-E sustainability can apply to healthcare.

- Environmental

Healthcare waste disposal can be a significant cost to the healthcare system, so sustainable practices can help reduce this.

- Social

Healthcare can focus on person- and community-centered approaches to address people's physical health needs and help them feel connected to their communities. This can lead to better outcomes for individuals, services, and communities.

- Economic

Healthcare companies can invest in E-S-G goals to continue and improve their sustainability, which can encourage others to do the same.

Intelligent Semantic Analysis for Healthcare
ISBN 978-0-443-33512-9
https://doi.org/10.1016/B978-0-443-33512-9.00013-5

2. Smart technologies for sustainable healthcare

One of the most prominent technology-driven shifts in healthcare is the rise of digital health and telemedicine. According to Schroeder et al. (2012). By enabling remote patient monitoring, virtual consultations, and connected care, these solutions reduce the need for in-person visits and the associated environmental footprint of patient transportation. They also expand access to healthcare, particularly in underserved or rural areas, promoting greater social equity.

For example, the use of wearable devices and remote monitoring tools allows patients with chronic conditions to manage their health from the comfort of their homes, reducing hospital admissions and the resource-intensive activities associated with on-site care. Similarly, telehealth platforms connect patients to healthcare providers through video calls, eliminating the need for physical travel. These innovations not only enhance sustainability but also improve patient convenience and satisfaction (Chiu and Fong, 2023).

Artificial intelligence (AI) and machine learning (ML) are transforming healthcare in myriad ways, with significant implications for sustainability. AI-powered clinical decision support systems can optimize treatment plans, streamline workflows, and reduce unnecessary tests and procedures, leading to more efficient use of healthcare resources.

In the realm of pharmaceutical development, (Knopp et al., 2023), AI is accelerating the drug discovery process, helping researchers identify promising drug candidates more quickly and effectively. This can shorten the timeline for bringing new medications to market, reducing the environmental impact of prolonged research and development cycles.

Furthermore, (Knopp et al., 2023), AI-driven predictive analytics are enabling proactive, personalized healthcare interventions that can prevent or mitigate the onset of chronic diseases. By empowering patients to manage their health more effectively, these technologies have the potential to reduce costly and resource-intensive hospital admissions and complications. Robotics and automation are also making significant inroads in healthcare, enhancing sustainability through improved efficiency and precision. Automated pharmacy systems, for instance, can reduce medication errors and minimize waste, while robotic surgical technologies enable more accurate and less invasive procedures, leading to shorter hospital stays and quicker recovery times for patients.

In the realm of hospital operations, autonomous delivery robots and drones can transport supplies, medications, and even food and linens between different facilities or floors, optimizing logistical workflows and reducing the carbon footprint associated with manual transportation.

Case studies highlighting successful implementation of smart technologies for sustainability in healthcare.

3. Case study: Healthcare & technology in the movie, "Big Hero 6"

Baird et al. (2014) in the city of San Fransokyo, 14-year-old robotics prodigy Hiro Hamada lives a life full of ingenuity and innovation, Scattered by the loss of his older brother, Tadashi. The city, a vivid blend of traditional Japanese aesthetics and futuristic technology, provides the perfect backdrop for a tale of discovery, redemption, and the essential role of technology in shaping the future of healthcare. The movie "Big Hero 6″ brings to life a story that intertwines human emotions with technological advancements, presenting us a glimpse of how smart choices in technology can lead to sustainable futures in healthcare.

At the heart of the movie is Baymax, a lovable, inflatable robot designed by Tadashi as a personal healthcare companion. Baymax is not just robot; he is an advanced piece of technology equipped with the capability to monitor and manage human health comprehensively. From physical injuries to mental health, Baymax's role in the story highlights the potential of technology to revolutionize healthcare. As we delve deeper into this fictional narrative, it becomes clear that the movie presents an inspiring vision of how technology can be harnessed to create a more sustainable and responsible healthcare system.

4. The power of innovation in healthcare

The core of the movie's plot revolves around Hiro's struggle for justice after his brother's untimely death. Tadashi's creation, Baymax, emerges out to be the Hiro's companion and protector, embodying the promise of technology to care for and improve human life. Baymax is designed to monitor various health parameters like physical and mental health, diagnose conditions, and even provide treatment or refer the patient to a healthcare professional when necessary. His non-threatening, approachable design also makes him an ideal companion for those who might be intimidated by traditional medical equipment. As well as he is a kind of robot who hugs the Hiro when he was feeling low by identifying his emotions.

This fictional technology reflects a real-world trend: the increasing integration of advanced robotics and artificial intelligence (AI) in healthcare (Knopp et al., 2023). Modern-Era healthcare systems are already seeing the benefits of AI in diagnostics, treatment planning, and patient management. Robotics, too, is playing a crucial role in surgery, rehabilitation, and elderly care. The case of Baymax serves as an imaginative extension of current technological trends, envisioning a future where healthcare is more accessible, personalized, responsive and available to individual needs.

5. Physical health: the first frontier

Baymax's primary function is to monitor and maintain the physical and mental health of his user. This involves measuring vital signs, diagnosing injuries, and administering appropriate care. In the real world, such technologies are becoming increasingly common. Rawas (2024). Wearable devices like smartwatches and fitness trackers have already started to provide continuous monitoring of physical health indicators such as heart rate, blood oxygen levels, and even sleep patterns. These devices empower individuals to take charge of their health by providing real-time data and insights that can prevent potential health issues.

However, Rawas (2024) the potential of technology in physical healthcare extends far beyond wearables. Robotics-assisted surgery is one area where the fusion of human expertise and machine precision is delivering unprecedented outcomes. Robots like the da vinci surgical system are being used in complex surgeries, offering greater precision and control than the human hand alone. These advancements not only improve surgical outcomes but also reduce recovery times and the risk of complications.

6. Mental health: The uncharted territory

While technologies for physical health are advancing rapidly, the same cannot be said for mental health care, which remains an area in dire need of innovation. Mental health issues such as depression, anxiety, and stress are becoming increasingly prevalent, exacerbated by the fast-paced, high-pressure environments in which many people live today. Unlike physical ailments, mental health conditions are often less visible and more difficult to diagnose and treat. This is where Baymax's fictional capabilities offer a glimpse into a future where technology could play a crucial role in mental health care.

Baymax is designed to provide holistic healthcare, which includes the ability to recognize and respond to mental health needs. This aspect of his design underscores the importance of considering mental health as a critical component of overall well-being. In today's world, (Knopp et al., 2023), we are seeing the beginnings of such technology with AI-driven mental health apps and chatbots that offer support and resources for those struggling with mental health issues. These tools can provide immediate assistance, help users track their mood and mental state, and even connect them with human therapists if needed.

7. The intersection of ethics and technology

As Hiro modifies Baymax to seek revenge, the narrative touches on the ethical dilemmas that accompany the development and deployment of advanced technologies. In Hiro's grief and anger, he momentarily forgets that Baymax was created with the purpose of helping people, not harming them. This moment in the story serves as a reminder of the ethical responsibilities that come with technological innovation.

In the real world, the integration of AI and robotics in healthcare must be guided by ethical principles that prioritize patient safety, privacy, and the equitable distribution of healthcare resources (Knopp et al., 2023). As technology becomes more embedded in healthcare, the potential for misuse or unintended consequences also grows. It is crucial that developers, healthcare providers, and policymakers work together to establish clear ethical guidelines and regulatory frameworks that ensure these technologies are used responsibly.

8. Toward a sustainable healthcare future

The journey of Hiro and Baymax in "Big Hero 6″ ultimately leads to a deeper understanding of the importance of using technology for the greater benefits. This lesson is particularly relevant in today's context, where the sustainability of healthcare systems is a growing concern. As populations age and the prevalence of chronic diseases rises, the demand for healthcare services is increasing, putting a strain on resources. Smart technology, when used responsibly, can help alleviate some of this burden by making healthcare more efficient, accessible, and personalized.

For example, telemedicine and remote monitoring technologies can reduce the need for physical visits to healthcare facilities, saving time and resources for both patients and providers. Rawas (2024) AI-driven diagnostic tools can assist healthcare professionals in making faster and more accurate diagnoses, leading to better patient outcomes. Furthermore, the use of big data and predictive analytics can help in identifying public health trends and preventing disease outbreaks, contributing to a healthier and more sustainable society.

9. Greening healthcare facilities

Beyond clinical applications, smart technologies are also transforming the physical infrastructure of healthcare institutions. Sustainable building design, integrated renewable energy systems, and intelligent building management technologies are helping healthcare facilities reduce their environmental impact and align with the principles of the circular economy (Abaku and Odimarha, 2024).

For example, the incorporation of solar panels, geothermal heat pumps, and energy-efficient lighting and HVAC systems can significantly lower a hospital's energy consumption and greenhouse gas emissions. Smart building management systems, in turn, can automate and optimize the operation of these sustainable technologies, further enhancing energy efficiency and resource conservation.

10. Examples of green healthcare facility initiatives around the world

Chiu and Fong (2023) Hospitals are increasingly prioritizing sustainability due to environmental concerns, regulatory obligations, and the opportunity for cost reduction. "Prominent healthcare institutions are adopting various sustainability measures to minimize their impact on the environment and adhere to global guidelines, such as the UN Sustainable Development Goals and the WHO's Global Green and Healthy Hospitals initiative (Table 18.1).

11. Conclusion: Embracing technology for a healthier tomorrow

"Big Hero six" is more than just a story about a boy and his robot; it is a narrative that encourages us to think about the future of healthcare and the role of technology in shaping it. Baymax represents the ideal of what healthcare technology can aspire to be: compassionate, responsive, and focused on the well-being of individuals. As we advance into a future where technology becomes increasingly integral to our lives, it is essential that we make smart choices that prioritize ethical considerations and sustainability.

In doing so, we can build a healthcare system that not only addresses the needs of today but also prepares us for the challenges of tomorrow. By advancing healthcare

Table 18.1 Prominent Healthcare Institutions adopting sustainability measures.

Hospital Name	Country	Sustainability Headline
Anadolu medical center	Turkey	Received LEED Gold certification for its sustainable design focused on energy, water, and waste management.
Chang Gung Memorial Hospital	Taiwan	Utilizes green building design, solar power, energy-efficient equipment, and environmentally preferred purchasing.
Singapore General Hospital	Singapore	One of the earliest hospitals in Singapore to adopt sustainable practices like energy audits, recycling, and green procurement policies.
University health Network	Canada	A leader in green healthcare building design, energy co-generation, and comprehensive sustainability reporting.
Barts health NHS Trust	UK	Has cut emissions by 27% since 2015 through LED lighting, better waste management, and energy efficiency initiatives.
Vancouver Coastal health	Canada	Aims for complete carbon neutrality by 2030 through energy retrofits, greener transportation, and sustainable procurement practices.
Western health	Australia	Implements holistic environmental management programs across energy, water, waste, procurement, and building design.
Kaiser Permanente	USA	Has committed to being carbon neutral by 2020 through energy efficiency, renewable power, and sustainable food programs.
Gundersen health system	USA	Became energy independent in 2014 through renewable energy sources like wind, waste management, and conservation efforts.
Heidelberg University Hospital	Germany	One of the largest hospitals in Europe working toward carbon neutrality, with extensive recycling, green building design, and staff engagement programs.

Source: healthcare-digital.com

responsibly, we can ensure that technology serves as a force for good, creating a world where everyone has the opportunity to lead a healthy and fulfilling life.

References

Abaku, E.A., Odimarha, A.C., 2024. Sustainable supply chain management in the medical industry: a theoretical and practical examination. Int. Med. Sci. Res. J. 4 (3), 319–340.

Alim, M., Sulley, S., 2024. Beyond healing: embracing the triple bottom line approach in post-pandemic healthcare. Cureus 16 (2).

Antonio Grasso. October 3, 2023. The seamless fusion of environmental, social, and economic sustainability in the digital age.

Chiu, W.K., Fong, B.Y.F., 2023. Recent trends of research and education in ESG and sustainability. In: Environmental, Social and Governance and Sustainable Development in Healthcare. Springer Nature Singapore, Singapore, pp. 99–112.

Baird, R. (Producer), Hall, D., & Williams, C. (Directors). (2014). *Big Hero 6*. Walt Disney Animation Studios.

Knopp, M.I., Warm, E.J., Weber, D., Kelleher, M., Kinnear, B., Schumacher, D.J., … Turner, L., 2023. AI-enabled medical education: threads of change, promising futures, and risky realities across four potential future worlds. JMIR Med. Educ. 9, e50373.

Mortimer, F., Isherwood, J., Wilkinson, A., Vaux, E., 2018. Sustainability in quality improvement: redefining value. Fut. Healthc. J. 5 (2), 88–93.

Rawas, S., 2024. AI: the future of humanity. Discov. Artif. Intell. 4 (1), 25.

Schroeder, K., Thompson, T., Frith, K., Pencheon, D., 2012. Sustainable Healthcare. John Wiley & Sons.

CHAPTER 19

Embedding sustainability into work-life balance practices with special reference to healthcare sector

Ulfat Andrabi[1], Priyanka Chhibber[1] and Aaliya Ashraf[2]

[1]Mittal School of Business, Lovely Professional University, Phagwara, Punjab, India; [2]School of Business, Chandigarh University, Chandigarh, Punjab, India

1. Introduction

To enhance productivity improvement, sustainable businesses need effective methods for managing their crucial human resources (Deery et al., 2004). To achieve this, sustainable enterprises and communities should place a high priority on ensuring that people have a healthy balance between their personal and professional life (Pfeffer and Fong, 2002). Work-life balance (WLB) practices are deliberate organizational actions that aim to lessen employee work-life conflict and boost their efficacy in both their professional and personal lives (Lazar et al., 2010). Now of globalization, it has grown to be recognized as an important challenge for any commercial enterprise to address the issue of staff turnover. According to Chan et al. (2016) employee turnover is a significant problem, particularly in terms of human resource management. The senior management does not put their focus on this significant problem because it receives less attention than other problems. It is possible that they are not fully capable of understanding the circumstances surrounding how the turnover of employees harms the productivity of the firm. As a result, it is more important than ever to carry out research on employee turnover to assist business organizations in recognizing their issues, conducting an analysis of the data, and making suggestions for potential remedies considering recent developments. The term WLB relates to the incapacity of experts to stabilize their schedules in such a way that they can concurrently achieve their personal as well as their professional goals. It frequently takes place when there are no clear boundaries between one's private life and professional life. Many people who work in professional settings find it difficult to draw a line between their requirements and those of their employees.

1.1 Factors affecting work-life balance

Being familiar with the factors that contribute to work-life imbalances. It may better assess whether it is time to adjust your daily routine and personal expectations. The following are some of the implications that can imbalance work-life balance.

Intelligent Semantic Analysis for Healthcare
ISBN 978-0-443-33512-9
https://doi.org/10.1016/B978-0-443-33512-9.00016-0

1. **Burnout:** Burnout harm one's mental health and can be caused by long working hours every day. The effect of fatigue can include a decrease in motivation and focus, disorientation an increase in the probability of making mistakes on the job, and a reduction in overall productivity.
2. **Decrease Free Time:** If the job excessively or exceedingly is difficult, then run the risk of missing significant family events like birthdays and anniversaries or other memorable time spent with loved ones because of the work responsibilities. When employees have constantly deal with the impact on how they connect with their family and friends. Even though they may have to make certain concessions in their personal life to meet the requirements of their job.
3. **High-Stress Level:** Work-life imbalance can rise employees' anxiety levels. Constant approaches to high stress can have bad effects on employees' health. Further, if employees' involvement stresses at any point, they may have to struggle to control or manage their emotions. It's significant to discover openings for their anxiety that permit them to live contentedly. Exercise thought, and other procedures of self-care can benefit them to decrease the anxiety they feel in their position.
4. **Reduced Activity:** An unbalanced work and personal life might make it more difficult to maintain an active lifestyle, particularly for people whose jobs do not require them to engage in physical labour, such as those who work in offices. If you spend most of your day sitting at a desk or in a chair, you should try to build more walks and other opportunities to move around into your routine. Moving about and maintaining an active lifestyle can assist in stress management and lower the risk of developing physical problems.

To fulfil necessities and give an attractive working environment, good compensation, and other perks in an economic strategy is highly complex and difficult for an individual to accomplish organization. Each organization has the goal of maximizing its production while simultaneously reducing the rate of staff turnover to achieve financial success. The control of worker turnover must achieve company objectives effectively. The importance of the work-life balance concept is steadily rising in the quickly, international society of today. The rising discussion regarding how to attain stability between job and private work and private life is due in part to 3 factors: 1) fast technical apprises, 2) extensive changes in opportunities for both workers and companies and 3) operative comprehension of the line between work and personal life is becoming more ambiguous because of emerging apps. Therefore, companies that appreciate their employees' diverse functions in addition to exploiting them as physical input to create output are highly valued by their workforce. These businesses routinely take steps to support a healthy balance between employees' personal and professional lives.

2. Research questions

RQ1: To study the HR practices followed by healthcare employees in WLB.
RQ2: To study HR practices Challenges of healthcare employees in WLB.

3. Review of literature

3.1 Conceptualization of work-life balance

The idea of WLB is becoming more and more common in the media. It is a statement that businesses, especially large ones, commonly employ and that is frequently touted as the cornerstone of their corporate welfare, for example. Nevertheless, despite the term's widespread use, the WLB idea is only partially understood by scholars. The notion of work-life balance extends beyond the family-friendly perspectives of previous HR practices, as it seeks to support all hired individuals, irrespective of marital or parental status, facilitating a more harmonious integration of professional and personal lives (White, 2003). More than ever before, people are worried about how job and family life interact and how they may affect one another. Work-life balance encompasses other aspects of life where people have duties and responsibilities to perform, in accumulation to the interaction among work and family responsibilities. These may consist of interests, pastimes, social activities, and other activities one engages in that enhance one's well-being apart from job and family obligations. Work-life balance was described by Gunz and Peiperl (2007) as the capacity to attain productivity and satisfaction in both work and personal life while reducing conflicts between the two (Felstead et al., 2002) went on to explain that, particularly in cultures where the labor market serves as the primary source of income, this balance is directly related to the cultural and institutional norms surrounding work and non-work domains. This term was developed Gelfand et al. (2007) by introducing the idea of "life balance," which includes both work and personal obligations. Overall, having a life balance involves being able to satisfactorily balance the demands of work, family, and personal life. This may require putting in long hours, working hard, and making consistent effort in all areas. The quality of domestic and familial life is adversely affected by extended work hours and elevated job intensity or pressure, leading to fatigue, stress, or other unfavorable psychological consequences. By "pleasure and excellent functioning at work and home with a minimum of role conflict," Clark (2000) defines balance. As a result, the phrase "work-life balance" (WLB) states the need for both employees and employers to strike stability between their personal and professional commitments. Doctors might be considered "the backbone of the medical system" in contemporary healthcare because they are the first line of patient treatment. Consequently, it is essential to safeguard their welfare and a healthy work-life balance to preserve the standard of doctors and, the standard of healthcare facilities. Working in the healthcare industry is undoubtedly stressful and demanding due to the long hours, unpredictable schedule, risks to health and safety, intense workload, patient-related stressors, poor pay, and disruptions to sleep. Job stress can be troublesome because overworked doctors may not be able to give adequate treatment, suffer from physical and mental health problems, get into arguments at work, take too many sick days, or quit their positions. To create an environment that will optimize healthcare employees' well-being and ability to maintain a healthy work-life balance, healthcare organizations must thoroughly assess their organizational culture. Changing the culture of the company will enable nurses to deliver superior medical care be more productive and raise patient happiness. Work-life balance does not mean dividing up one's time equally between work, family, and personal obligations. When it comes to the imbalance between work and life, the environment of the workplace has a stronger influence than the environment of the home. There are demands placed on a person's time, effort, and mental capacity by the organisation in which they work as well as the workplace itself. In their efforts to improve organizational effectiveness, managers place particular emphasis on raising employee organizational loyalty as one of their main areas of concentration.

3.2 COVID 19: Work-life balance

During the COVID-19 epidemic, healthcare workers come across considerable psychological problems that hurt their well-being. These problems include role conflicts, social isolation, anxiety, and worry (Chen et al., 2020). In previous pandemics, particularly the COVID-19 epidemic, the additional obligation of nurses because of an enhanced figure of circumstances and capability, as well as the inadequate figure of staff, may have initiated and may cause decreased work-life balance and mental complications (Karaarslan et al., 2021). This is especially true since there was an insufficient number of personnel. Magnavita et al. (2020) this study displays an unprotected exposure to infected patients is one of the variables that might lead to mental health issues like anxiety and sadness, as well as disrupted sleep patterns among healthcare professionals. Due to a lack of available personal protective equipment, those working in the healthcare industry are among the most likely to become sick. It is difficult for healthcare professionals to meet their own requirements, the requirements of their families, and the needs of their patients when they have a limited number of resources, which can have a severe impact on both their physical and mental health (Greenberg et al., 2020) found that a study that was carried out during the recent COVID 19 outbreak revealed that the epidemic generated a large deal of distress among health personnel. The epidemic encouraged front-line nurses to experience negative feelings (Sun, 2025). In the process of performing their jobs, nurses commit themselves to a heightened risk of mortality (Catton, 2020). Research from 2020, one of the factors that may contribute to a rise in the levels of depression, stress and anxiety experienced by healthcare professionals is a lack of faith in safety protocols (Magnavita et al., 2020). Fear of COVID-19 is one of the potential adverse effects that could hurt the psychological (mental) health of healthcare staff and nurses. An obsession is marked by an unusual dread and distress felt in response to a definite circumstance or object (going on a plane, heights, injections, etc.). This fear and anxiety can manifest in a variety of ways (Albashrawi et al., 2017). Individuals use fear as a form of defense when confronted with potentially harmful circumstances (Seçer et al., 2020). On the other hand, having a high level of anxiety may make an individual more susceptible to psychological issues (Shanafelt et al., 2015). During the COVID-19 epidemic, according to the findings of a study (Panisoara and Serban, 2013), medical professionals are concerned about passing on a sickness to their families and are emotionally distressed because of this anxiety. Nurses perform a vital role in the delivery of healthcare services, they must have a healthy balance of emotional comfort to ensure that the care they offer is beneficial, that excellence and execution are improved, and that efficiency is preserved (Aksu et al., 2020). According to Sexton et al. (2000) worsened psychological health decreases patient safety, the excellence of treatment, and nurse efficiency, and it also influences the profitability of institutions.

3.3 Conceptualization of sustainability

Research on sustainability consistently faces the challenge of overcoming the uncertainty and absence of simplicity that surrounds the idea of sustainability. Throughout the last few years, several studies that have been dedicated to the examination of the concept of sustainability have focused on the ambiguity that it produces as well as on the reasons for the lack of clarity regarding the concept, specifically its wide range of applications. According to the context or field (Cristina and Diana, 2014; Glavič and Lukman, 2007; Mebratu, 1998; Rassafi et al., 2006) in which the phrase is employed. Since there are so many various meanings of the idea, Mebratu (1998) asserts that there are over 80 diverse definitions of "sustainable development," while Ciegis assert that there are over 100 definitions of the term in the context of economics. However, none of the studies provides an explanation of these definitions. A lack of understanding about what "sustainability" means can cause problems for researchers because it can make it hard to put the idea into practice, lead to different ideas about the topic, and even make the studies less reliable. For example, say that many books, papers, and articles with the concept "sustainability" or "sustainable" in the title do not explain what it means. came to a similar conclusion. They looked at scientific research papers released in 2013 and found that 91.3% of those with the word "sustainability" in the theme did not explain what it meant. Particularly in research whose objective is to examine the sustainability of a specific framework, this could generate difficulties with the validity of constructs. This is due to research studies with this weakness being unable to exhibit that the factors and groups that were analysed are in line with a specific idea of sustainability. The lack of a definition raises the possibility that the researcher will focus their attention on something other than sustainability. According to researchers (Benito-Osorio et al., 2014), it is challenging to apply the concept of sustainability to research because it is not well defined. It can be challenging for a scholar to explain the set of classifications and/or variables that allow the opinion of sustainability in a particular structure, particularly when sustainability is the subject of the research. This is because an inexact explanation of the term or the use of a conventional title about it makes the concept challenging to elaborate. This is due to the concept's vague definition or the usage of a normative notion regarding it being a normative conception. When investigating sustainability as a topic for research, it is necessary to employ a scientific understanding of the term "sustainability."

3.4 Work-life balance practices

A requirement for establishing sustainability in healthcare has been recognized as the involvement of health professionals, which implies increased employee involvement in the administrative processes of healthcare organizations and maintenance of a stronger organizational commitment (Mohammad Mosadeghrad, 2013). Employee work involvement generally refers to the use of human resource management strategies that are specifically designed to encourage employee involvement in decision-making regarding their place of employment and provide them with the authority to improve the working environment (Cheesborough et al., 2017). Numerous researches have been done on the variables influencing work-life balance. The effect of demographic characteristics on the work-life balance among academic staff at the University of Kelaniya was examined by Mohammad Mosadeghrad (2013). They demonstrated that gender and marriage status as demographic determinants have a considerable impact on WLB. Korican Lajtman (2016) contributed to the body of knowledge on work-family issues that explain work-life problems. The findings of the research indicate that work has a greater magnitude on life than life has on work. Since married young professionals reported much more work-life conflict than other groups, marriage status arise as a strong interpreter of work-life confront. For women working in the IT sector (Vasumathi, 2018) identified the elements that may cause a work-life balance imbalance. She concluded that while all female employees look to their families for assistance, work-life balance is not much impacted by family structure. Apart from stage and conjugal status, other factors that were highlighted included company policy, which included paid maternity leave, vacations, religious leave, and leave collecting; public disposition (at home and work), which included spouse backing, family support, superior support, and colleague support; work flexibility; and transportation. Without these factors, it would be difficult for a woman to continue working. In the study on married working women. A successful WLB requires active engagement in tasks that result in beneficial results by directly transferring skills, abilities, affect, and beliefs from the workplace to the life sector. The fact that active participation in job responsibilities leads to realistic personal goals (such general life happiness) indicates the extent to which role engagement in one arena increases enjoyment and performance in another (Khare and Kapoor, 2019) found that the partner's occupation also had a major impact on work-life equilibrium. The people who were most impacted by this problem were those whose husbands worked in business and marketing. Sakthivel and Jayakrishnan conducted a study on nurses' organizational commitment and WLB in 2013. In this study, 328 nurses from both public and private hospitals were chosen and descriptive statistics and correlation analysis were used to conclude that professional life significantly interferes with family life, but less so with personal life. The study also revealed that nurses felt that having a better work-life balance inspired them to give more of themselves to their organizations and deliver greater results.,. WLB is the capacity of an individual to balance the requirements of their job with their personal and family requirements. Other researchers assert that other elements can be implemented through the provision of bonuses (benefits), health and fitness facilities, aid for the care of children and their old parents, and a strategy of leave for illness. It is desired that workers can achieve a balance between their obligations at home and work. Kim and Han (2014) completed a study on the link between employees' performance and work-life balance as well as the mediating effect of affective commitment. Understanding the multiple perspectives that can assist with maintaining an effective work-life balance has become highly prevalent as the vitality of WLB and stress at the personal and professional levels has increased. The study's major objective was to inspect the linkage between WLB and in-role accomplishment in the Korean scenario as well affective commitment acts as a mediating role. The study used structural equation modelling and regression analysis to evaluate the assumptions on a sample of 293 workers. Pennock (2016) investigated how women in senior management roles balance a successful job and family life in Austria, Denmark, and the UK. According to the results, women need flexibility from their employers as well as support from their families and partners to achieve a healthy work-life balance. The influence of functional periods on a worker's life-style balance is influenced by the total hours worked and the employee's perception of control over the timing and arrangement of their work schedule, as noted by Golden et al. (2011). A worker's perception of lacking discretion may be equally damaging to their ability to sustain a healthy work-life balance as working extra or extended hours. Research involving 100 respondents from a government execution agency in Malaysia identified work load as the primary reason influencing life-style balance, with role dispute as the secondary factor. Shaikh and Dange (2017) investigated employees' work-life balance in Mumbai IT corporates. The findings emphasized the importance of a workplace environment that facilitates a balance between work and personal life. It highlighted the importance of managers in fostering work-life balance. Factors influencing work-life balance in Jaipur, Rajasthan banking personnel, also include working circumstances, facilities, efficiency, support, freedom, and satisfaction (Mebratu, 1998). Research indicates that factors affecting the work-life balance (WLB) of female employees in educational institutions encompass marital status, duration of employment, the necessity for flexibility, extra working hours and overtime, the number of dependents, childcare requirements, the rationale for over shift, as well as the support from the functional head and the ability to openly address WLB concerns with the functional head. Sahrawat and Rawat (2021) conducted a study titled "Work-life and job satisfaction among corporate employees," which demonstrated a robust correlation between work-life balance and job satisfaction in this demographic. The results imply that a greater or adequate mark of job satisfaction influences corporate employees' ability to retain a fabulous work-life balance. Maintaining a positive work-life balance will make you more content in your employment. It has been discovered that job happiness is also correlated with work-life balance. WLB techniques should be put into operation by businesses to retain employees, reduce employee turnover, and improve employee job satisfaction, all of which contribute to higher efficiency and benefits. Work time features and WLB-health outcomes were evaluated in research on WLB-related factors for Korean Employees It was well known that Korea did not study the correlation between work-life balance (WLB) and health sufficiently. The information was gathered using the third Korean Working Conditions Survey, which looked at a sample of working people. This paper looked at overall characteristics, working time structure, work-related health, and WLB. Multivariate logistic regression was used to evaluate the relationships between work-life balance and health outcome variables such as health status, mental health, work-related health hazards, sick absenteeism, presentism, WMSD (work-related musculoskeletal disorder), headache/eye strain, and weariness. The general characteristics of the WLB were discovered to include gender, age, work sector, occupation, and job type. Working time characteristics were also significantly linked to WLB. WLB and several outcomes for well-being were also closely related. They showed the link between WLB's health results and Korean employment. The results suggested that therapies ought to include work-time metrics for better health.

4. Research methodology

This study's approach was taken from Manoharan and Singal (2017) work. The current study uses a five-step literature review methodology, including classifying the period, choosing databases, and journals, choosing articles, and classifying those papers.

4.1 Time frame

The analysis period included 19 years, from 1998 to 2021. To facilitate data gathering, the year 1998 was chosen as the starting point.

4.2 Collected information

The outline for this assessment is determined by the database choice, which is quite important. From some of the most popular databases, including Science Direct, Google Scholar, EBSCO, and Emerald Insight, pertinent articles are carefully selected.

4.3 Journals selected

Due to the interdisciplinary character of WLB, most publications are dispersed over numerous journals. The articles were first found using numerous databases to extend the search; after that, 50 journals in the hospitality industry, comprising the best journals for MDPI were found.

4.4 Article selection

The researchers utilized several general terms, including "work-life balance," "work-life satisfaction," "work-family satisfaction," "Sustainability," and "work-life conflicts," to make sure that all pertinent articles relating to the research of WLB were included. To avoid overlooking any important information, this was done.

4.5 Classification of articles

Articles were categorized into two areas to meet the review's goal: work-life balance and sustainability.

5. Findings

Because of the demanding and stressful nature of their work, those who work in the healthcare industry must prioritize set upon a healthy equilibrium between their professional and private lives. Some suggestions and methods supported by research for sustaining a healthy work-life balance in the healthcare industry are:

Establish limits: It is essential to set clear boundaries between one's personal life and one's professional life. Setting clear work hours, avoiding reading emails or taking work calls during personal time, and delegating responsibilities wherever it is possible are all examples of things that could fall under this category. According to a study, setting limits can improve work-life balance and reduce workplace stress.

Practice self-care: Self-care should be a priority for anybody working in the healthcare industry, and this includes making time for things like exercise, meditation, and hobbies outside of work. According to the findings of a study conducted by Sexton et al. (2000) healthcare professionals who participated in activities designed for their self-care reported greater levels of job satisfaction and lower rates of burnout.

Take breaks: Taking pauses, even brief ones, during the workday can help enhance productivity and reduce stress. According to the findings of a study conducted by Trougakos et al. (2015), taking brief breaks at various points throughout the workday helped improve the well-being of healthcare professionals and their level of job satisfaction.

Seek help: Individuals working in healthcare should look for assistance from their co-workers, managers, as well as family, and friends. According to the findings of a study healthcare personnel who received social support from their co-workers and superiors had much lower rates of burnout.

Prioritize Sleep: Sleep should be a priority for healthcare professionals because not getting enough sleep can lead to lower job performance and increased stress levels. Sleep deprivation should be a priority for healthcare practitioners. According to the findings of a study carried out by Geiger-Brown et al. (2015), healthcare professionals who got at least 7 hours of sleep each night had much-reduced rates of burnout.

5.1 Importance of sustainable work-life balance for healthcare professionals

For healthcare workers to give high-quality treatment while also preserving their own physical and mental well-being, a sustainable work-life balance is essential. Here are some methods and studies that back up a healthy work-life balance in the healthcare industry.

Flexible schedule: The use of flexible schedules can assist individuals working in the healthcare industry in striking a balance between their personal and professional lives. According to the findings of a study conducted by Sexton et al. (2000), employees who had more flexible schedules reported having a better work-life balance and were less likely to quit their jobs.

Self-Care: The act of providing for one's health and well-being is an important component of working in the healthcare field. This involves getting the recommended amount of sleep, engaging in regular physical activity, and cultivating a thoughtful attitude.

Supportive workplace: Stress reduction and increased job satisfaction can be aided by a supportive work environment that prioritizes finding a work-life balance. The study found that having a supportive work environment was associated with lower rates of burnout and better levels of satisfaction among physicians (Shanafelt et al., 2015).

Teamwork: Working together as a team can assist medical professionals in better managing their workload and providing support to one another. to increase work satisfaction and reduce burnout among healthcare professionals.

5.2 Need for bringing sustainability in WLB practices

WLB is a valuable factor in modern HRM and has received substantial consideration from academicians, the government, and employees. In this era of globalization, the HRM concept has grown from its original attention on benefits and management to an emphasis on employee satisfaction. Over the past few decades, the shift has been towards the employee-oriented aspects of High Involvement Management (HIM), Flexible management, Family friends HRM, and Work-Life balance. Due of the potential benefits it could provide both businesses and employees, this WLB concept is now widely accepted. Employees can benefit from increased well-being, less work-family conflicts, and more life satisfaction. Employees who have a healthy work-life balance are more creative and productive for their employers.

In addition to identifying practices that have the potential to improve work-life balance and negative consequences if not implemented. By offering flexible working options, firms can significantly aid all employees in reaching WLB. Through these flexible working hours, employees get the option to choose their time and place of work. As each employee may have his/her reasons to need a work-life balance: Some might require extra time to care for their kids, while others could require financial assistance. Because they are less worried about their personal affairs because of all these activities, employees might work harder. Organizations may, however, be unwilling to invest further funds in their workers if they are unsure that the investment would yield sufficient returns. Employees and others who do not require flexible work may abuse this situation and feel that it is unfair that they are not entitled to such assistance. Every obstacle people face increases the strain, which in turn causes tension, which causes emotional discord, and dissatisfaction. To maintain a healthy work-life balance, it is crucial to overcome obstacles. There are several causes of an unbalanced work-life balance, including:

Unreasonable demands at work: No doubt every organization is running for achieving its specific goals and objectives. For the attainment of these objectives, proper communication is required between employers and employees. Weekend meetings are arranged for the same, which might be tiring for the employees. Realistic demands from employees lead to conflict at work, such as one employee's urgent need for a weekend vacation or another's need for a holiday to visit his or her parents once a month. Employers shouldn't make demands like these because they disrupt the balance between work and life.

Turnover intentions:

Numerous work-related or non-work-related variable triggers intentions to quit; limited career opportunities, poor remuneration, unacknowledged hard work, and deficient flexible working policies. Work-family conflicts are one of the prominent causes behind the intention to leave the organization. Often, women compromise their careers if their spouse must relocate when relocation is important for career growth. On the part of women, expectations arising out of roles played by them such as "caretaker", "homemaker", "employee", "mother", "colleague", and "wife" make them overburdened and exhausted. Quitting their workplace responsibilities is the only way out left with them.

5.3 Sustainable human resource policies that indorse work-life balance

1. Adaptable Working Schedules:

Delayed hours, working from home, shortened workweeks, job sharing, and job exchanges are just a few of the adaptable work arrangements that enable people to combine work and family (Mohd Mahudin, Noor, Dzulkifli and Janon, 2016). Government sector personnel were made aware of the flexible start and finish time policy in June 2007 (Chandra, 2012). By this work schedule, nurses and physicians working in administration and management may pick between three different shift options: 7:30 a.m. to 4: 30 p.m.; (ii) 8:00:00 a.m. to 5:00:00 p.m.; and (iii) 8:30 a.m. to 5:30 p.m. The Human Resources Ministry introduced a policy allowing some personnel, exclusively the incapacitated or those with offspring's, to engage in remote work in October 2008. Part-time employment, which was introduced following the Employment Act 1955 revision in October 2010, is the third type of flexible working arrangement. According to this provision, part-time workers receive the same salary, assistance, and shields as full-time workers in terms of pay, vacations, public leave, and paid sick leave.

2. **Childcare Policy:**

   The Childcare Act of 1984, which puts forward a structure for debate on childcare provision, improves childcare policies. The introduction of daycare at the workplace is encouraged by this policy. It also highlights how crucial childcare assistance is in enabling parents, particularly mothers, to work. Subsidies and financial incentives are two government programs that are directly tied to this Act. These programs support childcare facilities and flexible work schedules. According to Mutter (2024), government personnel earning RM2,000 per month get an RM180 per kid rebate for subsidized daycare. In 2009 the expansion of this program to include households making less than RM3000 per month. To motivate firms, particularly those in the private sector, to establish their childcare conveniences.

3. **Job contribution and job separation:**

   Job sharing is a system in which a job is divided between two employees, each with predetermined working hours. A job-sharing relationship is believed and have occurred when two employees shared a full-time position. All aspects of employment, including compensation and benefits, are shared, claims the Tariq et al. (2012). The many job-sharing patterns include a Split week, which has a 5-day workweek and requires each employee to work for two and a half days, and a Split day, which is a division of the day. When one employee operates in the morning and the other in the afternoon, or the other way around, Week overlap: both employees work 3 days a week, creating an overlap of 1 day; different weeks: each worker works the entire week and receives the identical amount of time off from work. A job-sharing procedure can only be effective if each personnel can give it his all. When a job-sharing program is correctly handled, both the organization and the employees can profit since happier employees are more productive workers (Fong et al., 2011). The preferred technique for preventing layoffs is a job pooling program that is successful.

1. **Term-time working/developing:** Employees are hired under this system on long-term deals and are permitted to take time off during school breaks, either on a paid or unpaid basis. Working parents with school-aged children who want to

invest time with their families at home will benefit from such an approach (Tariq et al., 2012).

2. **Delayed hours:** The arrangement in which start time, lunchtime, and finish time is different for different people is also referred to as "Rota working". Employees in this system have set working hours, but they also have individual starting, finishing, and lunch/break times. Additionally, it aids in avoiding rush-hour traffic (Tariq et al., 2012). Delayed hours are a key method of reducing traffic, claims 2004 research by the International Labour Office Geneva. Workers must take care to set commencement and closing times between the suggested range.
3. **Job breaks and unpaid leaves:** This approach empowers the employees to take a fixed sum of time off from work to address personal or family matters. This Programme is a crucial tool for keeping on-board employees who will benefit the company. Voluntary leaves are brief absenteeism that generally lasts 1 day to 3 months. Breaks in the job, and extended leaves of absence, typically lasting three to 12 months (Tariq et al., 2012). Professionals normally take unpaid breaks from their jobs to address personal commitments (European Investment Bank, 2005).

5.4 Striking an equilibrium between work-life

Life and work are there to enjoy—not bringing you down! The choice is yours!

Work frequently takes priority over all other aspects of our existence. We could neglect our health to advance professionally. However, striking a healthy work-life balance is essential if you want to advance your career as well as your physical, emotional, and mental health. Work-life balance, to put it briefly, is a condition of equilibrium where a person can balance the demands of their personal and professional lives.

5.4.1 Employees can strike a balance

- By choosing between what your organization wants from you and what your family needs from you. Create a good schedule for yourself thinking of the best way to maintain an equilibrium among work and life.
- If you feel that your workplace has a toxic work environment and you do a job that you do not love or you could not be able to perform it in the best possible way, it's time to change to a new job. Do prioritize your choice of job, the job you dreamt to do.
- Your physical, mental, and emotional health should be your main concern. Prioritizing your health first and foremost will make you a better employee and person. You could be able to work more and be productive and innovative.
- Set limitations for yourself and for your co-worker's, to evade stress. When you leave your workplace, avoid thinking of workplace matters till night. Try to take different mobile phones for your workplace and shut them if when reached

home. In your workplace, your colleagues will come to realize that you are not accessible as you are engaged with your activities.

6. Conclusion

A crucial component of the formula for the long-term stability of healthcare organizations is internal organizational sustainability. For a business to be internally sustainable, human resource management strategies that support health professionals are essential. Hence, the healthcare industry must follow and implement practices that focus on the work-life balance of healthcare staff. The ongoing struggle to strike a balance between one's responsibilities to one's family and one's job can have major effects on a person's health and general quality of life. Health and wellness programs can, for sure help healthcare staff in balancing work and life imbalances. The key to effective work-life management; developing interactive and interesting training sessions for an employee on work-life balance, open communications is another way out, providing incentives who are overworking in the organization and embedding employee engagement practices.

References

Aksu, K., Naziroğlu, T., Özkan, P., 2020. Factors determining COVID-19 pneumonia severity in a country with routine BCG vaccination. Clin. Exp. Immunol. 202 (2), 220–225. https://doi.org/10.1111/cei.13507.

Albashrawi, M., Kartal, H., Oztekin, A., Motiwalla, L., 2017. The impact of subjective and objective experience on mobile banking usage: an analytical approach. In: Presented at the Hawaii International Conference on System Sciences. https://doi.org/10.24251/HICSS.2017.137.

Benito-Osorio, D., Muñoz-Aguado, L., Villar, C., 2014. The impact of family and work-life balance policies on the performance of Spanish listed companies. M@n@gement 17 (4), 214–236. https://doi.org/10.3917/mana.174.0214.

Catton, H., 2020. Nursing in the COVID-19 pandemic and beyond: protecting, saving, supporting and honouring nurses. Int. Nurs. Rev. 67 (2), 157–159. https://doi.org/10.1111/inr.12593.

Chan, K.M.A., Balvanera, P., Benessaiah, K., Chapman, M., Díaz, S., Gómez-Baggethun, E., , … Turner, N., 2016. Why protect nature? Rethinking values and the environment. Proc. Natl. Acad. Sci. 113 (6), 1462–1465. https://doi.org/10.1073/pnas.1525002113.

Chandra, V., 2012. Work–life balance: eastern and western perspectives. Int. J. Hum. Resour. Manag. 23 (5), 1040–1056. https://doi.org/10.1080/09585192.2012.651339.

Cheesborough, J.E., Gray, S.S., Bajaj, A.K., 2017. Striking a better integration of work and life: challenges and solutions. Plast. Reconstr. Surg. 139 (2), 495. https://doi.org/10.1097/PRS.0000000000002955.

Chen, N., Zhou, M., Dong, X., Qu, J., Gong, F., Han, Y., , … Zhang, L., 2020. Epidemiological and clinical characteristics of 99 cases of 2019 novel coronavirus pneumonia in Wuhan, China: a descriptive study. Lancet 395 (10223), 507–513. https://doi.org/10.1016/S0140-6736(20)30211-7.

Clark, S.C., 2000. Work/family border theory: a new theory of work/family balance - sue Campbell Clark, 2000. Hum. Relat. Retrieved from https://journals.sagepub.com/doi/abs/10.1177/0018726700536001.

Cristina, B., Diana, A., 2014. The perspective of concept sustainability. Proced. Soc. Behav. Sci. 116, 2257–2261. https://doi.org/10.1016/j.sbspro.2014.01.555.

Deery, M., Jago, L., Fredline, L., 2004. Sport tourism or event tourism: are they one and the same? J. Sport Tourism 9 (3), 235–245. https://doi.org/10.1080/1477508042000320250.
Felstead, A., Jewson, N., Phizacklea, A., Walters, S., 2002. Opportunities to work at home in the context of work-life balance. Hum. Resour. Manag. J. 12 (1), 54–76. https://doi.org/10.1111/j.1748-8583.2002.tb00057.x.
Fong, C., Ooi, K., Tan, B., Lee, V., Yee-Loong Chong, A., 2011. HRM practices and knowledge sharing: an empirical study. Int. J. Manpow. 32 (5–6), 704–723. https://doi.org/10.1108/01437721111158288.
Geiger-Brown, J.M., Rogers, V.E., Liu, W., Ludeman, E.M., Downton, K.D., Diaz-Abad, M., 2015. Cognitive behavioral therapy in persons with comorbid insomnia: a meta-analysis. Sleep Med. Rev. 23, 54–67. https://doi.org/10.1016/j.smrv.2014.11.007.
Gelfand, M.J., Erez, M., Aycan, Z., 2007. Cross-cultural organizational behavior. Annu. Rev. Psychol. 58 (1), 479–514.
Glavič, P., Lukman, R., 2007. Review of sustainability terms and their definitions. J. Clean. Prod. 15 (18), 1875–1885. https://doi.org/10.1016/j.jclepro.2006.12.006.
Golden, S.A., Covington, H.E., Berton, O., Russo, S.J., 2011. A standardized protocol for repeated social defeat stress in mice. Nat. Protoc. 6 (8), 1183–1191. https://doi.org/10.1038/nprot.2011.361.
Greenberg, N., Docherty, M., Gnanapragasam, S., Wessely, S., 2020. Managing mental health challenges faced by healthcare workers during covid-19 pandemic. BMJ 368, m1211. https://doi.org/10.1136/bmj.m1211.
Gunz, H.P., Peiperl, M., 2007. Handbook of Career Studies. SAGE Publications.
Karaarslan, F., Demircioğlu Güneri, F., Kardes, S., 2021. Postdischarge rheumatic and musculoskeletal symptoms following hospitalization for COVID-19: prospective follow-up by phone interviews. Rheumatol. Int. 41 (7), 1263–1271. https://doi.org/10.1007/s00296-021-04882-8.
Khare, A.K., Kapoor, A., 2019. A study on work life balance among women labor in garment units of NCR, India. Soc. Sci. Res. Netw. [SSRN Scholarly Paper] https://papers.ssrn.com/abstract=3526712.
Kim, Y.J., Han, J., 2014. Why smartphone advertising attracts customers: a model of Web advertising, flow, and personalization. Comput. Hum. Behav. 33, 256–269. https://doi.org/10.1016/j.chb.2014.01.015.
Korican Lajtman, M., 2016. Impact of Personal Factors on the Work Life Conflict and its Co-influence with Organizational Factors on Employee Commitment in Croatia.
Lazar, I., Osoian, C., Ratiu, P.I., 2010. The Role of Work-Life Balance Practices in Order to Improve Organizational Performance. https://www.um.edu.mt/library/oar/handle/123456789/31996.
Magnavita, N., Tripepi, G., Di Prinzio, R.R., 2020. Symptoms in health care workers during the COVID-19 epidemic. A cross-sectional Survey. Int. J. Environ. Res. Publ. Health 17 (14), 5218. https://doi.org/10.3390/ijerph17145218.
Manoharan, A., Singal, M., 2017. A systematic literature review of research on diversity and diversity management in the hospitality literature. Int. J. Hospit. Manag. 66, 77–91. https://doi.org/10.1016/j.ijhm.2017.07.002.
Mebratu, D., 1998. Sustainability and sustainable development: historical and conceptual review. Environ. Impact Assess. Rev. 18 (6), 493–520. https://doi.org/10.1016/S0195-9255(98)00019-5.
Mohammad Mosadeghrad, A., 2013. Obstacles to TQM success in health care systems. Int. J. Health Care Qual. Assur. 26 (2), 147–173. https://doi.org/10.1108/09526861311297352.
Mohd Mahudin, N.D., Noor, N., Dzulkifli, M., Janon, N., 2016. Religiosity among muslims: a scale development and validation study. Makara Hubs-Asia 20, 109–120. https://doi.org/10.7454/mssh.v20.
Mutter, J., 2024. Giving children their due voice in human resource management research. Lab. Ind. 34 (2), 132–154. https://doi.org/10.1080/10301763.2023.2244744.
Panisoara, G., Serban, M., 2013. Marital status and work-life balance. Proced. Soc. Behav. Sci. 78, 21–25. https://doi.org/10.1016/j.sbspro.2013.04.243.
Pennock, M., 2016. Wellbeing, economic development and economic growth: a literature synthesis and proposed theory. Int. J. Hydropower Dams 3 (1), 44–63. https://doi.org/10.1504/IJHD.2016.076196.
Pfeffer, J., Fong, C.T., 2002. The end of business schools? Less success than meets the eye. Acad. Manag. Learn. Educ. 1 (1), 78–95. https://doi.org/10.5465/amle.2002.7373679.

Rassafi, A.A., Poorzahedy, H., Vaziri, M., 2006. An alternative definition of sustainable development using stability and chaos theories. Sustain. Dev. 14 (1), 62–71. https://doi.org/10.1002/sd.268.

Sahrawat, P., Rawat, P., 2021. Work Life Balance and Job Satisfaction Among Corporate Employees. https://doi.org/10.13140/RG.2.2.26694.19522.

Seçer, İ., Ulas, S., Karaman-Özlü, Z., 2020. The effect of the fear of COVID-19 on healthcare professionals' psychological adjustment skills: mediating role of experiential avoidance and psychological resilience. Front. Psychol. 11. https://doi.org/10.3389/fpsyg.2020.561536.

Sexton, J.B., Thomas, E.J., Helmreich, R.L., 2000. Error, stress, and teamwork in medicine and aviation: cross sectional surveys. BMJ 320 (7237), 745–749. https://doi.org/10.1136/bmj.320.7237.745.

Shaikh, P., Dange, U., 2017. Antecedents of work life balance: a study of healthcare sector. IRA-Int. J. Manag. Soc. Sci. 8 (1), 1. https://doi.org/10.21013/jmss.v8.n1.p1 (ISSN 2455-2267).

Shanafelt, T.D., Hasan, O., Dyrbye, L.N., Sinsky, C., Satele, D., Sloan, J., West, C.P., 2015. Changes in burnout and satisfaction with work-life balance in physicians and the general US working population between 2011 and 2014. Mayo Clin. Proc. 90 (12), 1600–1613. https://doi.org/10.1016/j.mayocp.2015.08.023.

Sun, Z., 2025. Determining human resource management key indicators and their impact on organizational performance using deep reinforcement learning. Sci. Rep. 15 (1), 5690. https://doi.org/10.1038/s41598-025-86910-2.

Tariq, A., Aslam, H.D., Siddique, A., Tanveer, A., 2012. Work-life balance as a best practice model of human resource management: a win-win situational tool for the employees and organizations. Mediterr. J. Soc. Sci. 3 (1), 577–585.

Trougakos, J.P., Beal, D.J., Cheng, B.H., Hideg, I., Zweig, D., 2015. Too drained to help: a resource depletion perspective on daily interpersonal citizenship behaviors. J. Appl. Psychol. 100 (1), 227–236. https://doi.org/10.1037/a0038082.

Vasumathi, A., 2018. Work life balance of women employees: a literature review. Int. J. Serv. Oper. Manag. 29 (1), 100–146. https://doi.org/10.1504/IJSOM.2018.088477.

White, R.E., 2003. Soils for Fine Wines. OUP, USA.

CHAPTER 20

The role of the National Health Digitization Agency (NHDA) in digitizing the Algerian health sector: Between reality and prospects contributors

Yalouli Tarek[1], Bouguerra Mabrouk[1], Sonali Vyas[2], Grari Sabrina[1] and Ouali Khawla[1]
[1]Faculty of Economics, Commerce and Management Sciences Badji Mokhtar, Annaba, Algeria; [2]IET, JK Lakshmipat University, Jaipur, Rajasthan, India

1. Introduction

It is no secret to anyone that modern technologies in information and communication have become today one of the driving forces that depend on the development of social and economic programs, regionally or nationally, and in all sectors without exception. Comprehensive digital, taking into account the priorities and stakes of the country, and intended to build a digital society that elevates the national economy, and makes it based on science and knowledge, and all of this requires a beginning of commitment and reducing the digital divide by managing access to information and communication technology, and the comprehensive spread of digital services and networks, and also obliges everyone to promote Technological capabilities in their physical and cognitive form, and e-governance must be strengthened by providing an appropriate ecosystem for digitization, as well as establishing general principles of digital culture.

The Algeria Conference on Digitization was a prominent event and station to put the reality, challenges and prospects of digitization on the table for discussion between sector officials and the most important actors in it at the national level, between a clear political will, a committed government effort, and a diligent private sector. Available and huge, no matter how great the difficulties and challenges are, and whatever the situation is described, it is certain that the digitization train in Algeria has taken off, and is moving toward promising prospects.

In this context, and after the public authorities represented in the Ministry of Digitization and the Ministry of Health, Population and Hospital Reform announced the development of projects to digitize the health sector, where the medical files of Algerian patients are digitized, and a health database for these patients is created, provided that this

Intelligent Semantic Analysis for Healthcare
ISBN 978-0-443-33512-9
https://doi.org/10.1016/B978-0-443-33512-9.00004-4

step is a gateway to digitizing relations Contractual agreements with social security bodies according to the applicable cure card, which will link and document electronic pharmacy records as a base for consumer drug data.

The decision to digitize health gave a strong impetus for institutions to start work to implement and expand the project of creating and developing management and management approaches, and archiving medical files electronically, and this may provide a strong incentive to develop integrated hospital information systems, which serve digital transmission and storage, analysis and interpretation of available multimedia patient records.[1] This is embodied through the upgrade of the National Agency for Health Documents to the National Agency for the Digitization of Health Documents, where the latter will be supported with qualified and multidisciplinary human resources, especially those specialized in software and automated information, for the purpose of modernizing the health sector, and facilitating the process of accessing patients' health files, this agency. It constitutes an integral member of the newly developed health sector system, because of its great importance in giving a new soul by using communication techniques and databases.

1.1 Previous studies

Serai Oumessaad; Hadjab Nadjat: "Digital transformation in health institutions and its role in improving the quality of health care services in light of the Corona pandemic - the experience of the Kingdom of Saudi Arabia", (2022). This study targeted the role of digital transformation in health institutions in the experience of the Kingdom of Saudi Arabia to improve the quality of health services in light of the Corona pandemic. The descriptive method was used to collect and analyze information and data. The research reached a set of results, the most important of which is that the outbreak of the Corona epidemic had an important role in digitizing the health sector and activating email. This contributed to the improvement of health care services.

MEZIANE Tadj; TOUTA Nawel: The Algerian health system and its indicators: Reforms at the crossroads and their Challenges during the post-Corona covid19 pandemic period. (2022). This paper aims to clarify the most important reforms that made the Algerian health system adaptable to the new economic, social and political reality, as the stages of development of the Algerian health sector were revealed until 2020, and to monitor the impact of these changes on the development of public health indicators. Finally, a number of recommendations were set to help obtain a new vision for the development of a strong, comprehensive and sustainable health policy.

[1] Stelios C. Orphanoudakis, Eleni Kaldoudi, Manolis Tsiknakis, 1996, Technological Advances in Teleradiology, European Journal of Radiology, Special Issue: Towards a Digital Radiology Department - Technological Advances and Clinical Evaluations,vol. 22, pages 205–217.

Abdelmadjid SEDJEL; Faiçal MOKHTARI: Public-private partnerships as a tool for modern public management and improving the efficiency of Algeria's healthcare system, (2022). The study contributed to understanding how to reform the health system and increase its efficiency through the use of the method of cooperation between the public and private sectors, as well as the principles of modern public administration. Several problems faced by the current health system in Algeria due to the traditional model of public administration, which does not consider cost and quality as criteria for efficiency and effectiveness, were also highlighted. Finally, a third option was adopted that combines the financial, logistical and institutional capacities of the public sector with the rationality and competitiveness of the private sector.

The projects of modernizing the health sector have also been launched thanks to these completed studies targeting strategic plans, and awaiting the demarcation of the decision by the various ministries concerned with digitization, and the embodiment of the actual launch of the project in general, and for institutions in particular. The strength of these institutions lies in their human resources, so they must be given great attention because of their important sensitivity in the process of modernization. Which leads to the following main problem: How effective is the National Agency for the Digitization of Health Documents in digitizing the Algerian health sector? This main question is divided into sub-questions as follows:

- What is meant by health digitization?
- What is the National Agency for the Digitization of Health Documents?
- What are the prospects of the new agency in achieving the goals of the Algerian health sector in digitization?

The scientific importance of such topics is attracting thinkers and academics to research such future research, in order to facilitate the tasks of decision-makers in the practical implementation of these projects, which mainly aim to keep pace with the rapid digital modernization experienced by some advanced countries of the world at the regional and international levels, and to achieve prosperity Health justice for every citizen at the local and national levels. Reducing the bill of health costs, and its positive reflection on the economic reality of the country, while reducing the time factor in treatment and eliminating all forms of health bureaucracy, and other problems of the traditional health system.

The answer to the main problem and the sub-questions requires reliance on the descriptive approach in all locations of the study, and sometimes the comparative approach and the analytical approach are used to highlight some requirements that need to be clarified to bring the idea closer to the recipient.

2. Theoretical framework

In the era of rapid technological advancement, digitizing the health sector represents a transformative shift toward more efficient, accessible, and patient-centered services, leveraging information and communication technologies (ICT) to bridge gaps in traditional healthcare delivery. This theoretical framework elucidates core concepts of health digitization, including m-health, remote care, telehealth, and telemedicine, which form the foundational pillars for understanding the National Health Digitization Agency's (NHDA) role in Algeria. By defining these elements and their interrelations, the section provides the analytical lens to evaluate the agency's contributions amid Algeria's ongoing health modernization efforts.

2.1 Digitization concepts

1. **M-health:** It should be noted that medical practices in this framework are usually based on so-called "mobile" tools and services, which are accessed on smart devices: digital tablets, smartphones, and personal digital assistants (PDAs).), other wireless devices, etc.) in promoting the use of health applications, connected health objects, monitoring devices, social networks for medical purposes, (Emission ADN, 2020) or health information platforms;[2] (E-santé, 2005)
2. **Remote care:** This consists in escorting or monitoring a particular patient remotely by a medical professional, (Emission ADN, 2020) or a pharmacist in contact with a patient;[3]
3. **Telehealth:** The intent is telemedicine, mobile health and telecare, which makes it possible to follow as much as possible the actors in the path of their care, from prevention to follow-up, thanks to the use of tools to create, transmit, share and manage digital information, but Also thanks to training, online health services or social networks (Executive Decree No. 22-251 of June 30, 2022);[4]
4. **Telemedicine:** that is, keeping the patient in constant contact with health professionals from a distance, for the purpose of prevention, diagnosis, description or monitoring. Telemedicine in this regard relies on five medical practices that lie in: (Halim Ammar khodja & Autres, 2021) [5]

[2] Williams SY, Adeyemi SO, Eyitayo JO, Odeyemi OE, Dada OE, Adesina MA, Akintayo AD. Mobile Health Technology (Mhealth) in Combating COVID-19 Pandemic: Use, Challenges and Recommendations. European Journal of Medical and Educational Technologies (2020); 13(4): em2018. https://doi.org/10.30935/ejmets/8572

[3] Alexandra Queirós, Joaquim Alvarelhão, Margarida Cerqueira, Nelson Rocha, Remote Care Technology: A Systematic Review of Reviews and Meta-Analyses (2018); https://www.researchgate.net/publication/323135789_Remote_Care_Technology_A_Systematic_Review_of_Reviews_and_Meta-Analyses (Visited in 30/06/2023).

[4] Omolayo M. Ikumapayi, Joseph F. Kayode, Adeniran Sunday Afolalu, Telehealth and Telemedicine -An Overview, Conference: third African International Conference on Industrial Engineering and Operations Management, At: Nsukka, Nigeria, April 5–7, 2022, (Visited in 30/06/2023).

[5] Stelios C. Orphanoudakis, Eleni Kaldoudi, Manolis Tsiknakis, op., cit.

5. **Remote consultations:** Medical advice in this regard takes place between the doctor and his patient remotely, and sometimes this is done in the presence of a third-party care provider;[6]
   a. **Telemedicine:** that is, remotely assisting a healthcare professional to another counterpart in the conduct and implementation of the respective medical practices; (https://esante.gouv.fr/les-acteurs-de-terrainfr)
   b. **Remote medical monitoring:** In this context, the health professional interprets the health data that has been monitored remotely, these data are necessary for the medical monitoring of the patient, and for making decisions regarding his care, (Acher, n.d.) using the necessary medical tools. With the recording and transmission of health data prepared by the health care provider or from the patient himself; [7]
   c. **Tele expertise:** by means of which experiences are exchanged remotely between health professionals that, whereby these professionals establish another medical diagnosis by examining the medical information of the patient, within the training or experience practices of the concerned professional (https://www.emro.who.int/health-topics/ehealth/);[8]
   d. **Medical regulation:** It is a medical response conducted over the phone by a health professional who is assigned to the emergency call center, which guides the patient in the best way, in line with his health condition.

2.2 E-health actors

The most important active pillars in the digital health sector in general can be divided into two important categories, one of which relates to the active representatives in

[6] Rebecca Rosen, Sietse Wieringa, Trisha Greenhalgh, Claudia Leone, Sarah Rybczynska-Bunt, Gemma Hughes, Lucy Moore, Sara E Shaw, Joseph Wherton, Richard Byng. Clinical risk in remote consultations in general practice: findings from in- COVID- 19 pandemic qualitative research, BJGP; DOI: 10.3399/BJGPO.2021.0204, Published: 29 June 2022, in: https://creativecommons.org/licenses/by/4.0

[7] NHS, Clinical guide for the management of remote consultations and remote working in secondary care during the coronavirus pandemic. Specialty guides for patient management during the coronavirus pandemic, November 2020, https://www.rcslt.org/wp-content/uploads/2021/10/NHS-England-clinical-guide-for-the-management-of-remote-consultations-and-remote-working-in-secondary-care-during-the-coronavirus-pandemic.pdf, (Visited in 30/06/2023).

[8] MedIsi, La Télé-Expertise C'est Quoi? URPS ML Nouvelle-Aquitaine, Fiche 1, Mars 2023, https://www.urpsml-na.org/uploads/images/1677698307_Fiche%20g%C3%A9n%C3%A9rale%20TLX_010322.pdf, (Visited in 30/06/2023).

the field, and the other relates to the active representatives at the institutional level. In this context, this can be clarified more according to the following: (Acher, n.d.).[9]

2.2.1 Field actors

The most important players represented in digital health at the field level can be summarized in five main actors as follows (Fig. 20.1):

The previous figure shows representatives of the digital health field as follows:

- Health users: it includes both doctors, nursing employees from the paramedical sector, employees of the health sector administration, and others;
- Health institutions: These are hospitals, clinics of various specializations, health laboratories, pharmacies, etc.;
- Social medical institutions: This is related to insurance agencies, geriatric homes, social medicine institutions, and other institutions;
- Industries: It means the various laboratories and factories specialized in the production of medicines and various other health supplies, and other health products required in the health market;
- Patients: This means all beneficiaries of medical services, hospitalization, treatment, and other services.

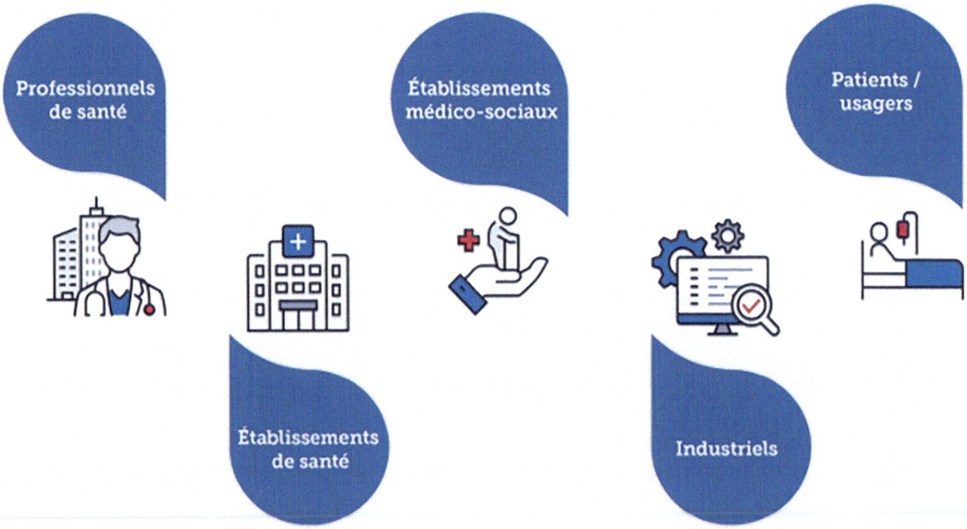

Figure 20.1 Field actors. *(https://esante.gouv.fr/les-acteurs-de-terrain.)*

[9] World Health Organization, Global strategy on digital health 2020–25, 2021, https://www.who.int/docs/default-source/documents/gs4dhdaa2a9f352b0445bafbc79ca799dce4d.pdf, (Visited in 30/06/2023).

2.2.2 Institutional actors

In order to uncover the institutional players in the digital health ecosystem, the e-health sector is supposed to benefit from a rich as well as complex institutional ecosystem, where the skills and tasks of its players interact: digital players, healthcare, health actors, social medicine, And public and private actors, national, regional and international players., where the following figure shows the most important actors at the institutional level as follows (Orphanoudakis et al., 1996) (Fig. 20.2):

The above figure shows the most important active players as follows:[10]

1. *The Ministry of Health and Hospital Reform:* It is considered the pioneer of digital transformation, as it defines the national strategy to be followed, and works to provide the necessary resources for the dissemination of e-health. Among its most important players are the Ministerial Delegation for Digital Health, the General Directorate of Health Care, the Directorate of Social Security, the General Directorate of Health, the General Directorate of Social Cohesion, the Department of Research, Studies, Evaluation and Statistics.

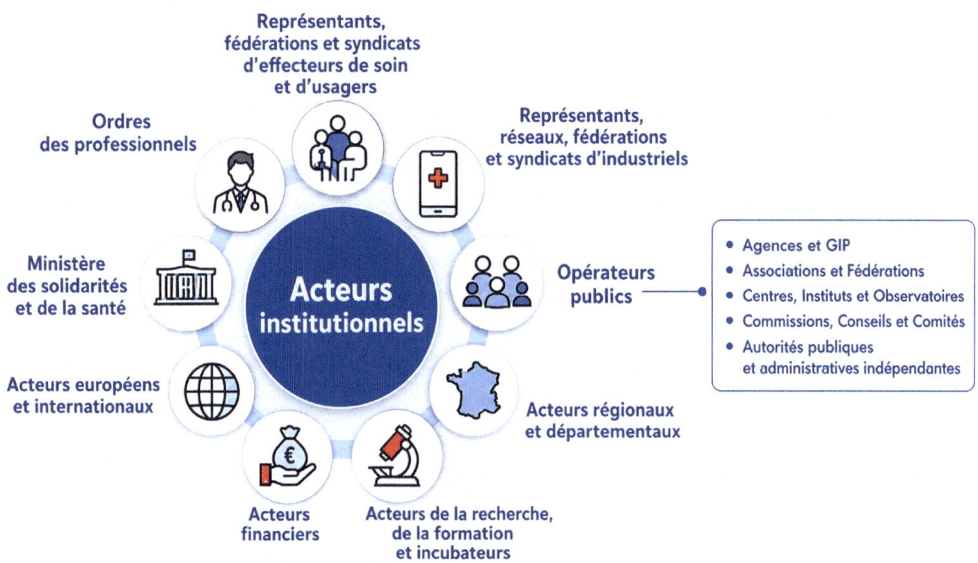

Figure 20.2 Actors at the institutional level: Institutional actors. *(https://esante.gouv.fr/les-acteurs-de-terrain.)*

[10] https://esante.gouv.fr/les-acteurs-de-terrain

2. *Professional demands:* Forming the organizational structures of health professions that guarantee the professionalism of their workers, the principle of medical ethics, and thus qualitative and reliable patient care. Among its most important representatives are the National Council of the Medical Syndicate, the National Pharmacists Syndicate, the National Nurses Syndicate, and the National Surgeons-Dentists Syndicate (Fig. 20.3).
3. *Representatives, federations and unions of care providers and users:* their goal is to raise the voice and defend the interests of patients and those who care for them. These include patient associations, medical associations, dentists, pharmacists for public and private health and social institutions, clinical research companies, and others.
4. *Industry representatives, networks, federations, and unions:* their mission is to carry a voice and defend the interests of e-health manufacturers, and to create networks of entrepreneurs to stimulate collective intelligence. Among the most important players within this framework are pharmaceutical companies, health and social information systems companies, the Federation of Electrical, Electronic and Communication Industries, the Industries Group, the Federation of Health Industries, etc.
5. *Public operators:* any institutional bodies of public or private status that supervise the e-health sector and implement state policies by fulfilling the public service mission, with the aim of organizing, informing, supporting, securing, motivating and coordinating within the framework of the digital transformation of the health system.
6. *Regional and administrative players:* They are concerned with the task of supervising digital transformation, and supporting actors in the field at the regional and administrative levels. They are regional health agencies, regional association of health professionals, regional health insurance funds, regional delegations, regional health observatories, regional emergency observatories, regional council for quality and coordination of care, etc.

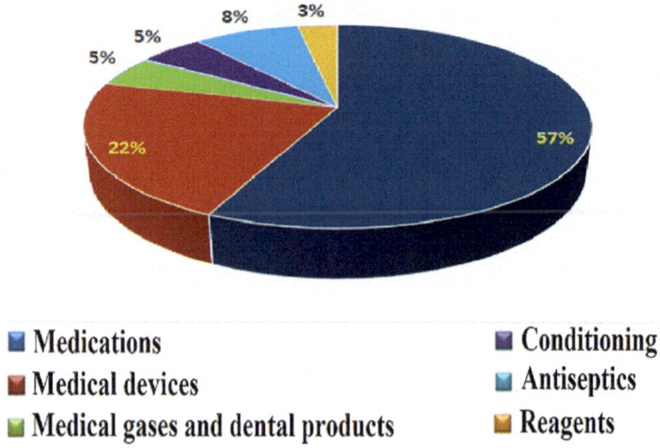

Figure 20.3 Pharma, equipment and medical devices market data. *(Halim Ammar Khodja & Autres. Reformes et perspectives du secteur de la santé et de la E-Santé en Algérie et au Maroc. Bretagne Commerce International, Webinaire, 21 Janvier 2021. P. 15. In: www.BretabneCommercsInternational.com.)*

7. *Research and training actors and incubators*: Their mission is to activate the promotion of research in the still-nascent e-health sector, and to raise awareness of digital health challenges through training and stimulating innovation. They can be listed in: The Ministry of Higher Education and Scientific Research, the National Center for Scientific Research, national institutes related to research in digital science and technology, national institutes related to health and medical research, research and documentation institutes in health economics, national institutes for statistics and economic studies, the School of Advanced Studies in Public health, national bodies for continuous training of hospital staff, postgraduate studies center in economics and hospital management, national social security bodies, live laboratories in health, . and other bodies.
8. *Financial entities:* These entities provide basic financial resources for health, medical and social care to people, as well as the will to do so in the e-health sector. It consists of the following categories: the National Health Insurance Fund, the National Solidarity Fund, health insurance, pension funds, social security funds, the National Office for Compensation for Medical Accidents, the National Health Insurance Fund for wage workers, the National Old-Age Insurance Fund for wage workers, . and others. That of other bodies.
9. *European and international actors:* European and international e-health auditors, multilateral guarantors of interoperability between countries, and vectors for establishing the global e-health market. The most important players in these entities are: European Medicines Agency, World Health Organization, European Commission, European Medical Informatics Federation, World Medical Association, Health Care Information and Management Systems Association (European Hospital and healthcare associations), etc.

3. The reality and prospects of the health sector and digital health in Algeria

3.1 The reality of the health sector in Algeria

The Algerian government passed a new health law in 2018, to reinforce the principle of free health care enshrined in the constitution, in order to introduce a new dynamism in the sector. So that the law applies new measures, including the promotion of electronic health, and the adoption of new technologies, especially computerized medical files in the national system (Orphanoudakis et al., 1996) (Fig. 20.4).

So the Algerian authorities launched an ambitious program to build five large hospitals nationwide in partnership with foreign companies. However, due to the economic crisis, the projects remained frozen until 2019. The government has also completed 34 small hospitals, plans to launch 26 more hospitals, and create 27,000 new jobs as part of the 2021 budget. It will continue to renovate, modernize and acquire new equipment

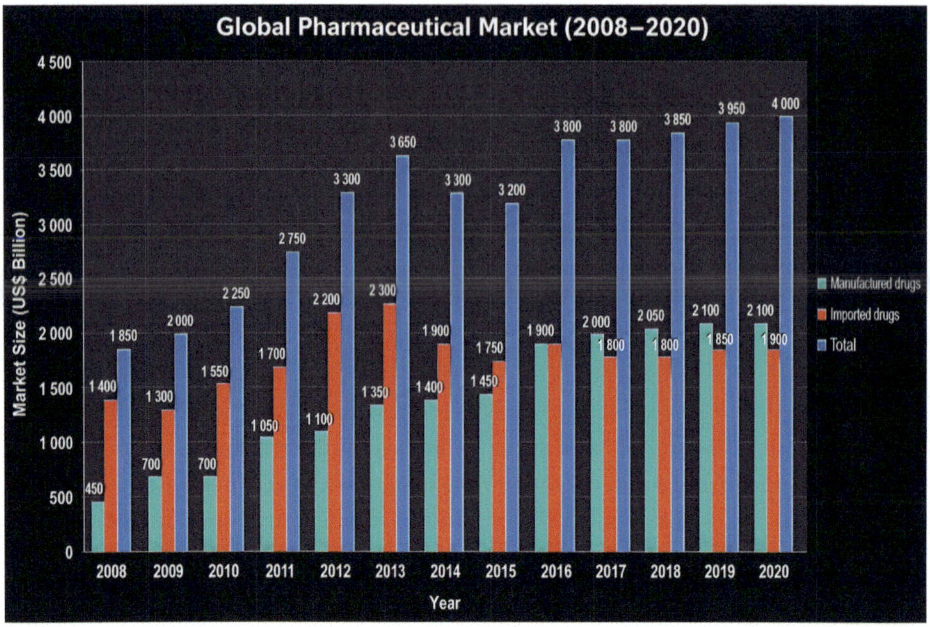

Figure 20.4 Evolution of the pharma market from 2008 to 2020 in millions of $. *(Halim Ammar Khodja & Autres. Reformes et perspectives du secteur de la santé et de la E-Santé en Algérie et au Maroc. Bretagne Commerce International, Webinaire, 21 Janvier 2021. P.18. In: www. BretabneCommercsInternational.com.)*

for 304 existing public hospitals. , especially with the pandemic that has revealed the main shortcomings of the sector. In the same vein, the private sector is investing in health care to respond to the increasing demand and shortcomings of the public sector. In Algeria, there are 208 private clinics in almost all specialties, and four to five new clinics are opened every year in addition to doctors' offices. (The Algerian health market is, n.d.)

Therefore, Algeria witnessed a remarkable improvement in health indicators thanks to an increase in the share of the state budget allocated to the health sector, which represents 5.2% of GDP. It has allocated $28 billion over the past decade, including $3 billion in the state budget for 2020. In order for the population to continue to benefit from free public health care and the social security system.

The health system is organized around the Department of Health, Population and Hospital Reform (MSPRH) and has five regional health boards, regional health observatories, and 48 health and population directorates (one direction per state) across the country. According to the Ministry of Health, Algeria has 586 public healthcare facilities including 15 university hospital centers (CHU) and 83 specialized hospitals (EHS). (webtv.univbejaia.dz.)

Algeria is experiencing an epidemiological shift and chronic diseases associated with lifestyle, such as diabetes, cancer and hypertension, require more complex and expensive treatment that has replaced infectious diseases. Life expectancy has increased to 76.5 years, and the population is expected to reach 50 million by 2030 from 44.7 million in January 2021.[11]

In general, the health sector in Algeria witnessed a high dynamic and rapid development, as this sector achieved a quantitative and qualitative leap that brought it, during the year 2021, to the following.

3.1.1 Presentation of the algerian health sector

In 2021, Algeria's health coverage indicators reflected significant infrastructural and human resource development in the public health sector (Williams et al., 2020). Specifically, there was one treatment room for every 6886 inhabitants, one general practitioner per 1118 citizens, one multidisciplinary clinic per 23,917 citizens, and one specialist doctor per 1413 citizens. Additionally, the country maintained 1.76 beds per 1000 residents, one dental surgeon per 2972 citizens, one drugstore per 3467 citizens, one midwife per 1203 citizens, and one paramedical aid per 325 citizens.[12]

These figures indicate a robust expansion in healthcare accessibility, particularly in paramedical support and general practitioner availability, though ratios for specialists, dental surgeons, and multidisciplinary clinics suggest areas for further improvement to meet growing demands amid population growth and epidemiological shifts. Overall, they underscore Algeria's quantitative achievements in health infrastructure, aligning with national reforms aimed at enhancing service delivery efficiency (Williams et al., 2020).

3.1.2 Distribution of private healthcare institutions by type of structure

The distribution of Algeria's private healthcare infrastructure across two main categories: individual offices/clinics on the one hand, and group practices/centers on the other. At the individual level, there are 9042 specialist doctors' offices, 7298 general-practitioner offices, 6514 dental surgeries and 10,260 pharmacies. These figures place pharmacies first in terms of density, followed by specialist offices, general-practice offices and finally dental surgeries, reflecting a demand-driven investment pattern that prioritizes medicines and specialized care.

[11] Aissaoui Nasreddine, The Reform of Primary Health Care/PHC in Algeria: A First Step Towards a More Equitable and Efficient Health System, HEALTH SCIENCESVolume 2021, ID 366, in.

[12] Halim AMMAR KHODJA & Autres, « Reformes et perspectives du secteur de la santé et de la E-Santé en Algérie et au Maroc », Bretagne Commerce International, Webinaire, 21 Janvier 2021. In: www.BretabneCommercsInternational.com.

Turning to group facilities, numbers are far smaller: 249 medical and surgical clinics, 36 diagnostic clinics, 151 hemodialysis centers equipped with 1780 dialysis machines, 709 group-practice clusters and only eight medical clinics. The data reveal a clear concentration of private investment in high-tech or specialized services (Queirós et al., 2018), while collective infrastructure remains limited compared with individual units. This suggests a private-sector model that favors small, scattered investments over larger collective ones, probably because of higher equipment and management costs or because financing and reimbursement schemes are structured to reward individual service delivery.[13]

3.2 Breakdown of private healthcare establishments by type of structure

Private healthcare infrastructures now occupy an important place in the healthcare sector. These infrastructures are distributed as follows:[14] (Table 20.1).

3.2.1 Distribution of clinics

This section presents the distribution of private healthcare establishments by facility type, providing insight into the structural composition of the private healthcare sector in the studied region. The data reveals a significant concentration of surgical clinics compared to other types of medical facilities.

The table offers a revealing snapshot of Algeria's private-health-care landscape, where the distribution of facilities is sharply polarized between high-volume individual offices and far scarcer collective structures. With 10,260 pharmacies leading the field, followed by 9042 specialist offices, 7298 general-practice offices and 6514 dental surgeries, the private sector clearly privileges entry-level, outpatient-focused investments that require limited capital and offer rapid turnover. This pattern mirrors consumer behavior: medicines and specialist consultations are the first port of call, while dental care remains a lower-priority expenditure.

Table 20.1 Breakdown of private healthcare establishments by type of structure.

Structures	NBRE	%
Medical clinic	28	8,95
Surgical Clinic	249	79,55
Diagnostic clinic	36	11,5
TOTAL	313	100

Halim Ammar Khodja & Autres. Reformes et perspectives du secteur de la santé et de la E-Santé en Algérie et au Maroc. Bretagne Commerce International, Webinaire, 21 Janvier 2021. P. 15. In: www.BretabneCommercsInternational.com.

[13] Halim AMMAR KHODJA & Autres, op., cit.

[14] BENHAMOUDA Youcef, KHELIFA Hadj, Measuring the quality of health services in Algeria, Journal of Contemporary Business and Economic Studies Vol.(05) No.(1) (2022). In: https://www.asjp.cerist.dz/en/downArticle/617/5/1/176391, (Visited in 30/06/2025).

In contrast, group infrastructure is thin: only 249 medical-surgical clinics, 36 diagnostic clinics, 151 dialysis centers (1780 machines) and 709 group-practice clusters exist nationwide, alongside a token eight general medical clinics. The ratio of individual to collective units is therefore almost 20:1, signaling a market that avoids the heavy fixed costs, complex accreditation and longer pay-back periods associated with multi-specialty facilities. Dialysis centers are the exception, their numbers driven by chronic-disease demand and predictable reimbursement schedules; even here, however, bed capacity ($\approx$11.8 machines per center) suggests modest-scale units rather than regional hubs.

Overall, the data expose a two-tier system: a dense, atomized front line of solo practices and retail pharmacies that eases geographical access, but a bottleneck in intermediate-level, capital-intensive services (imaging, day surgery, short-stay diagnostics) where private risk capital is reluctant to invest. Policy implications are clear: incentives must shift from simple licensing to targeted subsidies, concessional financing or PPP models that de-risk larger facilities, integrate care pathways and prevent costly duplication of high-tech equipment. Without such levers, Algeria's private sector will continue to excel in dispensing pills and consultations while remaining chronically under-represented in the very settings—cohesive clinics and diagnostic centers—capable of relieving pressure on overcrowded public hospitals (Ikumapayi and Kayode, 2022).

3.2.2 Breakdown of private practices

This section presents the distribution of private medical practices across different categories, providing a comprehensive overview of the healthcare service landscape. The following table details the numerical breakdown of various types of private healthcare providers, highlighting the relative prominence of each category within the private healthcare sector.

The table offers a striking snapshot of Algeria's private clinic landscape: out of 313 licensed facilities, almost four-fifths (249 units, 79.6%) are surgical clinics, while only 28 (8.9%) are general medical clinics and 36 (11.5%) are diagnostic centers. This distribution signals a market that has chosen to concentrate capital, skills and equipment on the most remunerative segment—interventions and day surgery—rather than on primary-level ambulatory care or preventive diagnostics. The scarcity of general medical clinics (fewer than one for every nine surgical sites) suggests that investors perceive lower margins in basic consultations and routine follow-ups, preferring instead the fee-for-service revenues generated by operating theaters and short-stay procedures. Diagnostic clinics, although twice as numerous as general clinics, still represent barely one in 10 facilities, indicating that high-tech imaging and laboratory work are regarded as ancillary profit centers rather than as standalone businesses. Overall, the data reveal a two-tier private system: a plethora of individual doctors' offices (evident in earlier tables) that handle the bulk of outpatient demand, overlaid by a thin but lucrative upper tier of surgical clinics whose business model relies on procedures reimbursed at premium rates. Policy implications are clear: unless reimbursement schedules or concessional financing are re-calibrated to make primary-level clinics and diagnostic hubs financially attractive, the private sector will continue to under-supply the very settings—non-surgical, intermediate-complexity, capital-intensive—that could relieve pressure on over-crowded public hospitals and complete the continuum of care (Rosen et al., 2022) (Table 20.2).

3.3 Pharmaceutical industry in Algeria

The health sector is considered a strategic sector in Algeria, because its general budget is ranked fourth, and because of the growing health needs in the country. To this end, preparations are underway for a comprehensive reform of the health system, within the framework of defining the new health policy. So the volume of health structures being established at the end of 2020 is approximately 40 hospitals of medium capacity (60–240 beds), some of which are specialized (burn victims, psychiatric, anticancer, mother-child, etc.). NHS (2020) The volume of production was also expected to be similar to 2021. The medical equipment and devices sector generates annual sales of about 700 million US dollars, with a growth rate of 7% annually.[15]

In another context, thanks to the new government policy, the Algerian pharmaceutical industry has witnessed an important recovery in recent years, reaching 175 pharmaceutical production units operating in 2020 (MedIsi et al., 2023). These units ensure the production of pharmaceutical products as shown in the following figure:[16]

Table 20.2 Distribution of private practices.

Cabinets	NBRE
Specialists	9042
Generalists	7298
Oral surgeon	6514
Drugstores	10,260
Of group	709

Halim Ammar Khodja & Autres. Reformes et perspectives du secteur de la santé et de la E-Santé en Algérie et au Maroc. Bretagne Commerce International, Webinaire, 21 Janvier 2021. P. 15. In: www.BretabneCommercsInternational.com.

[15] The Algerian health market is growing, Tuesday 7 April 2020 - Healthy ambition - Algeria - from two to 3 December 2020, https://www.businessfrance.fr/le-marche-algerien-de-la-sante-est-en-croissance

[16] The Algerian health market is growing, op., cit.

From this figure, it appears that the large proportion of the pharmaceutical industry in the Algerian market (domestic) belongs to pharmaceutical products at a rate of 57%, as the Algerian pharmaceutical industry achieves an annual sales volume of more than 3.5 billion US dollars (a third market in the Middle East and Africa), 45% of which are imported, with more than 80 production units open to technology partnerships (World Health Organization, 2021). As for other pharmaceutical products, they vary in rates, according to the production capacity of this young and fertile market. Perhaps the following figure highlights the following:[17]

It is clear from this figure, that the pharmaceutical industry in the Algerian market is witnessing an increasing growth during the period (2008–2020), with the continued dependence on imports. Also, this market is distinguished by the following:[18]

- 40 countries, 300 suppliers and 50 local manufacturers: this has been the stated objective of the authorities for decades, regardless of the country's financial health;
- Strategic market in permanent expansion;
- Market dependent on imports;
- A balance of security funds;
- A 2009 local production recovery plan;
- A reference tariff;
- Power of substitution in pharmacy.

In this context, the sales volume of the Algerian health market is increasing by 11% annually. As for the pharmaceutical industry, where the sales volume is estimated at more than three billion euros, despite import restrictions and an increase in local production, with the stability of drug imports remarkably during the period (2016–2020).

As for medical devices, Algeria's imports of medical devices amounted to about $770 million in 2018, which puts Algeria in the list of the top five countries in the African continent, according to Algerian customs. Despite the economic crisis in the country, this sector is still growing and 90% depends on imports mainly from the United States (18%), China (14.5%), Germany (12.5%), France (11%), and Japan (10.5%). (Nasreddine, n.d.)

The sector also relies on public tenders launched by CHP (Central Hospitals Pharmacy) to meet the requirements of public hospitals with about 40%, and about 60% for the private sector (clinics and direct purchases). There are more than 200 importers of medical devices in Algeria and more than 10,000 pharmacies. The local production of the devices is very minimal, and it focuses on low-tech products, but it is currently in line with the new government policy that encourages local production, benefiting from a 25% bonus on the prices of local products compared to the importer in all public

[17] Ibid.
[18] Halim AMMAR KHODJA & Autres, op., cit.

tenders. In this context, the government imposes restrictions or sometimes suspends the import of a product in order to satisfy the local producer. To contribute to the national efforts to combat the epidemic, for example, there are two leading private companies; Vital Care and Salem Diagnostics launched local production of rapid diagnostic tests. Vital Care, which already produces self-monitoring blood glucose strips in Algeria, is a semi-manufacturer of the COVID-19 IgG/IgM Rapid Diagnostic Tests with the Canadian partner that has registered market success.[19]

As for the pharmaceutical market in Algeria, which is considered the second largest market in the African continent after South Africa, the Algerian pharmaceutical market recorded a double-digit recovery during the past decade (at a rate of 17%), that is, about $4 billion, according to the Algerian Minister of Pharmaceutical Industry. The import bill for pharmaceuticals recorded $2 billion in 2019, of which $600 million were drugs to treat cancer and $400 million was insulin for diabetics. Canadian pharmaceutical exports to Algeria increased from $4.4 million in 2016 to $11 million in 2017. In 2018, they registered more than $24 million, according to Algerian Customs.[20]

Algeria in particular is also attractive to major international pharmaceutical groups that have set up pharmaceutical factories with the French dominating the market. Leading companies include Sanofi, Novo Nordisk, Pfizer, Al Kindi, Hikma, GSK, and local companies. Groupe Saidal, Merinal Laboratories and Biopharm. Through a cabinet reshuffle in June 2020, a new Ministry of Pharmaceutical Industries was created as a strong signal to boost the sector. For the purpose of supervising the regulatory aspects that govern the pharmaceutical industry, the medical device market, and ensures accessibility, availability, registration, import and price control of medicines. The newly established National Medicines Agency will be strengthened to facilitate and simplify the registration process. Furthermore, a new Minister Plenipotentiary in charge of hospital reform has been appointed to speed up the reform process, particularly the digitization of health care.

In this context, the policy of the Algerian government aims to gradually reduce imports and enhance local production, especially in generic medicines, to reach the desired rate of 70% of local manufacturing by 2022. The government imposed a ban on some imported medicines that are produced locally, in return for progress Tax incentives, which contributed to a fivefold increase in the market share of domestic drug production in the past 5 years, reaching 52%, or about $2 billion in 2018. According to the ministry, there are 97 drug manufacturers spread across 21 states, including 49 For the central region, 36 for the east, and 12 for the west.[21]

[19] The Algerian health market is growing, op., cit.

[20] The Algerian health market is growing, op., cit.

[21] Ibid.

Therefore, the pharmaceutical industries can be considered as a "strategic sector" in the supplementary financing bill for 2020, and as such remain subject to the cap at 49% of the participation of any foreign investor in the capital of the Algerian company (called rule 51/49). It is possible to exclude from this rule the investments related to the manufacture of basic products that are innovative and have a high added value, that require complex and protected technology, and are intended for the local market and for export.[22]

4. Prospects for digital health in Algeria

Algeria is the largest African country in terms of area. Its population of 44.7 million lives mostly in the hills and green northern strip with a beach on the Mediterranean. While 80% of the land is made up of desert with an arid climate which presents a great challenge to the living conditions. Ironically, the Sahara is the country's most generous as it brings in oil and gas revenues that particularly help fund a free health care system for Algeria's residents.

On paper, the coverage of the current health system appears comprehensive, but in reality it is somewhat dysfunctional as a whole with some reasons arising from challenges related to the geographical extent of the territory it is supposed to cover.

In another related context, the use of the mobile phone and the Internet was as much a phenomenon in Algeria as anywhere else. This has greatly contributed to enabling solutions in the health sector that he only dreamed of a few years ago. Many of the initiatives and projects that have been implemented so far, have been launched through specialized conferences held since 2006. Where the various efforts aroused great interest, and a serious discussion about the importance of their unification through the National Association of Telemedicine in Algeria. This national scientific association is not for profit, and its mission is to promote and develop competencies in the field of telemedicine and digital health in Algeria within a national legal and regulatory framework. It also strives to achieve these goals by designing and implementing practical medical projects and organizing local and regional seminars, such as the one that was held in Bejaia in December 2015.[23]

As a result, the dynamics toward achieving this goal were initially slow, starting with the launch of the Ministry of Health for the first time a national promotion agency, which contributed significantly to the holding of teleconferencing between hospitals in the south and north. This was followed by the launch of the Ministry of Higher Education of scientific centers to enhance scientific support for e-health such as the Center

[22] Ibid.

[23] webtv.univbejaia.dz.

for Biomedical Engineering at the University of Tlemcen in the West, and the Research and Development Center for Advanced Technologies in Algiers. These initiatives have also boosted interest in the benefits of telemedicine and e-health in Algeria in the scientific and medical community and among administrative bodies. Finally, this led to the emergence of a government project to create a national e-health system and the establishment of the Algerian Society of Telemedicine and eHealth, (ASTeH).

4.1 Presentation du secteur de la e-sante

The objective of the implementation of the national e-Health strategy was set during the year 2019 by Mohamed El Amine DJAKER, specialist in information and communication technologies at the World Health Organization (WHO) in Algeria:

- Health law 18–08 has dedicated a provision to it;
- Creation and installation of the national digitization agency in January 2021;
- Several innovative and active companies in this field, particularly in telemedicine;
- An activity that is developing in the private sector;
- An activity that is developing more in medical devices than in the pharmaceutical sector.

4.2 Opportunities and challenges of health digitization in Algeria

4.2.1 Opportunities

Various opportunities exist in terms of funding and readiness from providers, healthcare professionals and patients alike. Also, the proven successes of some e-health initiatives in some developing countries will allow this success to be extended to Algeria through various partnerships. This will greatly allow the evaluation of the characteristics of successful Algerian e-health applications in the future. It is also worth noting those lessons that can be learned from failed international e-health activities and that will also help in the success of future Algerian projects.

The new health law emphasized free health care in public structures and the authorities' obligation to provide an adequate supply of health care. For this purpose, projects appeared to build 160 public and private health facilities from 2019 to 2021. It was planned to build 24 hospitals and in 2019, two cancer centers affiliated with the military institution are about to be completed, and the private sector expects to establish three to four clinics annually, and it has verified more health From 50 projects for the production of medical devices.

In another context, priority goes to the care of chronic diseases, such as: diabetes, high blood pressure, and cancer in particular. A new plan to combat cancer is being prepared during the period (2020–25). Where the costs of treatment for patients are partially covered by the state. This situation identifies very important short- and long-term needs in terms of reagents, imaging, and prevention. This made the country's authorities research electronic health, cheaper medical devices, etc., as part of their commitment to rationalizing health budgets. Algeria also has huge needs for innovative molecules and knowledge expertise in the pharmaceutical industry, because the Algerian pharmaceutical industry aims to meet 70% of the needs of its population by 2020, compared to 55% in 2018 (Youcef and Hadj, 2022).

4.2.2 Challenges

However, like other middle-income countries, Algeria faces a number of challenges, including growing inequality in access to health care, and increasing economic and social vulnerabilities. Despite the huge expenditures, the public health system suffers from poor management, poor quality of services, chaos, and heavy bureaucracy. The recent health crisis also exposed systemic turmoil, which led the government to further engage in various reforms to modernize the sector.

Despite the challenges of primarily market access, the bureaucratic registration process (but less severe than previously), the complex regulatory environment, and the specter of pre-existing risks to the country, Algeria's large and growing health sector will continue to generate sales opportunities and multiple investments over the 10 years coming. The favorable demographics and healthcare infrastructure projects will continue to support the increasing demand for medical treatment. Algeria's strategic geographic location, government incentives for investors, and the country's ambitious long-term plan will also support Algeria's making of the country as an export base toward the African market.

Therefore, successful companies in the Algerian market adopt a long-term strategy, dedicate time to market development, and build strong partnerships with local players who are equipped to deal with local regulations. Among the promising opportunities that Canadian companies include: production of pharmaceuticals, nutrients, medical devices, digital health development, telemedicine, health care service on the one hand, and continuing medical training on the one hand. Products of importance in the medical system: such as medical imaging, in vitro diagnostics, radiology, surgery, orthopedics, prosthetics, respiratory system, X-rays, laboratory equipment and reagents, on the other hand (Youcef and Hadj, 2022).

4.3 The National Health Digitization Agency (NHDA)

The (NHDA) is a new institutional entity specialized in promoting the digitization process in the Algerian health sector. The Algerian legislator has assigned a number of tasks to it, which can be clarified according to the following (Fig. 20.5).

4.3.1 The creation of the (NHDA)

The (NHDA) was created on the ruins of the National Agency for Bankrupt Health Documents, which was created by the requirements of Executive Decree No.: 95–319 of October 14, 1995. The new agency has taken over as a public enterprise

Figure 20.5 Structural organization of the (NHDA). *(Prepared by researchers based on the requirements of the executive decree establishing the agency.)*

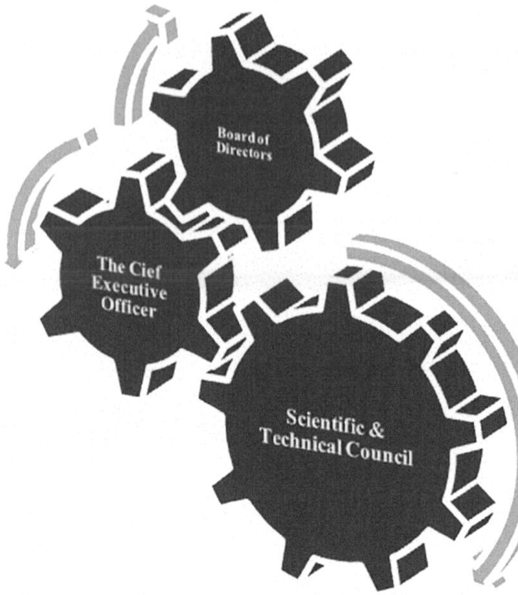

of an industrial and commercial character (EPIC).[24] by Executive Decree No. 22-251 of June 30 (2022).

4.3.2 Organizational structure of the (NHDA)

The organizational structure of this agency is based on three main elements, which can be illustrated according to the following figure:

The above figure highlights the agency's structure, which is composed of:

- **a.** *Board of Directors:* This council is responsible for running of the (NHDA), [24] within the periodic deliberations (twice a year).[25]They are appointed for a period of 3 years, renewable once by the Minister of this guardianship.[26]
- **b.** *The Director General:* He works on the good running of the Agency within his duties authorized in the text of Article 21 of the aforementioned decree, and the latter is assisted by an Assistant Director General, whom the Minister of Health proposes to appoint, [27]
- **c.** *Scientific and Technical Council:* This advisory body provides the board of directors and the general manager with opinions, suggestions and recommendations related to

[24] Article 08 of the aforementioned executive decree.
[25] Article 09 and Article 13 of the same executive decree.
[26] Article 11 of the same executive decree.
[27] Article 19 of the same executive decree.

medical, scientific and technical issues working to frame the use of information and communication technology at the level of the health sector.[28] This scientific and technical council is headed by one of its members who is elected by his peers for a renewable 3-year term.[29]

4.3.3 The role of the (NHDA) in the digitization of the health sector

The (NHDA) can be considered as the backbone of the digital health desired by decision-makers in the health sector in Algeria, and this can be justified by addressing the following.

4.3.3.1 Tasks of the (NHDA)

In order to achieve the goals of the health sector in health digitization, the Algerian legislator singled out a number of tasks assigned to this agency, these tasks are as follows:[30]

- Setting the digital hospital across a patient-centered floor, enabling health professionals to safely view the course of treatment in real time, allowing for a significant improvement in the quality of treatment by integrating, in particular, data on admissions, clinical data, treatment protocols, units, medical grounds and technical interests for the management of intensive treatment (Medical analysis laboratory, medical imaging and hospital pharmacy) and the measurement and maintenance of medical equipment;
- Developing the shared medical file, available to health professionals, through an identifier given to each patient and containing all data that encourage the exchange, sharing and security of data of a personal nature necessary for the patient's medical care and coordination of medical treatment,
- Implementation of telemedicine, telemedicine, tele-radiology, and tele-visual lectures for health professionals enables remote medical care of patients and the safe exchange and sharing of data, taking into account, in particular, the shortcomings of the treatment offer in areas with insufficient health coverage;
- Developing medical training via Internet platforms, in cooperation with relevant institutions and bodies, was able to ensure medical and quasi-medical training and encourage distance education;
- Creating interoperable networks and treatment poles at the local, regional and national levels that facilitate directing patients to health structures and institutions that are most suitable for their hospitalization and care, especially in the case of urgency;

[28] Article 23 of the same executive decree.
[29] Article 25 of the same executive decree.
[30] Article 06 of the same executive decree.

- Developing and establishing the coding of diseases and occupational work (medical and paramedical), across a platform, in coordination with the relevant sectors, health professionals and social security bodies;
- A ground delegation to the health vigilance and the health of health is directed to health controls and health professionals and employees to obtain reliable information on health, especially on the remedial diseases, the mission, the conversion, the environment, the remedy, the remedy, the remedy, the remedy, the remedies, the remedies, the remedies, the remedies, the remedies, the remedial,
- Developing a portal for reporting unwanted health events that enables people to report to the health authorities every undesirable event or any unusual adverse health effect in order to avoid epidemics, pandemics and health disasters;
- Digitizing the management and preservation of medical archives, as well as all other health archives;
- Contribute to the design and preparation of legislative and regulatory measures that enable the framing of the use of information and communication technologies in the health sector, especially those related to the practice of the health profession, the preservation of medical secrets, the exchange and sharing of health data of a personal nature;
- Ensuring every scientific and technical information and every document necessary for training in the field of health;
- Design, preparation, production, acquisition and publication on all supports of all documents, literature, publications, information and educational, medical, scientific and technical means.

The agency is also tasked with investigating:[31]

- "Maintaining the platform of the national health information system that lists references and standards as well as all health data and ensures interoperability with the information systems of other sectors of activity, in accordance with the legislation and regulation governing information and communication technology;
- Providing and securing digital identification of health structures, institutions and health professionals by creating an identification reference that allows confidentiality and security of equipment, housing and storage of data and periodic audits;
- Preparing the books of conditions related to the studies and achievements to be carried out in the field of developing digitization in the field of health;
- Preparing periodic reports on the progress of the works that have been initiated and on the development of indicators for the development of digitization and proposing every measure aimed at improving it and submitting it to the Minister in charge of Health.

[31] Article 07 of the same executive decree.

The main objective of assigning the new agency is to carry out tasks of a strategic nature with the aim of modernizing the health sector as a whole. Where will the computerization and digitization of health activities be relied upon, and the management of human resources, financial means and equipment, as previously mentioned?

4.3.3.2 Applications of the (NHDA) in the field of digital health

As part of the country's endeavor with the substantive measure that falls within the framework of the government's strategy for digitization and improving the performance of the public health facility. This endeavor resulted directly in the establishment of the National Agency for the Digitization of the Health Sector, as a strategic tool of the Ministry of Health, Population and Hospital Reform, with the aim of initiating the embodiment of the following exemplary projects:

- Digital management of patient care at the level of multi-service clinics;
- Digital management of radiology sessions schedules;
- Digital management of appointments for vaccination against the Covid-19 virus;
- Development of the digital ground for the spatial identification of health structures.

Through the Health Template Management System (THSM) platform shown in the following figure:[32] (Fig. 20.6)

It is clear that the (TMSS) Template management système platform aims to manage the sector's business, especially with regard to the organization and management of work teams at the hospital level, and at the management level.etc. In this case, he means: (Template) (Le Canevas), where all audit procedures are included, for the purpose of creating a clear vision of reality.

As for the hospital, it contains two systems:[33]

- Hospital Information system
- ERP: It is the part dedicated to management

It has begun to digitize the part dedicated to the application as a first stage, to be able to cover all stages with digitization according to a well-studied plan within a general strategy dedicated to the digitization of the health sector in general (Fig. 20.7).

Therefore, the health aspect was divided within the previous structure of the sector into: a reception floor; internal network; and an external network. The interaction between all these three networks, that is, the two networks and the receiving floor, creates what is known as the hopeful visual for the health department.

[32] Emission ADN, 2020, La Numérisation du secteur sanitaire de l'Algérie: Réalité et Perspective, Canal El Hayat TV, le: 28/12/2020, dans: https://www.youtube.com/watch?v=qmlEFrgOzu4&t=2547s. (Visite le 20/08/2024).

[33] Emission ADN, 2020, La Numérisation du secteur sanitaire de l'Algérie: Réalité et Perspective, Canal El Hayat TV, le: 28/12/2020, dans: https://www.youtube.com/watch?v=qmlEFrgOzu4&t=2547s. (Visite le 20/08/2024).

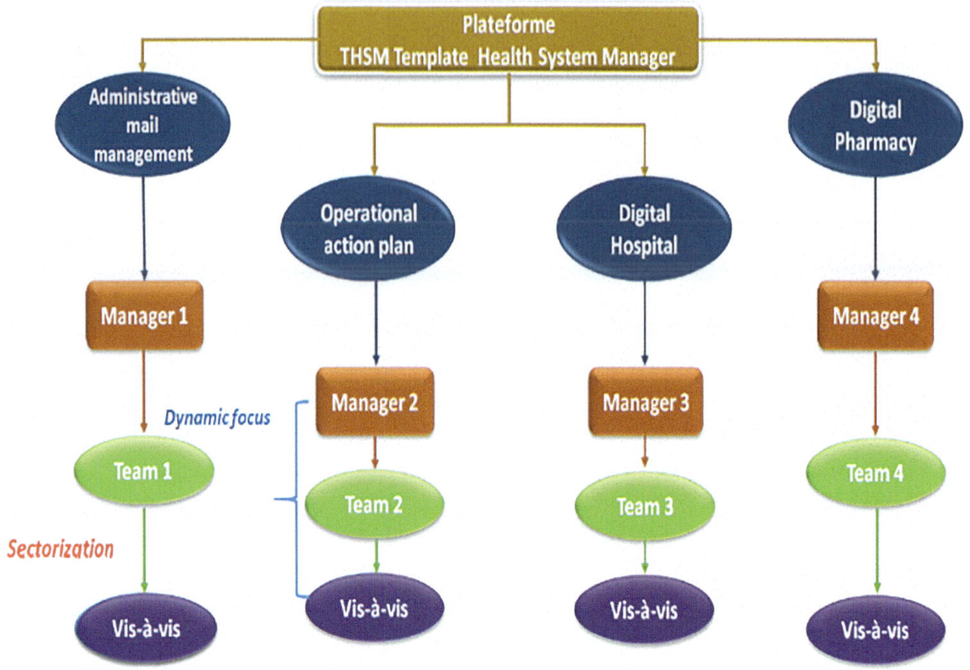

Figure 20.6 Health template management system platform (THSM). *(Prepared by researchers based on.)*

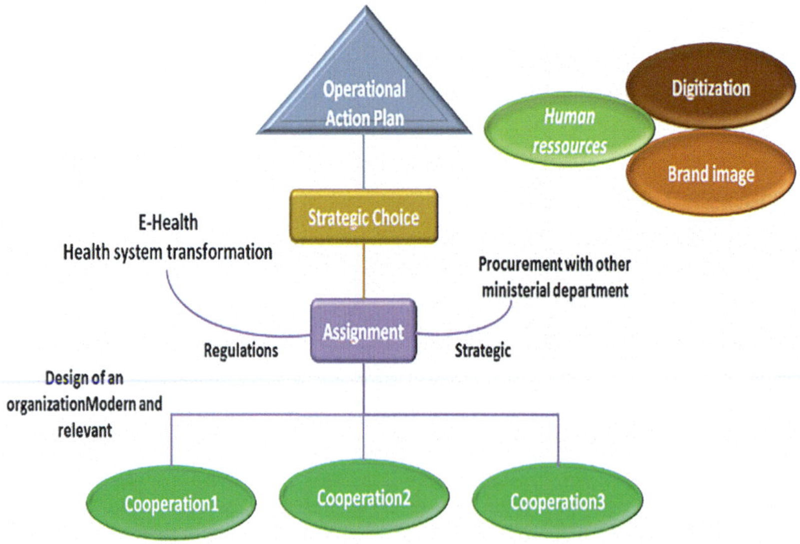

Figure 20.7 Operational action plan. *(Prepared by researchers based on.)*

At the level of the reception floor, 03 Attrebus is placed: the first is the Social Security number, which opens to the contract with the Social Security Agency; And the National Identification Number (NIN), the latter was included for the first time, because not all patients or those seeking treatment are enrolled in the Social Security Fund, but there is a significant segment of citizens who are not involved in that. This requires tracking all these patients by placing an electronic bracelet containing (Barcode) in the emergency department, which represents the digital hospital center, and so that they are followed up during the treatment process, after transferring them and knowing their consumables of m This will enable us to know the cost of treating each patient, of course, after tracking the medical file transmitted with the joint (Partage) technology. Social security. So when the diagnosis and medical monitoring process is completed at the level of the emergency department, this patient may be transferred to the hospital to continue his treatment, and in this case, each of the treatment departments in this hospital is informed in advance of the arrival of this patient through its connection to the digital floor. With the patient's stay in the hospital, and his continued consumption of a certain amount of medication designated for his treatment, and his conduct of various analyzes, and other treatment facilities, he will eventually be able to know all the costs of treatment for the patient until he is discharged from the hospital and the end of his treatment. Including the possibility of adjusting the number of patients treated in the hospital daily, and evaluating the daily cost of hospital treatment, while also identifying insured and uninsured patients. This will allow rationalizing expenditures at the level of these hospitals, in order to avoid the accumulated losses over the years within the budgets of these hospital authorities.

The ultimate goal of all of this remains to provide the best care for the patient and the optimum quality of treatment within the framework of the digital hospital. In this regard, as a preliminary experiment, a digital hospital was chosen, which was not disclosed (for certain reasons), to be taken as a model hospital (a pioneer) within the components of the internal network, which also contains the medical file of the joint patient; And the extranet, which is one of the most important components of the treating doctor, whether it is a private doctor, a private medical clinic, a private pharmacy, and a hospital pharmacy. All of these bodies are directly linked to the Social Security Fund and the national identity number, thus achieving the desired transparency and visibility for all of these bodies, that is, being able to address the waste of annual state budget funds allocated to the health sector.[34]

[34] Emission ADN, 2020, La Numérisation du secteur sanitaire de l'Algérie: Réalité et Perspective, Canal El Hayat TV, le: 28/12/2020, dans: https://www.youtube.com/watch?v=qmlEFrgOzu4&t=2547s. (Visite le 20/08/2024).

As for the Operational Action Plan (PAO), this process is considered a recent application in Algeria, and it mainly aims to give visibility and transparency to the regular daily activities of the administration within the health sector, with the various other guardianship bodies. The following figure shows the following:[35]

Through this form, this is done depending on the method of work (Bound Down House) that contains the template (Template), and also contains the budget of the sector, where competencies are divided to enable encouraging workers in the public administration of the sector and motivating them to do more competencies and energies latent they have. This will make it possible to know the operational costs in managing the health sector, and the bodies affiliated with the trusteeship. Note that the digitization of the administration is completely independent of the digitization of the hospital.

The operational work plan (PAO), according to the above-mentioned form, is in line with the will of the ministry, which consists of a group of directorates and bodies affiliated with it, so that each directorate or body is placed separately, and the work of each body is drawn up, then the work is divided into (Les Taches) tasks, and each A task that connects with others. This will make it easier to know the cost of each task, while identifying the person assigned to do it, and this is called the exact work of all these bodies (similar to the work of an ant cell), to be drawn within a control panel for analysis and decision-making. At the end of the year, the budget of each directorate is adjusted in a real and accurate way, and the various human energies that gave a lot of their efforts during that year are known, and thus the possibility of correcting all errors, and engaging in perfection and excellence at work.

This will also achieve the sector (La Dématérialisation) or non-material, that is, enabling access to zero paper in the administrative transactions of the health sector, and then real access to the developments of digitization.

As for (Bureau d'ordre général BOG), which can be described within what is shown in Figure No. (09) the following:[36] (Fig. 20.8)

Through this figure, it is clear that when the correspondence is received, it is transferred to the Postal Processing Authority (BOG), which ensures that it is received, read, recorded, and contained through the electronic digitization system to the General Secretariat of the concerned health authority, to be subsequently directed to the various subsidiary bodies The aim is to trace (La Trassabilité) the incoming correspondence, within the various processes related to archiving, storage, etc. Through the visible control panel, it is possible to analyze the quantity and movement of incoming and outgoing

[35] Ibid.

[36] Emission ADN, 2020, La Numérisation du secteur sanitaire de l'Algérie: Réalité et Perspective, Canal El Hayat TV, le: 28/12/2020, dans: https://www.youtube.com/watch?v=qmlEFrgOzu4&t=2547s. (Visite le 20/08/2024).

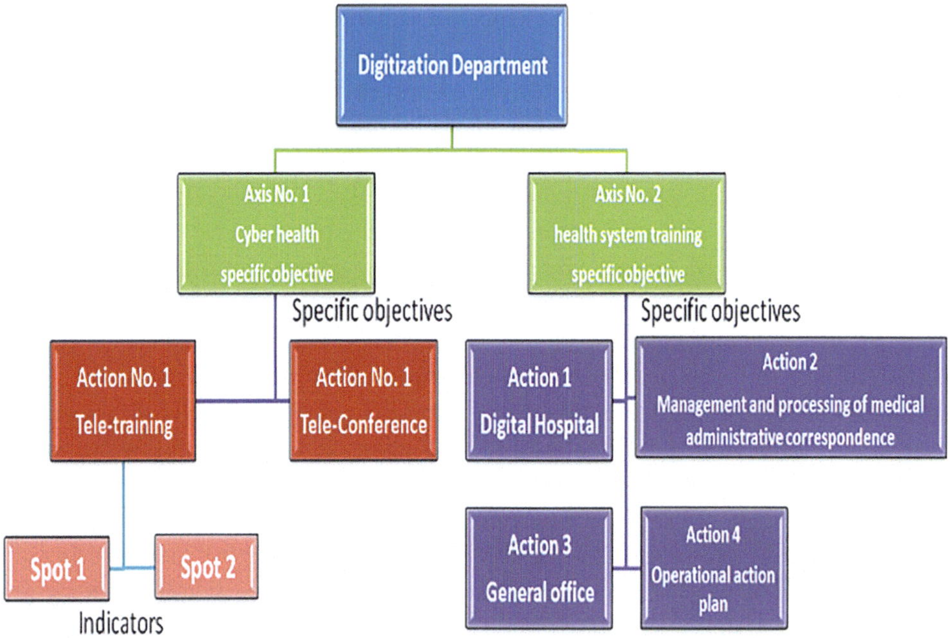

Figure 20.8 Health template management system platform (THSM). *(Prepared by researchers based on.)*

correspondence, and to identify the broken correspondence, which has not yet been processed.

Finally, and in general, the application and embodiment of digitization in this health sector will enable the achievement of a set of benefits that can be summarized as follows:

- Achieving speed in treatment;
- Keeping patient appointments;
- Profit in efficiency and profit in money;
- Protecting doctors from infectious diseases (such as what happened with the Corona pandemic);
- Improving internet flow in Algeria.

It is assumed that the digitization of medical records for patients in hospitals will be generalized before June 2023. This was announced by the Algerian Minister of Health, as he confirmed that "the era of paper is over" and that the exorbitant costs of using papers will end. Through his inspection visit to the University Hospital Center in Beni Messous, Algiers, in order to inquire about the effective application of digitization, and to achieve this goal, the state is striving in this direction.

Inspection committees have been formed at a high level of management to roam all regions of the country with the aim of inquiring about the state of hospitals in digitization. In addition to hygiene, medical emergencies, financial management, improving staff performance, and the quality of services provided to citizens, in addition to meeting staff needs and listening to their complaints.[37]

It should also be noted that Prof. Lyce Rahal, Director General of Health Services, recently referred to the digitization strategy adopted by the Ministry of Health, which is based on modernizing health structures and activities. According to him, the first step in this process is to enter an identification number into the patient's electronic medical record, in order to improve the effectiveness, efficiency, quality and safety of health care.

In this context, on the website of the University Hospital of Oran, it was noted that the Electronic Medical File (DEM) collects medical information related to the patient, necessary for coordinating care, prescriptions, medical summaries, hospitalization reports, test results, and indication of allergies. He added, "DEM is a computerized patient file management system that aims to replace traditional paper files." According to the same source, the digitization of a patient's medical file will ensure that any access by certified professionals can be traced, in real time, to any patient's DEM consultation procedure or supply. It was also a matter of confidentiality, the website said, noting that for users, maintaining the confidentiality of a patient's personal medical data is an ethical and legal obligation. He indicated that medical confidentiality is an obligation aimed at protecting the health of people who can trust the doctor. The physician's ethical obligations regarding maintaining a record of electronic medical records are essentially the same obligations that apply to a paper record. He stressed the need to take deterrent technical and legal measures to prevent the risks of misuse.[38]

According to the same site, data security, particularly DEM, is ensured via a secure computer platform, allowing professionals to manage user identity and access control, data integrity and journaling.

5. Conclusion

E-Health has made significant progress in Algeria. Indicators of this progress are evident from the specific national examples and activities covered. Especially what has been dealt with at the level of the health sector, and this progress can be better controlled when legal frameworks are introduced in regulating this, and the development of national policies related to e-health, with training and qualification of more human resources, commitment to regular funding, and respect for the long-term plans established for this.

Telemedicine, for example, allows canceling distances, reducing hospitalization time, and reducing mobility (travel), while providing a means for specialists to move

[37] Samia Acher, Santé: Généralisation de la numérisation avant juin 2023, in Le jeune INDEPENDANT (in 22 0.12. 2022 at 22:45) https://www.jeune-independant.net/sante-generalisation-de-la-numerisation-avant-juin-2023/(Visited in 30/06/2025).

[38] Ibid.

electronically to various regions and virtually. Despite the movement of doctors or patients between different medical sites, the National Agency for Health Digitization can, through its applications, transfer patient information through internal and external network connections quickly and simply.

5.1 Research results

The preliminary results of the evaluation of the e-health initiative in Algeria reveal the desire and enthusiasm of its actors to use information and communication technology to offer quality health care remotely. However, lack of resources, lack of coordination, fragile and unqualified infrastructure, and insufficient funding prevent the adoption of these technologies. Numerous Western country reports have documented the intrinsic benefits of e-health in delivering optimal healthcare. This evaluation revealed Algeria's initiative in this regard. By adopting a number of projects in progress, and due to their lack of appropriate documentation and evaluation, the current evaluation is limited, and the overall success of e-health projects is uncertain.

In this context, and given the recent applications of this agency, and the beginnings of the spread of digitization platforms in the health sector, and despite some digital platforms that have been operated by private individuals, this is considered insufficient to answer the main problematic of this topic, which expresses the effectiveness of this agency in digitizing this sector. The beginnings suggest the success of the first stages or the first steps of digitization, and the evidence for this lies in the great interest that the Algerian authorities attach to the issue of digitization in general, and the digitization of the health sector in particular, as they have harnessed very important material and human capabilities for the success of the digitization of the health sector through the efforts and available human energies The National Agency for Health Digitization. Perhaps the beginning of the operation of digitization components in this sector, especially the operation and linking of the digital hospital, digital pharmacy, and digital medicine, with the digital file of the treated patient, within internal and external networks, and the association of this with the National Social Security Fund and the national identification number, for non-believers, are all signs of practical success. This ambitious digitization.

However, efforts remain lacking in rigor and high seriousness to make this happen, by decision-makers in the country, because the issue of digitization, especially in the health field, is linked to other necessary requirements, such as the issue of delaying the realization of the high flow of the Internet in the country, (despite the promises related to improving the flow of the Internet in hospitals), And the lack of partnership with highly experienced foreign dealers in this regard, with the low rate of training of health sector users, to enable the smooth and simple implementation of digitization requirements, and in record time.

5.2 Recommendations

This agency constitutes the digital locomotive in the health sector, thus creating a kind of competition with the rest of the other sectors, especially the vital ones. In addition, the exchange of medical documents through the digital platform will enable a number of benefits to be achieved, and in order to achieve this, the following recommendations must be taken into account:

- Improving the flow of the Internet in the country in general and hospitals in particular;
- Activating the partnership with foreign specialists with high experience in the digitalization of the health sector;
- Training of all health sector users in the field of digitalization according to the competence of each user;
- Opening university specializations in the field of health digitization, and the digitization of health sector management;
- Allocating more funds necessary to accelerate the realization of digitization in this sector;
- Stimulating other sectors parallel to the health sector to digitize to facilitate the electronic linking of these sectors to the e-health sector;
- Supporting health digitization in various pillars of the Fourth Industrial Revolution, such as augmented (virtual) reality, big data, 3D printing. etc.

References

Acher, S. Santé : Généralisation de la numérisation avant juin 2023, in Le Jeune Independant. (Visited in 22. 12. 2024 at 22:45). https://www.jeune-independant.net/sante-generalisation-de-la-numerisation-avant-juin-2023/.

Emission ADN. 2020. La Numérisation du secteur sanitaire de l'Algérie : Réalité et Perspective, Canal El Hayat TV, le : 28/12/2020, dans. https://www.youtube.com/watch?v=qmlEFrgOzu4&t=2547s. (Visite le 20/August/2024).

E-santé, 2005. une révolution pour l'extension des couvertures sante en Afrique », AIM, Healthcare and social benefits fir all, ResoDoc by Doc&You. In: E-Health in the Eastern Mediterranean Region: What Is E-Health? Cairo. World Health Organization Regional Office for the Eastern Mediterranean. https://www.emro.who.int/HIS/ehealth/AboutEhealth.htm. (Accessed 8 August 2024). Accessed.

Executive Decree No. 22-251 ofJune 30, 2022. Relating to the National Agency for Digitization in Health.

Halim Ammar Khodja & Autres. Reformes et perspectives du secteur de la santé et de la E-Santé en Algérie et au Maroc, Bretagne Commerce International, Webinaire, 21 Janvier 2021. www.BretabneCommercsInternational.com.

https://esante.gouv.fr/les-acteurs-de-terrain.

https://www.emro.who.int/health-topics/ehealth/. (Accessed 09 August 2024).

Ikumapayi, O.M., Kayode, J.F., 2022. Adeniran sunday Afolalu, telehealth and telemedicine -an overview. In: Conference: 3rd African International Conference on Industrial Engineering and Operations Management.

MedIsi, La Télé-Expertise C'est Quoi?. URPS ML Nouvelle-Aquitaine. Fiche 1, Mars 2023. https://www.urpsml-na.org/uploads/images/1677698307_Fiche%20g%C3%A9n%C3%A9rale%20TLX_010322.pdf.

Nasreddine, A. The reform of primary health care/PHC in Algeria : a first step towards a more Equitable and efficient health system, HEALTH SCIENCESVolume 2021, ID 366. in: https://www.researchgate.net/publication/348899727_The_Reform_of_Primary_Health_Care_PHC_in_Algeria_A_First_Step_Towards_a_More_Equitable_and_Efficient_Health_System#fullTextFileContent.

NHS, 2020. Clinical guide for the management of remote consultations and remote working in secondary care during the coronavirus pandemic. In: Specialty Guides for Patient Management During the Coronavirus Pandemic. https://www.rcslt.org/wp-content/uploads/2021/10/NHS-England-clinical-guide-for-the-management-of-remote-consultations-and-remote-working-in-secondary-care-during-the-coronavirus-pandemic.pdf.

Orphanoudakis, S.C., Kaldoudi, E., Tsiknakis, M., 1996. Technological advances in Teleradiology. Eur. J. Radiol. 22, 205–217.

Queirós, A., Alvarelhão, J., Cerqueira, M., Rocha, N., 2018. In: Remote Care Technology: A Systematic Review of Reviews and Meta-Analyses. https://www.researchgate.net/publication/323135789_Remote_Care_Technology_A_Systematic_Review_of_Reviews_and_Meta-Analyses.

Rosen, R., Wieringa, S., Greenhalgh, T., Leone, C., Rybczynska-Bunt, S., Hughes, G., Moore, L., Shaw, S.E., Wherton, J., Byng, R., 2022. Clinical risk in remote consultations in general practice: findings from in- COVID- 19 pandemic qualitative research. BJGP. https://doi.org/10.3399/BJGPO.2021.0204. (Accessed 29 June 2022).

The Algerian health market is growing, Tuesday 7 April 2020 - Healthy ambition - Algeria - from 2 to 3 December 2020. https://www.businessfrance.fr/le-marche-algerien-de-la-sante-est-en-croissance.

webtv.univbejaia.dz.

Williams, S.Y., Adeyemi, S.O., Eyitayo, J.O., Odeyemi, O.E., Dada, O.E., Adesina, M.A., Akintayo, A.D., 2020. Mobile health technology (Mhealth) in combating COVID-19 pandemic: use, challenges and recommendations. Eur. J. Med. Educ. Technol. 13 (4), em2018. https://doi.org/10.30935/ejmets/8572.

World Health Organization, 2021. Global Strategy on Digital Health 2020-2025. https://www.who.int/docs/default-source/documents/gs4dhdaa2a9f352b0445bafbc79ca799dce4d.pdf.

Youcef, B., Hadj, K., 2022. Measuring the quality of health services in Algeria. J. Contempor. Business Econ. Stud. 05 (1). https://www.asjp.cerist.dz/en/downArticle/617/5/1/176391.

CHAPTER 21

Smart healthcare systems: In the era of metaverse, artificial intelligence, and data science

Ahmed Imran Kabir[1], Md Warif Islam Mahim[1], Dewan Ariyan Mehdi[1] and Sonali Vyas[2]
[1]Department of Bachelor of Business Administration, School of Business and Economics, United International University, Dhaka, Bangladesh; [2]Department of Computer Science, JK Lakshmipat University, Jaipur, Rajasthan, India

1. Introduction

The landscape of healthcare is undergoing great transformations driven by advancements in artificial intelligence (AI) and the burgeoning concept of Metaverse (Yang et al., 2022). These technologies are becoming more ingrained into healthcare systems; they offer unusual opportunities to enhance patient care and medical training. This paper contains an exhaustive analysis of the future of AI and Metaverse technologies, which are going to take the healthcare system to another level with much quality and reliable experience. In this smart world, AI technologies are already fulfilling the responsibility of assistance, quick replies, and in many other positions by intelligent chatbots and (DSS) decision support systems. On the other hand, Metaverse offers a virtual platform to create a more connecting experience (Ali et al., 2023). Metaverse can be a solution to health care education more precisely in the future, where students can experience surgical procedures, practices, therapy sessions, medics, etc. It will give them a chance to work and experience as professionals, develop their skills, and receive patients' quality care. This development comes with challenges and opportunities, which this article will explore by analyzing the technology acceptance model, circulating the innovation theory, and other related frameworks.

In addition, implementing these technologies, there will be challenges in the ability of AI, reliable use of data, and also the engagement emotionality with patients of the AI system, which will impact patient trust in the healthcare providers. We conducted a survey with the healthcare professionals and patient consumers, and with the collected data, we analyzed using statistical methods to provide deeper insight and develop a hypothesis about the effectiveness of this development or integration of AI and Metaverse technologies in the healthcare system.

Intelligent Semantic Analysis for Healthcare
ISBN 978-0-443-33512-9,
https://doi.org/10.1016/B978-0-443-33512-9.00020-2

2. Literature survey & hypothesis development

2.1 When integrating AI and metaverse technologies into healthcare

The field of healthcare is developing day by day, greeting the large advancements of the emerging concepts of AI and Metaverse (Tyagi and Saxena, 2022; Dang et al., 2023). By using intelligent chatbots and a Decision Support System, the healthcare system is developing for patient interactions by offering 24/7 healthcare assistance and support (Rachmad, 2022; Al Kuwaiti et al., 2023). These technologies are constantly developing and learning from the conversation (Ali et al., 2023; Bashir et al., 2023). These technologies not only respond to patient queries but also educate themselves about every massage interaction and inquiry, which improves patient experience and care quality (Lee, 2022; Jamshidi et al., 2022).

Moreover, Metaverse is a whole new visual experience of surgical training and therapy sessions in healthcare services (Wang et al., 2022). This technology will let you enter a fake world with the feel of real experience, where surgeries can be practiced with zero risks and improving skills (Tan et al., 2022; Bhattacharya et al., 2022; Moztarzadeh et al., 2023; Mozumder et al., 2023). Without risking any real situation and outcomes, the metaverse is allowing the healthcare system to improve safely and professionally.

2.2 Organizational culture and smart healthcare adoption

The healthcare system and its organizational cultures show the success of smart healthcare systems within healthcare institutions. By using advanced technology, smart healthcare system is continuously learning and is open to change, facilitating the integration and effective uses of it (Islam, 2024; Moodley, 2024). Smart healthcare is a culture that promotes collaboration affiliation across multitasking teams, essentials for developing and innovative healthcare solutions (Bibri and Jagatheesaperumal, 2023).

Moreover, the interest of the staff in new technologies plays a crucial part in manipulating the fullest potential of a smart healthcare system, as utilizing these technologies directly.

2.3 Theoretical frameworks supporting smart healthcare systems

Understanding and analyzing the dynamics of smart healthcare systems within healthcare organizations shows many relevant theoretical frameworks and models:

2.4 Hypothesis development

Based on the research, the study introduces theories to see the outcome of artificial intelligence (AI) and patient reliability in the healthcare environment:

Hypothesis 1 (H1): Explain the ability of AI Systems to improve patients' Trust in Healthcare Services (Ali et al., 2023).

This hypothesis proposes that the simplicity and opportunities of AI systems greatly add to creating patient reliability. If AI systems look more explainable, patients might be more certain about the decisions suggested by these systems.

Hypothesis 2 (H2): Using Intelligent Chatbot Positively Influences Patients' Trust in Healthcare Services.

This hypothesis shows the application of chatbots, recognized for their ability to provide reactions based on patients' perceptions, which increases the reliability of the healthcare system to patients. By providing instant help, these chatbots develop the patient experience and also increase reliability in smart healthcare systems.

Hypothesis 3 (H3): Presence of Human Emotions in AI Systems Increases Patients' Trust in Healthcare Providers.

This hypothesis analyses the impact of human-like sentiments in AI systems on patient reliability. The existence of human emotions in AI systems takes interaction to another level, and it looks less robotic and more professional. It increases the reliability of smart healthcare systems more with its realistic humor.

Hypothesis 4 (H4): Data Security estimates build up Patients' Trust in Healthcare Services (Al Kuwaiti et al., 2023).

This hypothesis analyses the examination of how data security strength within AI systems may increase the trust of patients. Ensuring the security and confidentiality of patients' data are important, and it is proven in theory that implementing effective security estimation could possibly augment the willingness of patients to participate in AI-increased healthcare services.

Hypothesis 5 (H5): Appropriate use of AI and Data Science showcases the relationship between AI explain ability and patients' Trust (Letafati and Otoum, 2023).

This hypothesis asses if the appropriate use of AI and data science mediating the impacts of AI explains the ability of patients to trust. It shows those ethical considerations in AI usage not only keep transparency but also make the trust stronger to patients in smart healthcare systems.

Hypothesis 6 (H6): The Ethical Use of AI and Data Science Mediates the Impact of Intelligent Chatbot Usage on Patients' Trust (Dang et al., 2023).

This final hypothesis shows how, ethically, AI impacts the connection between intelligent chatbot usage and patient trust; it also suggests that ethical practices in deploying chatbots, including privacy considerations and user profiles, play a vital role in preserving and boosting trust.

Fig. 21.1 shows the structure of research and study in this paper. Figure 1explains the hypothesis that happens to be a significant relationship as well as positiveness between the adoption of patients' trust in AI systems and Healthcare Providers and the explainability of AI systems, intelligent chatbot usage, the presence of human emotions in AI systems, and data security measures (H1, H2, H3, H4).

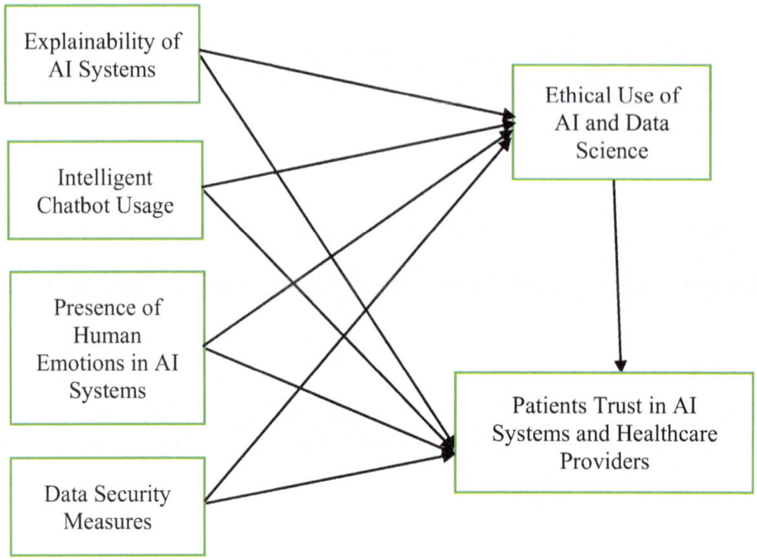

Figure 21.1 Network diagram of the hypothesis.

Moreover, its breakdowns the hypotheses that the goal to use ethical AI and data science indicates the association between the adoption of patients' trust in AI systems and the explainability of AI systems, intelligent chatbot usage, presence of human emotions in AI systems, and data security measures (H4, H5, H6). We created these hypotheses for testing relationships between using AI technologies and their joint impacts on patients' reliability of healthcare services. Every hypothesis goal is to reveal the essential perspectives on how technology with ethical approaches all together impact patient evaluations and trust in the smart healthcare system.

3. Methodology

3.1 Study context

In this paper, the network diagram serves as a visualization of all the hypotheses and the relationship between AI technologies within intelligent healthcare systems and their impact on patient trust. The IV, which shows the explainability of AI systems, chatbot usage, human emotions, DSS in AI systems, and data security, are advanced to influence both mediating variables, organizational culture, and directly affect the DV, patients' trust in AI systems, and healthcare Services. The diagram describes direct route from each IV to the DV to show direct effects that are not mediated.

Additionally, organizational culture is represented as a central mediating variable that potentially channels the effects of this AI technology on patient trust. This pattern allows extensive analysis of both direct and indirect influences, showing a clear map for understanding how improvement in AI technology might increase or decrease trust within healthcare contexts.

3.2 Research design

A survey design is employed to investigate the impact of an intelligent healthcare system within Metaverse and Artificial Intelligence (AI) on healthcare outcomes. This research approach is suitable for understanding broad trends and correlations within a large population, given its efficiency in data collection and the breadth of data it can cover (Ullah et al., 2023).

3.3 Questionnaire design

There were 25 questions in the final survey and five variables were constructed. Prior to the execution of the final survey, we executed a pretest to confirm the reliability of the questionnaire we prepared. Likert scales were used to measure the construct in research ranging from 1 being "strongly disagree" to 5 being "strongly agree".

3.4 Sampling and data collection

Participants were selected using stratified random sampling from various healthcare institutions that have adopted intelligent systems (Islam, 2024). Sample size was determined based on the population size of healthcare professionals in these institutions, aiming for a representative distribution across different roles and experience levels. This method ensures diversity and enhances the generalizability of the findings.

3.4.1 Operationalization of variables

- Independent Variables (IVs): Explainability of AI systems, intelligent chatbot usage, presence of human emotions in AI systems, and data security measures.
- Dependent Variable (DV): Patients' trust in AI systems and healthcare providers.
- Mediating Variable: Ethical use of AI and data science.
- Data Collection Methods and Tools
- Using social media platforms, email marketing, and personal contacts, the data was collected through Google Forms. This strategy of a multi-channel approach helped us to reach the maximum number of respondents. All the data were managed in Microsoft Excel, where working with huge data is comparatively easier.

3.5 Data analysis

3.5.1 Presentation of descriptive statistics

With the help of descriptive statistics as means, standard deviation, and ranges, we summarized the key variables. By this summarization, we got to know the basic understanding of the data set and its distribution and tendency of the responses.

3.5.2 Analysis of the correlation between independent and dependent factors

Regression analysis was done to understand the correlation between the decision factors (e.g., explainability of AI systems and intelligent chatbot usage) and the dependent variable (patients' trust in AI systems). This method helps in understanding how changes are associated between IVs and DVs.

3.6 Mediation analysis

To inspect the effect of the purpose to use in this research, a two-step method was utilized by observing at the impact of H1, H2, H3, and H4 on H6 acceptance through intent to use was utilized in the first phase. The implied effect of H3, H5, H6 and H2, H5, H6 was similar in this study. Using H1, H2, H3, and H4 without ignoring the mediator (H5), the study prioritized observing the effects on the H6 for H5. Moreover, the mediating variable is the "Ethical use of AI and data science," and it is used to examine how it impacts the relationship between various IVs (such as explainability of AI systems, intelligent chatbot usage, etc.) and the DV.

4. Results and analysis

These methodologies and analytical techniques provide an understanding of the dynamics at play in smart healthcare systems within the Metaverse and AI era, offering insights that can guide future research and practical applications in the field.

4.1 Profile of Respondent's

Based on cross-sectional data collected from the medical university students the study was done. Table 21.1 shows the demographics of gender, and Table 21.2 shows the age demographics who were a part of the survey. Men made up 70.3%, and women made up 29.2% of the population; 86.3% were between the ages of 20–25, 10.4% were between the ages of 26–30, 2.4% were between 31–35, and 0.9% were between the age of 41–45. The total number of respondents was 212, and the respondents were a mix of medical students, professionals, regular patients, and students from different universities and different programs. The cumulative percentage of gender demographics in Table 21.1 is 29.2 for females and 99.5 for male and the cumulative percent of age demographics in Table 21.2 is 86.3% for ages between 20–25, 96.7% for age between 26–30, 99.1% for age between 31–35, and lastly 100% for age between 41–45.

4.2 Validity of discriminant

4.2.1 Criterion analysis by Fornell–Larcker

Measures the average variance received by the items of each set up in relation to the variance due to measurement error. Values are above 0.50 for all constructs, which is the recommended threshold, signifying satisfactory convergent validity.

Table 21.1 Theoretical foundations.

Topic	Theory
Technological acceptance model (TAM):	This model was designed by Davis in 1989; this model helps to understand the acceptance and rejection of the technology, which is crucial in developing the smart healthcare system.
Diffusion of innovation theorem:	This theorem gives a deep analysis of how old technologies spread within an organization and healthcare at large. It assists in identifying categories and adapters and the factors that affect the adoption rate, providing a structured way to mitigate barriers to development or adaptation.
Organizational readiness for changing:	This framework helps to show if the healthcare organization is ready to develop new technologies. It focuses on organizational factors like the existing infrastructure, employee incompetence, and the overall culture, which are necessary for a very cooperative transition from the healthcare system to a smart healthcare system.
Social-technical systems hypothesis (STS):	STS shows the importance of addressing both social and technical aspects when authorizing technology in the healthcare background. It highlights the connection between technology, healthcare professionals, and patients as a crucial element for failing.
Meager supervision:	Obtaining from manufacturing, lean principles in healthcare concentrate on improving inefficiency, increasing waste, and increasing patient scare. This principle can help facilitate the ineffective implementation of smart technologies by changing the healthcare system and promoting an environment of frequent decline.

Table 21.2 Gender demographics.

		Frequency	Percent	Valid percent	Cumulative percent
Valid	Female	62	29.2	29.2	29.2
	Male	149	70.3	70.3	99.5
	Other	1	0.5	0.5	100
	Total	212	100	100	

Therefore, our research model demonstrates robust reliability and good convergent validity across all constructs, indicating strong internal consistency and that the indicators used explain a significant part of the variance in the constructs Table 21.3.

This table distinguishes the square root of the Average Variance Extracted (AVE) for each concept with the correlations between constructs. For valid discriminant validity, the sloped elements (which are the square roots of AVE) should be greater than the off-sloped elements in their own rows and columns. The provided values, such as 0.724 for

Table 21.3 Age demographics.

		Frequency	Percent	Valid percent	Cumulative percent
Valid	20–25	183	86.3	86.3	86.3
	26–30	22	10.4	10.4	96.7
	31–35	5	2.4	2.4	99.1
	41–45	2	0.9	0.9	100
	Total	212	100	100	

Table 21.4 : The Fornell-Larcker criterion analysis for discriminant validity.

	1	2	3	4	5	6
Data security measures	0.724					
Ethical use of AI and data science	0.506	0.743				
Explainability of AI systems	0.540	0.560	0.803			
Intelligent chatbot usage	0.388	0.621	0.693	0.775		
Patients trust in AI systems and healthcare providers	0.508	0.633	0.732	0.784	0.729	
Presence of human emotions in AI systems	0.531	0.602	0.425	0.409	0.551	0.829

"Data security," indicate that each construct divides more variance with its indicators than other constructs, declaring discriminant validity.

4.2.2 Heterotrait-Monotrait (HTMT) analysis

The validity of discriminant analysis provided checks if constructs that are supposed to be unrelated are indeed showing low correlations, ensuring that they measure different concepts Table 21.4, Fig. 21.2.

Note: Validity of Discriminant exists if the HTMT <0.85 (Henseler et al., 2005); Validity of Discriminant exists if the HTMT <0.90 (Gold et al., 2001).

A newer criterion where values less than 0.90 suggests adequate discriminant validity. Most HTMT values like 0.409 between "Patients' trust" and "Presence of." are well below 0.90, supporting distinctiveness between constructs (Table 21.5).

4.2.3 Cross-loads

Items should be loaded higher on their respective constructs than on other constructs. The numbers, like 0.718 for an item under "Data security," should be higher than any cross-loadings for the same item on other constructs (e.g., 0.550 on "Ethical use."), confirming distinctiveness Table 21.6.

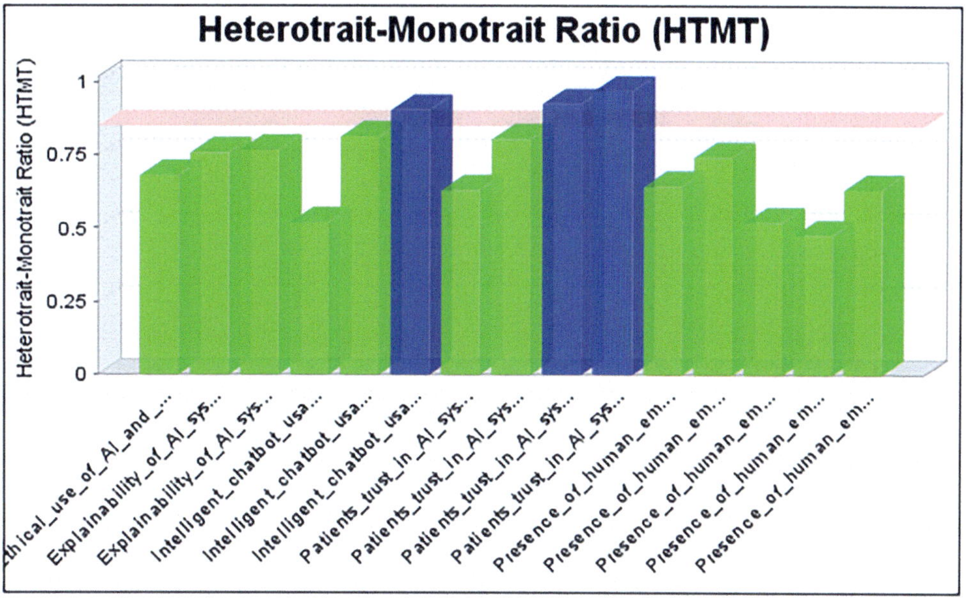

Figure 21.2 Heterotrait-monotrait (HTMT) bar chart analysis.

Table 21.5 The Heterotrait-Monotrait (HTMT) analysis for discriminant validity.

	1	2	3	4	5	6
Data security measures						
Ethical use of AI and data science	0.682					
Explainability of AI systems	0.758	0.768				
Intelligent chatbot usage	0.521	0.813	0.904			
Patients trust in AI systems and healthcare providers	0.631	0.804	0.926	0.970		
Presence of human emotions in AI systems	0.645	0.744	0.522	0.478	0.634	

4.3 SEM structural model study

Fig. 21.3 represents the results of all analyses in the SEM model for our study. Using a 5000-sample bootstrapping approach, the t-values and R squares were calculated.

To develop the standardized structural equational model (SEM) outputs displayed in Fig. 21.3, Smart PLS 3.2.9 was used. The path coefficients of all variables seem to be in decent condition.

Table 21.6 Values of the cross loads of individual items in the SEM.

	1	2	3	4	5	6
Doyoubelieveintelligentchatbotscaneffectivelybridgethegapinunder	0.410	0.450	0.550	0.720	0.573	0.267
DoyoubelieveitisimportantforAIanddatasciencetechnologiestoadhere	0.480	0.730	0.400	0.310	0.309	0.526
DoyoufeelcomfortablediscussingsensitivemedicalinformationwithAId	0.290	0.400	0.480	0.630	0.687	0.228
DoyouthinkAIsystemsshouldbedesignedtoexhibithumanlikeemotionssuc	0.460	0.480	0.310	0.330	0.41	0.821
Haveyoueverinteractedwithanintelligentchatbotforhealthcarerelate	0.170	0.470	0.490	0.780	0.59	0.235
Howconcernedareyouaboutthesecurityofourpersonalhealthdatawhenus	0.700	0.330	0.450	0.410	0.364	0.22
HowconfidentareyouintheabilityofAIsystemstoprotectourpersonalhe	0.350	0.500	0.430	0.530	0.751	0.425
HowconfidentareyouinusingAIdrivenhealthcaresolutionsthatoffertra	0.420	0.420	0.810	0.650	0.712	0.386
HowimportantdoyouconsideritforAIsystemstoprovideexplanationsfort	0.390	0.410	0.800	0.550	0.537	0.228
HowimportantisitforyouthatAIsystemsinhealthcaredemonstrateempath	0.460	0.460	0.230	0.250	0.392	0.905
Howimportantisitforyoutobeinformedaboutthemeasurestakentoensuret	0.820	0.48	0.41	0.280	0.468	0.544
Howmuchconfidencedoyouhaveintheethicalstandardsfollowedbydevelop	0.220	0.700	0.390	0.480	0.508	0.312
HowmuchtrustdoyouplaceinAIdrivenhealthcaresystemstoprovideaccura	0.400	0.410	0.57	0.540	0.699	0.417
Howsatisfiedareyouwiththeexplanationsprovidedbyintelligentchatbo	0.250	0.400	0.4800	0.780	0.527	0.17
Howsatisfiedareyouwiththeleveloftransparencyandexplainabilitypro	0.300	0.380	0.500	0.610	0.719	0.366
Inouropinionhowmuchdointelligentchatbotscontributetoimprovingpa	0.360	0.570	0.620	0.820	0.712	0.534
Inouropinionwhatroleshouldgovernmentregulationsplayinensuringda	0.650	0.280	0.360	0.230	0.203	0.233
Overallonascaleof1to10howmuchtrustdoyouhaveintheintegrationofAIa	0.410	0.480	0.600	0.550	0.758	0.269
TowhatextentdoyouagreethatunderstandinghowAIarrivesatadiagnosisw	0.490	0.520	0.800	0.450	0.491	0.397
TowhatextentdoyoubelieveAIsystemscanimprovethequalityofhealthcar	0.450	0.580	0.610	0.590	0.759	0.641
TowhatextentdoyouthinkethicalconsiderationsinfluenceourtrustinA	0.380	0.700	0.340	0.560	0.497	0.384
TowhatextentdoyouthinkthepresenceofhumanemotionsinAIsystemscould	0.360	0.380	0.380	0.43	0.485	0.745
WouldyoubemorelikelytouseAIdrivenhealthcareservicesifyouwereassu	0.410	0.840	0.520	0.49	0.552	0.550
Wouldyoubewillingtosharemorepersonalhealthinformationifyouwereas	0.710	0.330	0.350	0.2	0.374	0.460
WouldyoutrustAIdrivenhealthcaresolutionsmoreiftheyshowedempathya	0.460	0.630	0.450	0.35	0.517	0.835

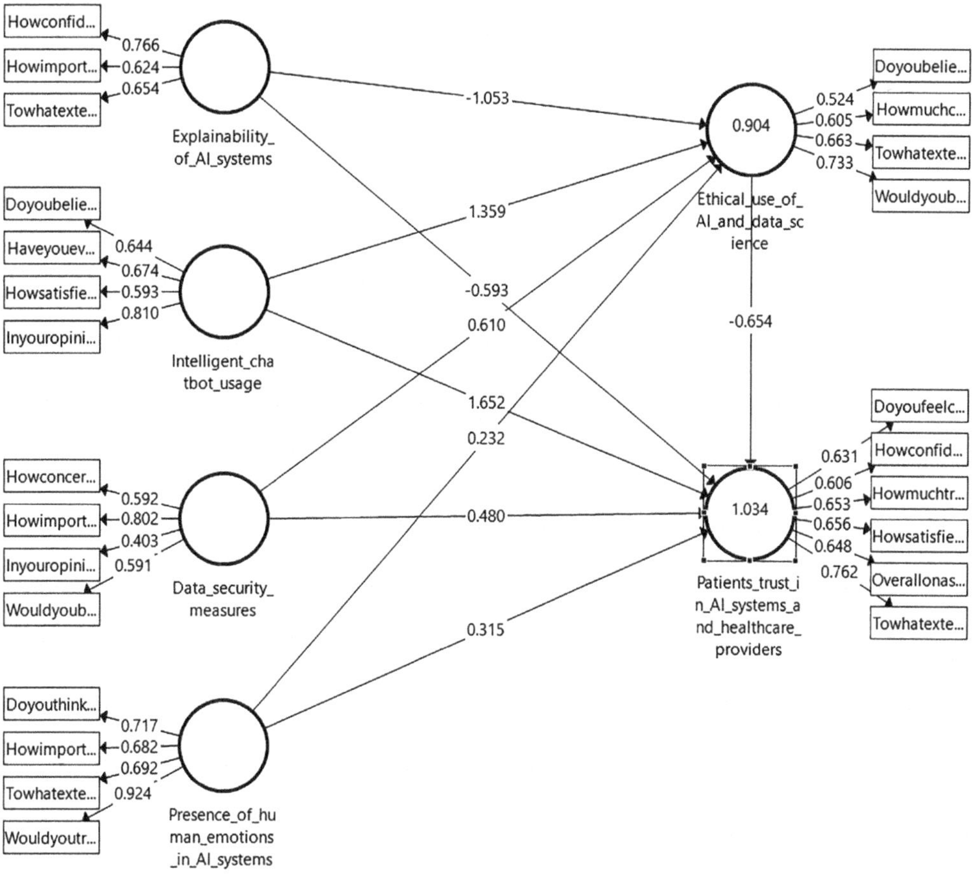

Figure 21.3 Standardized results of SEM calculations.

4.4 Hypotheses testing by direct and indirect effects

Table 21.7 shows the outcomes of direct and indirect hypotheses. To measure statistical t-values, Bootstrapping was used. Based on the *P*-value of Smart PLS 3 3.2.9, a 95% confidence interval can be utilized in social science research.

In Table 21.7, Data Security Measures - > Ethical Use of AI - > Patient Trust: The original indirect effect here is 0.008, with a bootstrap mean of 0.010 and a standard deviation of 0.014. The t-statistic of 0.572 and a *P*-value of 0.567 suggest that this pathway's influence is not statistically significant. The numbers indicate a weak and non-significant indirect effect of data security measures on patient trust through ethical AI usage.

Explainability of AI Systems - > Ethical Use of AI - > Patient Trust: This pathway shows an even smaller original effect of 0.005, with a bootstrap mean of 0.002 and a

Table 21.7 Result of direct and indirect effect hypotheses.

	Original sample (O)	Sample mean (M)	Standard deviation (STDEV)	T statistics (|O/STDEV|)	*P* values	Decisions
Data security measures - > ethical use of AI and data science - > patients trust in AI systems and healthcare providers	0.008	0.01	0.014	0.572	0.567	Rejected
Explainability of AI systems - > ethical use of AI and data science - > patients trust in AI systems and healthcare providers	0.005	0.000	0.010	0.490	0.624	Rejected
Intelligent chatbot usage - > ethical use of AI and data science - > patients trust in AI systems and healthcare providers	0.023	0.03	0.032	0.720	0.471	Rejected
Presence of human emotions in AI systems - > ethical use of AI and data science - > patients trust in AI systems and healthcare providers	0.020	0.02	0.025	0.804	0.421	Rejected

standard deviation of 0.010. The t-statistics are 0.490, and the *P*-value is 0.624, further confirming the non-significance of this effect. This suggests that AI explainability does not significantly affect patient trust through ethical practices.

Intelligent Chatbot Usage - > Ethical Use of AI - > Patient Trust: The observed effect here is somewhat larger at 0.023, with a bootstrap mean of 0.027 and a standard deviation of 0.032. However, the t-statistic of 0.720 and a *P*-value of 0.471 still do not indicate a statistically significant effect. While the numbers are higher compared to other pathways, they do not support a significant mediation of trust through ethical AI via intelligent chatbot usage. Presence of Human Emotions in AI Systems - > Ethical Use of AI - > Patient Trust: This pathway shows an original indirect effect of 0.020, with a bootstrap mean of 0.019 and a standard deviation of 0.025. The t-statistic of 0.804 and a *P*-value of 0.421 are the highest among the pathways but still fall short of significance. This suggests a moderate but not significant effect of human emotions in AI systems influencing patient trust through ethical practices Table 21.8.

Data Security Measures - > Ethical Use of AI - > Patient Trust: the Original Effect: 0.008, Bootstrap Mean: 0.010, Bias: 0.002, Confidence Interval: 0.000 to 0.049 Since the confidence interval includes zero, the effect of data security measures on patient trust, mediated by ethical AI usage, is not statistically significant. This

Table 21.8 Confidence intervals bias corrected.

	Original sample (O)	Sample mean (M)	Bias	2.50%	97.50%
Data security measures - > ethical use of AI and data science - > patients trust in AI systems and healthcare providers	0.008	0.01	0.002	−0.008	0.049
Explainability of AI systems - > ethical use of AI and data science - > patients trust in AI system and healthcare providers	0.005	0.002	−0.002	−0.005	0.035
Intelligent chatbot usage - > ethical use of AI and data science - > patients trust in AI systems and healthcare providers	0.023	0.027	0.003	−0.025	0.098
Presence of human emotions in AI systems - > ethical use of AI and data science - > patients trust in AI systems and healthcare providers	0.02	0.019	−0.001	−0.031	0.07

pathway does not support a meaningful impact on patient trust. Explainability of AI Systems - > Ethical Use of AI - > Patient Trust: the Original Effect: 0.005, Bootstrap Mean: 0.002, Bias: 0.002, Confidence Interval: 0.000 to 0.035 the confidence interval again includes zero, indicating that the indirect effect of AI explainability on patient trust through ethical usage is non-significant. This pathway also does not support a substantial impact. Intelligent Chatbot Usage - > Ethical Use of AI - > Patient Trust: the Original Effect: 0.023, Bootstrap Mean: 0.027, Bias: 0.003, Confidence Interval: 0.002 to 0.098, although this interval is broader and the effect size larger, the inclusion of zero in the interval means that the effect of intelligent chatbot usage on patient trust through ethical AI practices does not reach statistical significance. Presence of Human Emotions in AI Systems - > Ethical Use of AI - > Patient Trust: the Original Effect: 0.020, Bootstrap Mean: 0.019, Bias: 0.001, Confidence Interval: 0.031 to 0.070, this confidence interval does not include zero, suggesting a statistically significant effect. The presence of human emotions in AI systems positively influences patient trust through ethical AI usage. Among the pathways analyzed, only the influence of human emotions in AI systems on patient trust via ethical practices is supported by the data as statistically significant. The other pathways—concerning data security, explainability, and chatbot usage—do not show significant impacts on patient trust. This indicates the selective significance of human-centric AI features in enhancing trust within intelligent healthcare systems.

5. Conclusions

In conclusion, the paper emphasizes the potential of implementing artificial intelligence (AI) and metaverse technologies into healthcare systems (Yang et al., 2022; Tyagi and Saxena, 2022). It emphasizes that these advanced technologies are not just reshaping patient interactions and experiences but also transforming medical training and healing practices by offering a virtual, risk-free environment. Findings while doing this research highlight adapting technologies in the healthcare system. Developing and implementing these technologies fully was done only because of this supporting culture, which always values learning and adaptability.

Moreover, the paper and study hypothesis prove that H1, H2, H3, and H4 significantly increase patients' trust and reliability over smart healthcare systems. In implying these technologies successfully, trust and reliability were very crucial and important (Petrigna and Musumeci, 2022). Overall, the paper and its methodology suggest smart healthcare outcomes are influenced by the ethical use (Siau and Wang, 2018; Letafati and Otoum, 2023) and the transparent communication skills artificial intelligence and metaverse have (Wang et al., 2022), moving toward a future where technology and healthcare are continuously affiliated (Dang et al., 2023).

6. Implementation of the study

To collect the data systematically and to assess transparently the impact of Ai and metaverse, we employed a survey design. This survey targeted young medical students and patients to ensure a comprehensive representation of the sample. The survey contains 25 questions with four independent variables and one mediating and one dependent variable to identify the human perception and acceptance of AI and metaverse technologies in the smart healthcare system. The survey link was marketed in different possible channels and personal contacts to reach the maximum response possible:

a) Social media platforms like Facebook, linked-in, and Instagram.
b) email marketing targeting specific groups like medical students, professionals, regular students, and university faculty members.
c) Personal contacts in hospitals or medical students helped more to get higher responses.

For a meaningful statistical sample size, the population of healthcare workers in the questioned organization was determined to ensure a mix of positions and experience levels. This technique helped us to make the study findings more accurate and generalized. All the collected data was stored and analyzed in Microsoft Excel. By using SPSS, we analyzed the demographics of the data, and by using PLS 3, we determined the descriptive statistics, which show mean scores, standard deviation, and ranges, which show the tendency of the data and the variety of responses in the data. In addition, regression analyses were done to see the relationship between IVs and DVs. This detailed, methodical approach concluded with the implantation of artificial intelligence AI and Metaverse in smart healthcare systems.

7. Limitations and directions for further research

This study is not beyond limitations like other studies, and it seems that we need a few areas for improvement in this study and future research suggestions in further exploration.

7.1 Limitations

Table 21.9.

Table 21.9 Limitations of the study.

Sample diversity:	We collected survey data from different demographics, but there was a gap in collecting data from different regions. Not every region and demographic have the same opinion and response as ours. This limitation can affect the study's understanding of the usefulness and effectiveness of implementing smart healthcare services.
Survey methodology:	The research and study relied on surveys created by us, which can introduce bias, such as respondents may answer desirable answers rather than reading out the question and giving honest answers.
Technological changes:	Constant evolution of AI and metaverse technologies means that findings may become outdated quickly as new technology develops.

7.2 Directions for future research

Table 21.10.

Table 21.10 Directions for future research.

Extensive studies:	In future research, extensive studies should be implemented to keep track of changes over time; deeper analysis can be possible on trust in AI and metaverse in smart healthcare systems.
Geographical coverage:	By collecting data from different regions, language may develop a deeper analysis and more accurate success in the field of smart healthcare systems.
Qualitative methods:	Implementing interviews or focusing on specific groups may provide deeper insights into individual experiences and perceptions of the data from surveys.
Ethical consideration:	A deeper analysis and study are needed in ethical considerations in AI and metaverse in smart healthcare systems. Also, data security and patient autonomy are well maintained with the advancement of technology.

References

Al Kuwaiti, A., Nazer, K., Al-Reedy, A., Al-Shehri, S., Al-Muhanna, A., Subbarayalu, A.V., , … Al-Muhanna, F.A., 2023. A review of the role of artificial intelligence in healthcare. J. Personalized Med. 13 (6), 951. https://doi.org/10.3390/jpm13060951.

Ali, S., Abdullah, Armand, T.P.T., Athar, A., Hussain, A., Ali, M., , … Kim, H.C., 2023. Metaverse in healthcare integrated with explainable AI and blockchain: enabling immersiveness, ensuring trust, and providing patient data security. Sensors 23 (2), 565. https://doi.org/10.3390/s23020565.

Bashir, A.K., Victor, N., Bhattacharya, S., Huynh-The, T., Chengoden, R., Yenduri, G., , … Liyanage, M., 2023. Federated learning for the healthcare metaverse: concepts, applications, challenges, and future directions. IEEE Internet Things J. 10 (24), 21873–21891. https://doi.org/10.1109/JIOT.2023.3304790.

Bhattacharya, P., Obaidat, M.S., Savaliya, D., Sanghavi, S., Tanwar, S., Sadaun, B., 2022. Metaverse assisted telesurgery in healthcare 5.0: an interplay of blockchain and explainable AI. In: 2022 International Conference on Computer, Information and Telecommunication Systems (CITS). IEEE, pp. 1–5. https://doi.org/10.1109/CITS55221.2022.9832978.

Bibri, S.E., Jagatheesaperumal, S.K., 2023. Harnessing the potential of the metaverse and artificial intelligence for the internet of city things: cost-effective XReality and synergistic AIoT technologies. Smart Cities 6 (5), 2397–2429. https://doi.org/10.3390/smartcities6050109.

Dang, V.A., Vu Khanh, Q., Nguyen, V.H., Nguyen, T., Nguyen, D.C., 2023. Intelligent healthcare: integration of emerging technologies and Internet of Things for humanity. Sensors 23 (9), 4200. https://doi.org/10.3390/s23094200.

Gold, A.H., Malhotra, A., Segars, A.H., 2001. Knowledge management: an organizational capabilities perspective. J. Manag. Inf. Syst. 18 (1), 185–214.

Henseler, J., Ringle, C.M., Sinkovics, R.R., 2005. The use of partial least squares path modeling in international marketing. Adv. Int. Mark. 20, 277–319.

Islam, M.M., 2024. The impact of transfer learning on AI performance across domains. J. Artif. Intell. Gen. Sci. (JAIGS) 1 (1).

Jamshidi, M.B., Ebadpour, M., Moghani, M.M., 2022. Cancer digital twins in metaverse. In: 2022 20th International Conference on Mechatronics-Mechatronika (ME). IEEE, pp. 1–6.

Lee, C.W., 2022. Application of metaverse service to healthcare industry: a strategic perspective. Int. J. Environ. Res. Publ. Health 19 (20), 13038. https://doi.org/10.3390/ijerph192013038.

Letafati, M., Otoum, S., 2023. Digital healthcare in the metaverse: insights into privacy and security. IEEE Consum. Electron. Mag. 13 (3), 80–89. https://doi.org/10.1109/MCE.2023.3334494.

Moodley, K., 2024. Artificial intelligence (AI) or augmented intelligence? How big data and AI are transforming healthcare: challenges and opportunities. S. Afr. General Practitioner 5 (4), 170–175. https://hdl.handle.net/10520/ejc-mp_sagp_v5_n4_a6.

Moztarzadeh, O., Jamshidi, M., Sargolzaei, S., Jamshidi, A., Baghalipour, N., Malekzadeh Moghani, M., Hauer, L., 2023. Metaverse and healthcare: machine learning-enabled digital twins of cancer. Bioengineering 10 (4), 455. https://doi.org/10.3390/bioengineering10040455.

Mozumder, M.A.I., Armand, T.P.T., Imtiyaj Uddin, S.M., Athar, A., Sumon, R.I., Hussain, A., Kim, H. C., 2023. Metaverse for digital anti-aging healthcare: an overview of potential use cases based on artificial intelligence, blockchain, IoT technologies, its challenges, and future directions. Appl. Sci. 13 (8), 5127. https://doi.org/10.3390/app13085127.

Petrigna, L., Musumeci, G., 2022. The metaverse: a new challenge for the healthcare system: a scoping review. J. Funct. Morphol. Kinesiol. 7 (3), 63. https://doi.org/10.3390/jfmk7030063.

Rachmad, Y.E., 2022. MediVerse: challenges and development of digital health transformation towards metaverse in medicine. J. Eng. Electr. Inform. 2 (2), 72–90.

Siau, K., Wang, W., 2018. Building trust in artificial intelligence, machine learning, and robotics. Cutter Bus. Technol. J. 31 (2), 47. https://ink.library.smu.edu.sg/sis_research/9371.

Tan, T.F., Li, Y., Lim, J.S., Gunasekeran, D.V., Teo, Z.L., Ng, W.Y., Ting, D.S., 2022. Metaverse and virtual health care in ophthalmology: opportunities and challenges. Asia-Pac. J. Ophthalmol. 11 (3), 237–246. https://doi.org/10.1097/APO.0000000000000537.

Tyagi, Y., Saxena, S., 2022. Applications of artificial intelligence, metaverse and data science for intelligent healthcare. AAYAM: AKGIM J. Manag. 12 (2), 141–149. https://doi.org/10.4018/979-8-3693-0928-5.ch004.

Ullah, H., Manickam, S., Obaidat, M., Laghari, S.U.A., Uddin, M., 2023. Exploring the potential of metaverse technology in healthcare: applications, challenges, and future directions. IEEE Access 11, 69686–69707. https://ieeexplore.ieee.org/abstract/document/10153590.

Wang, G., Badal, A., Jia, X., Maltz, J.S., Mueller, K., Myers, K.J., , … Zeng, R., 2022. Development of metaverse for intelligent healthcare. Nat. Mach. Intell. 4 (11), 922–929. https://www.nature.com/articles/s42256-022-00549-6.

Yang, Y., Siau, K., Xie, W., Sun, Y., 2022. Smart health: intelligent healthcare systems in the metaverse, artificial intelligence, and data science era. J. Organ. End User Comput. 34 (1), 1–14. https://doi.org/10.4018/JOEUC.308814.

Further reading

Sun, T.Q., Medaglia, R., 2019. Mapping the challenges of Artificial Intelligence in the public sector: evidence from public healthcare. Gov. Inf. Q. 36 (2), 368–383. https://doi.org/10.1016/j.giq.2018.09.008.

CHAPTER 22

Challenges and future scope of intelligent semantic healthcare

Pranali Pangam[1], Sonali Nikam[2], Swapnali Patil[3], Shreyasi Kadam[5] and Pournima Sankpal[4]

[1]Department of Pharmaceutical Chemistry, Dr. N. J. Paulbudhe College of Pharmacy, Ahmednagar, Maharashtra, India; [2]Department of Pharmaceutical Chemistry, D. Y. Patil Education Society, D. Y. Patil College of Pharmacy, Kolhapur, Maharashtra, India; [3]Department of Pharmaceutics, Vijayrao Naik College of Pharmacy, Shirwal, Maharashtra, India; [4]Department of Pharmaceutical Chemistry, Bharati Vidyapeeth College of Pharmacy, Palus, Sangli, Maharashtra, India; [5]Department of Pharmaceutics, Genesis Institute of Pharmacy, Kolhapur, Maharashtra, India

1. Introduction

Intelligent Semantic Healthcare refers to the integration of semantic technologies with intelligent systems to improve healthcare delivery, management, and outcomes. Semantic technologies enable machines to understand and interpret the meaning of information, facilitating better data integration, retrieval, and usage. Intelligent systems, powered by AI and machine learning, leverage this semantically rich data to provide advanced analytics, decision support, and personalized healthcare solutions (Sarker, 2022).

Semantic technologies encompass a range of tools and standards that allow for the representation, sharing, and linking of data across diverse systems. These include:

1.1 Ontologies

Ontologies are structured frameworks that define the relationships between concepts within a specific domain, providing a common language for consistent data interpretation and integration. In healthcare, ontologies like SNOMED CT (Systematized Nomenclature of Medicine—Clinical Terms) and the Unified Medical Language System (UMLS) play a crucial role. SNOMED CT offers a comprehensive clinical vocabulary that covers a wide range of medical concepts, including diseases, symptoms, procedures, and findings, and establishes relationships among these concepts. This standardization allows for accurate data exchange and integration across various healthcare systems and applications, ensuring that clinical data is consistently recorded and interpreted (Sachdeva and Bhalla, 1936). It includes a Metathesaurus, which is a large thesaurus of biomedical and health-related concepts and their relationships, and a Semantic Network that categorizes these concepts and their interrelationships. Additionally, UMLS provides lexical tools for natural language processing, which help extract and standardize information from textual sources (Torab-Miandoab et al., 2023).

Intelligent Semantic Analysis for Healthcare
ISBN 978-0-443-33512-9,
https://doi.org/10.1016/B978-0-443-33512-9.00010-X

1.2 Semantic web technologies

Semantic Web technologies, such as RDF (Resource Description Framework) and OWL (Web Ontology Language), are pivotal in creating interconnected data networks that facilitate the integration and linking of disparate data sources. RDF provides a framework for representing information about resources in a graph form, using triples composed of a subject, predicate, and object. This structure allows data to be interlinked and queried efficiently, enabling the retrieval of relevant information across different systems. OWL, on the other hand, is a language designed for defining and instantiating Web ontologies. It provides more expressive power than RDF Schema (RDFS) for specifying complex relationships between concepts and for creating rich, semantically meaningful data models. By using these technologies, disparate data sources can be seamlessly integrated, making it easier to access and combine information from multiple origins. This capability is especially valuable in fields like healthcare, where data from various sources, such as electronic health records, research databases, and clinical trials, need to be linked and analyzed together. Semantic Web technologies thus enhance data interoperability and enable more comprehensive and meaningful data analysis, driving advancements in knowledge discovery and decision-making (Bose and Sugumaran, 2007).

1.3 Natural language processing (NLP)

Natural Language Processing (NLP) encompasses techniques that enable machines to understand and process human language, facilitating the extraction of meaningful information from text. In the healthcare sector, NLP is particularly valuable as it can analyze unstructured data sources such as clinical notes, patient records, and research articles. These unstructured texts contain a wealth of information that is often not easily accessible through traditional data processing methods (Chowdhary and Chowdhary, 2020). This structured data can then be analyzed and integrated with other health information systems, enabling healthcare providers to gain actionable insights. Additionally, NLP can assist in literature review by summarizing research articles and identifying relevant studies, which can be particularly useful for keeping up with the vast amount of new medical research (Williams and Petrovich, 2023).

1.3.1 Applications of semantic technologies in healthcare

Semantic technologies are revolutionizing various aspects of healthcare by enhancing data integration, improving decision-making, and accelerating research. Here are some of the key applications:

Electronic Health Records (EHRs): Semantic technologies significantly enhance the interoperability of Electronic Health Records (EHRs) by providing a standardized framework for integrating data from various sources. This framework ensures that information from different healthcare systems—such as hospitals, clinics, and laboratories—can be accurately combined and understood, addressing the challenges posed by disparate data formats and terminologies (Panahiazar et al., 2014). This enriched record supports better-coordinated care and facilitates more informed clinical decision-making. Furthermore, the use of standardized data formats and terminologies promotes seamless data exchange between healthcare providers, leading to more efficient communication and collaboration. This improved data sharing is essential for delivering high-quality care and ensuring that all members of a patient's healthcare team have access to relevant and accurate information (Peng et al., 2020).

Clinical Decision Support Systems (CDSS): Clinical Decision Support Systems (CDSS) utilize semantic technologies to deliver evidence-based recommendations that significantly enhance healthcare practices. By leveraging ontologies and structured data, CDSS offer relevant suggestions for diagnosis and treatment, which improves clinical outcomes and supports more effective healthcare interventions. The application of semantic data within these systems refines diagnostic accuracy and helps optimize treatment plans by analyzing patient data in the context of comprehensive medical knowledge (Wagholikar et al., 2012). This analysis provides valuable insights that enable precise and tailored healthcare decisions. Furthermore, ontologies ensure consistent interpretation of medical data across various systems, which enhances the reliability and effectiveness of decision support tools. This consistency is crucial for maintaining the integrity of recommendations and ensuring that healthcare professionals can trust the information provided by these systems (Dissanyake et al., 2020).

Research and Development: Semantic technologies play a pivotal role in research and development by facilitating the integration of data from diverse research sources, such as clinical trials, observational studies, and laboratory results. This integration supports a more comprehensive analysis of medical data, allowing researchers to synthesize information from various studies and gain a holistic view of their research domain. By linking and standardizing research data, semantic technologies enhance the effectiveness of data analysis, enabling the identification of patterns and correlations that might not be evident in isolated datasets (Sim et al., 2014). This comprehensive and cohesive approach to data analysis accelerates the discovery of new medical insights and innovations, significantly contributing to the development of novel treatments and therapies. The ability to integrate and analyze diverse datasets not only streamlines research processes but also fosters breakthroughs that advance medical science and improve patient care (Ohmann and Kuchinke, 2018).

Personalized Medicine: Semantic technologies significantly advance personalized medicine by enabling the customization of treatment plans based on a comprehensive analysis of individual patient data. This includes genetic information, medical history, and lifestyle factors, allowing for tailored treatment strategies that are specifically designed to meet the unique needs of each patient. Such personalization not only enhances the effectiveness of treatments but also minimizes the risk of adverse effects by considering the patient's individual characteristics and health conditions. By integrating and analyzing patient-specific data, semantic technologies facilitate the development of more targeted and effective treatment strategies, leading to improved patient outcomes. This approach ensures that treatments are not only more precise but also more aligned with each patient's distinct health profile, ultimately contributing to better overall healthcare and quality of life (Zhang et al., 2018).

Public Health Surveillance: Semantic technologies greatly enhance public health surveillance by integrating data from diverse sources, including health records, epidemiological studies, and sensor data. This integration enables more accurate and timely monitoring of disease outbreaks and public health trends (Savel and Foldy, 2012). By correlating data from different reports and studies, semantic technologies provide a comprehensive view of public health issues, which is crucial for understanding complex health patterns and trends. This holistic analysis supports more informed and effective responses to public health challenges, allowing for targeted interventions and resource allocation. Ultimately, the use of semantic technologies in public health surveillance enhances the ability to manage and mitigate health risks, contributing to better population health and more efficient public health strategies (Mirhaji, 2009).

Medical Imaging: Integrating semantic technologies with medical imaging significantly enhances the analysis of imaging data, leading to more advanced and accurate interpretations. This integration improves the detection of anomalies and allows for a more nuanced understanding of imaging results by contextualizing them with patient records. Additionally, semantic technologies facilitate the linkage of imaging data with other relevant patient information, such as clinical notes and Electronic Health Records (EHRs). This comprehensive approach provides a holistic view of the patient's health, supporting more precise diagnoses and informed treatment planning. By combining imaging data with broader health information, semantic technologies enable a more integrated and accurate assessment of patient conditions, ultimately contributing to better clinical outcomes and more effective healthcare delivery (Panayides et al., 2020).

2. Overview of intelligent systems in healthcare

Intelligent systems in healthcare leverage advancements in Artificial Intelligence (AI) and machine learning (ML) to analyze large volumes of data, identify patterns, and generate predictive models. These systems are instrumental in enhancing various aspects of healthcare, including diagnosis, treatment planning, and patient management (Ahmed et al., 2020) (Table 22.1).

Table 22.1 Key components of intelligent systems in healthcare (Zafar et al., 2023).

Key component	Description	Applications in healthcare
Artificial intelligence (AI)	Broad range of technologies designed to perform tasks requiring human intelligence.	**Image recognition:** Analyzes medical images (X-rays, MRIs, CT scans) for accurate detection of abnormalities. **Predictive analytics:** Predicts patient outcomes, disease progression, and complications. **Virtual health assistants:** Provides information, manages appointments, and assists in health monitoring.
Machine learning (ML)	A subset of AI that involves training algorithms on large datasets to recognize patterns and make predictions.	**Disease progression prediction:** Forecasts the progression of diseases to enable timely interventions. **High-risk patient identification:** Identifies patients at high risk of certain conditions for proactive care. **Personalized treatment suggestions:** Recommends tailored treatments based on patient data and historical outcomes.
Deep learning	Specialized branch of machine learning using neural networks with many layers to analyze complex data.	**Medical imaging:** Enhances image analysis for improved detection and classification of conditions. **Genomics:** Analyzes genetic data to identify mutations and disease associations. **Drug discovery:** Predicts interactions of compounds with biological targets to accelerate drug development.

3. Current state of intelligent semantic healthcare

3.1 Existing technologies and applications

3.1.1 Natural language processing (NLP) for medical records

NLP technologies are used to analyze and interpret unstructured data from medical records, such as clinical notes and patient histories. NLP converts this textual information into structured data that can be integrated with other health records and used for further analysis (Sheikhalishahi et al., 2019).

Applications:

Information Extraction: NLP extracts pertinent medical information from unstructured text, including details about symptoms, diagnoses, and treatment plans. This extraction process ensures that crucial data points are identified and made available for use in clinical decision-making and research.

Clinical Documentation Improvement: NLP improves the quality and completeness of clinical documentation by identifying gaps or inconsistencies in the recorded data. By suggesting enhancements and corrections, NLP contributes to more accurate and comprehensive patient records.

Patient Risk Assessment: NLP analyzes free-text data in medical records to assess patient risk factors and identify individuals who may be at risk for various conditions. This analysis supports early intervention and preventive measures, leading to better patient outcomes and more effective management of chronic diseases (Locke et al., 2021).

3.1.2 Semantic web technologies in health informatics

Semantic Web technologies, including the Resource Description Framework (RDF) and the Web Ontology Language (OWL), facilitate the creation of interconnected data networks by employing standardized vocabularies and ontologies. These technologies enhance data integration and interoperability across various health information systems.

Applications:

Data Integration: Semantic Web technologies enable the consolidation of data from diverse sources, such as Electronic Health Records (EHRs), research databases, and public health records. By providing a unified view of patient information, these technologies support comprehensive data analysis and improve the continuity of care.

Interoperability: These technologies ensure that different health information systems can effectively communicate and understand each other's data. This interoperability is crucial for seamless data exchange and coordination among various healthcare providers and systems.

Knowledge Representation: Semantic Web technologies allow for the representation of complex medical knowledge in a structured, queryable format. This structured representation enables advanced querying and analysis of medical data, facilitating better decision-making and knowledge sharing across healthcare systems (Kumar et al., 2017).

3.1.3 AI-driven diagnostic tools

AI-driven diagnostic tools utilize advanced algorithms to analyze medical images and data, providing accurate and timely diagnoses. These tools leverage machine learning and deep learning techniques to enhance diagnostic capabilities and support healthcare professionals in their decision-making processes (Yeasmin et al., 2024).

Applications:

Medical Imaging: AI algorithms are employed to analyze medical images, such as X-rays and MRIs, with high precision. These tools detect abnormalities, including tumors, fractures, and other conditions, facilitating early diagnosis and improving the accuracy of image interpretation.

Predictive Diagnostics: AI models predict the likelihood of diseases by analyzing various types of patient data, including imaging results and genetic information. These predictive capabilities enable healthcare providers to anticipate potential health issues and implement preventive measures or early interventions.

Clinical Decision Support: AI-driven diagnostic tools offer real-time support to healthcare professionals by providing evidence-based recommendations and alerts. These systems analyze patient data and medical knowledge to assist in making informed decisions, enhancing diagnostic accuracy and treatment planning (Abdel galil Abdel galil, 2023).

3.2 Case studies and real-world examples

3.2.1 Implementations in electronic health records (EHRs)

The integration of Natural Language Processing (NLP) and semantic technologies into Electronic Health Records (EHRs) significantly enhances data extraction and integration processes. By employing these technologies, systems like IBM Watson Health's NLP capabilities can transform unstructured data from clinical notes and patient histories into structured information that is seamlessly incorporated into EHRs. This integration improves the accuracy of patient records and facilitates better care coordination by providing a comprehensive view of patient information. The result is increased data accuracy and usability, which in turn leads to more informed clinical decision-making and more efficient patient care. Enhanced data quality enables healthcare professionals to make better decisions, streamline workflows, and ultimately improve patient outcomes (Savova et al., 2019).

3.2.2 Decision Support Systems (DSS)

Clinical Decision Support Systems (DSS), such as IBM Watson for Oncology, harness the power of AI and semantic technologies to analyze patient data and offer evidence-based treatment recommendations. By integrating vast medical knowledge databases and clinical guidelines, these systems provide healthcare professionals with valuable insights to support decision-making. The use of DSS improves diagnostic accuracy and optimizes treatment planning, leading to more personalized and effective patient care. This integration of advanced technologies ensures that treatment decisions are informed by comprehensive and up-to-date medical information, ultimately enhancing patient outcomes and contributing to more tailored healthcare solutions (El Naqa et al., 2018).

3.2.3 Personalized medicine

The integration of AI and semantic technologies in personalized medicine enables the development of tailored treatment plans based on individual patient profiles. For instance, platforms like Tempus utilize AI to analyze a combination of genetic and clinical data to recommend personalized cancer treatments. This approach allows for treatments that are specifically designed to match the unique characteristics of each patient's condition. The impact of such personalized strategies includes more effective treatments with reduced side effects, leading to improved patient outcomes and more optimized therapeutic strategies. By aligning treatment plans with the individual genetic and clinical profiles of patients, healthcare providers can offer care that is both targeted and efficient (Vidal et al., 2019).

4. Challenges in intelligent semantic healthcare

4.1 Data integration and interoperability

4.1.1 Heterogeneity of data sources

Healthcare data is produced from a range of sources, including electronic health records (EHRs), laboratory results, imaging systems, and wearable devices. Integrating these diverse types of data into a cohesive system poses a significant challenge. Each source employs its own format and structure, which complicates the process of consolidating and analyzing the information in a seamless manner. The varied nature of these data sources means that they must be harmonized to provide a unified view of patient information, a task that requires overcoming substantial technical and organizational hurdles (Huang et al., 2018).

4.1.2 Standardization issues

In healthcare, standards such as HL7 (Health Level Seven) and FHIR (Fast Healthcare Interoperability Resources) are used for data exchange and integration. However, ensuring consistency and compatibility across these standards presents a significant challenge. Variations in implementation and interpretation can lead to discrepancies, complicating efforts to create systems that can effectively communicate and exchange data. These inconsistencies can hinder the development of truly interoperable systems, making it difficult for healthcare providers to achieve seamless data integration and efficient information sharing across different platforms and organizations (Vorisek et al., 2022).

4.1.3 Semantic interoperability

Beyond the technical aspects of data integration, ensuring that information from different sources maintains a consistent meaning across various systems is essential. Semantic interoperability focuses on aligning the definitions and relationships of medical terms and concepts to ensure that data is accurately interpreted and utilized. This involves creating a shared understanding of medical terminology and ontologies, which can be complex and challenging to standardize across diverse healthcare environments. Achieving semantic interoperability requires harmonizing how medical concepts are represented and understood across different systems, facilitating accurate data exchange and meaningful insights (Bittner et al., 2005).

4.2 Data privacy and security

4.2.1 Ensuring confidentiality and integrity of health data

In healthcare, protecting sensitive health information from unauthorized access while maintaining its integrity is of utmost importance. Implementing effective security measures is crucial to safeguard patient privacy and prevent potential data breaches. These measures must be robust enough to prevent unauthorized access yet flexible enough to support necessary data sharing and analysis. Balancing stringent security protocols with the operational requirements of healthcare systems ensures that patient data remains confidential and accurate, enabling both secure handling and effective use in clinical settings (Thapa and Camtepe, 2021).

4.2.2 Compliance with regulations

Adhering to regulatory frameworks such as HIPAA (Health Insurance Portability and Accountability Act) in the U.S. and GDPR (General Data Protection Regulation) in Europe is essential for upholding data privacy and security in healthcare. These regulations impose stringent requirements on how data is handled, emphasizing the importance of privacy and consent. Healthcare organizations must comply with these regulations to protect patient information, ensuring that data practices are both lawful and secure. Rigorous adherence to HIPAA and GDPR helps maintain trust and integrity within healthcare systems by safeguarding sensitive health data and respecting patient rights (Bakare et al., 2024).

4.3 Scalability and performance

4.3.1 Handling large volumes of data

The healthcare industry generates vast amounts of data daily, including patient records, diagnostic results, imaging data, and other critical information. Intelligent systems must be engineered to efficiently manage and process this extensive volume of data. It is essential that these systems handle large data sets without compromising performance to ensure operational effectiveness. Effective data management is crucial for maintaining high-quality patient care and supporting the smooth functioning of healthcare operations (Dash et al., 2019).

4.3.2 *Real-time processing and response times*

In numerous healthcare applications, such as clinical decision support systems, real-time data processing is vital. These systems must analyze incoming data and provide immediate responses to facilitate effective decision-making and patient management. The ability to process information swiftly and deliver timely feedback is crucial for enhancing the quality of care and optimizing treatment outcomes. Ensuring that intelligent systems maintain high performance and low latency is essential for supporting critical clinical decisions and improving patient care efficiency (Abirami and Chitra, 2020).

4.4 Accuracy and reliability

4.4.1 *Ensuring high accuracy in diagnostic tools*

For AI-driven diagnostic tools to be effective, they must provide highly accurate results. Achieving and maintaining rigorous accuracy standards is essential for ensuring the reliability of these tools in clinical settings. High accuracy is crucial for making correct diagnoses and delivering appropriate treatment recommendations, which directly influences patient outcomes and fosters trust in the technology. Reliable diagnostic tools help clinicians make informed decisions, enhancing the overall quality of care and reinforcing confidence in AI applications within healthcare (Riaz, 2024).

4.4.2 *Dealing with biases in AI models*

AI models can inherit biases from their training data, potentially leading to skewed or unfair outcomes. Addressing and mitigating these biases is crucial for ensuring that diagnostic and treatment recommendations are equitable and reliable. Identifying and correcting biases helps improve the fairness and effectiveness of AI-driven healthcare solutions, fostering more accurate and just outcomes for all patients. By actively working to reduce these biases, we can enhance the overall performance of AI tools and promote better, more equitable patient care (González Sendino et al., 2024).

4.5 Ethical and legal issues

4.5.1 *Ethical considerations in AI decision-making*

The integration of AI into healthcare raises important ethical questions about how decisions are made. Understanding the decision-making processes of AI systems is essential to ensure they adhere to ethical standards. This involves examining the fairness, transparency, and accountability of AI-driven decisions. Ensuring that these processes align with established ethical principles helps prevent potential harm to patients and avoids unjust outcomes, maintaining trust in AI technologies while safeguarding the integrity of patient care (Gabriel, 2023).

4.5.2 Liability and accountability in automated systems

Determining liability for errors or adverse outcomes resulting from automated systems is a complex issue that intertwines legal and ethical concerns. Establishing clear mechanisms for accountability is crucial to address these challenges effectively. This involves defining the roles and responsibilities of all parties involved, including developers, healthcare providers, and the AI systems themselves. It is important to ensure that there are fair and effective processes in place for managing and resolving errors, which helps in addressing legal claims and maintaining trust in automated healthcare solutions. Clear accountability frameworks are necessary to navigate the intricacies of liability and to ensure that stakeholders are appropriately responsible for the outcomes of AI-driven systems (Schönberger, 2019).

4.6 User acceptance and trust

4.6.1 Gaining trust from healthcare professionals

For intelligent systems to be successfully integrated into clinical practice, it is crucial to gain the trust of healthcare professionals. Demonstrating the reliability and transparency of these systems is essential for securing their acceptance. Ensuring that the technology delivers accurate results and enhances the quality of care without compromising patient safety can build confidence among practitioners. Engaging with healthcare professionals and addressing their concerns about the technology's performance and utility helps in fostering trust and facilitating smoother integration into clinical workflows (Unertl et al., 2020).

4.6.2 Addressing fears and resistance to new technologies

Healthcare professionals may resist adopting new technologies due to concerns about their reliability or potential disruptions to established workflows. To overcome these fears, it is important to address these concerns directly and provide comprehensive training and support. Clear communication about the benefits of the technology, coupled with practical demonstrations and ongoing assistance, can help alleviate resistance and promote a positive reception. By actively engaging with users and offering resources to help them adapt, the transition to new technologies can be made more seamless and effective (Ngafeeson, 2015).

5. Future scope of intelligent semantic healthcare

5.1 Advancements in technology

The future of intelligent semantic healthcare is on the brink of significant breakthroughs driven by continuous advancements in technology. AI and machine learning are expected to enhance the accuracy and capabilities of diagnostic and predictive tools, enabling more precise and effective healthcare solutions. Improved semantic web technologies will play a crucial role in advancing data integration and interoperability, ensuring seamless information exchange across various platforms. Additionally, the integration of Internet of Things (IoT) and wearable devices will provide real-time health monitoring and deliver personalized insights, pushing the boundaries of proactive and tailored healthcare. These technological advancements will collectively push the field forward, offering innovative solutions and improving overall patient care (Sikdar and Guha, 2020).

5.2 Personalized medicine

Semantic technologies are set to significantly advance the field of precision medicine. By leveraging detailed semantic analysis of patient data, these technologies enable a more nuanced understanding of individual health profiles. This allows for the development of treatment plans that are finely tuned to each patient's unique needs, taking into account genetic, environmental, and lifestyle factors. The use of semantic technologies in analyzing patient data facilitates the identification of the most effective therapies and potential adverse effects specific to each individual. As a result, personalized medicine can enhance treatment efficacy, reduce side effects, and lead to more targeted and individualized care strategies. This shift toward a more personalized approach promises to improve patient outcomes and revolutionize healthcare delivery by aligning treatments more closely with each patient's distinct health characteristics (Amir-Aslani and Mangematin, 2010).

5.3 Enhanced decision support systems

The future of clinical decision support systems (CDSS) is set to advance significantly with the integration of next-generation capabilities. Emerging technologies will enable CDSS to incorporate advanced predictive analytics and proactive healthcare management features. These systems will be capable of providing more precise and timely recommendations by analyzing vast amounts of patient data and relevant medical information. As a result, healthcare professionals will be better supported in making informed decisions, leading to improved patient outcomes and more effective treatment strategies. The continued evolution of CDSS promises to enhance clinical decision-making processes, contributing to more efficient and personalized patient care (Gabriel, 2023).

5.4 Interoperability and standardization

The future of intelligent semantic healthcare will heavily rely on the establishment of global standards for health data to ensure seamless data exchange and integration. Developing improved frameworks for interoperability will facilitate more effective sharing of health information across diverse systems and geographic regions. This will enable better coordination of care by allowing healthcare providers to access and utilize comprehensive patient data regardless of its source. Enhanced interoperability will support more cohesive health management, leading to improved patient outcomes and a more integrated healthcare ecosystem (Riaz, 2024).

5.5 Ethical AI in healthcare

As intelligent systems become increasingly integral to healthcare, developing robust ethical guidelines and frameworks for AI will be crucial. These guidelines will address critical issues related to fairness, accountability, and transparency, ensuring that AI systems operate within ethical boundaries. Establishing these standards will help maintain trust and reliability in healthcare applications, ensuring that AI technologies are used responsibly and ethically. By upholding high standards of integrity, the healthcare industry can foster confidence in AI-driven solutions, ultimately enhancing their effectiveness and acceptance in clinical practice.

5.6 Patient-centric care

Future advancements in intelligent semantic healthcare will increasingly focus on empowering patients with enhanced access to their health data. Semantic technologies will play a pivotal role in facilitating personalized interactions and streamlined information sharing, enabling patients to make more informed decisions about their health. By improving patient engagement and education, these technologies will support individuals in taking an active role in managing their care. As a result, patients will be better equipped to participate in their treatment plans and overall health management, leading to a more collaborative and effective approach to healthcare (Bittner et al., 2005).

6. Conclusion

Intelligent semantic healthcare is poised to revolutionize the management, analysis, and utilization of healthcare data by integrating advanced technologies such as artificial intelligence (AI), machine learning, and semantic technologies. Despite its promising potential, several challenges must be addressed. Data integration and interoperability issues stem from the heterogeneity of healthcare data sources and the need for standardized frameworks. Privacy and security concerns are paramount, requiring stringent measures to protect sensitive health information while ensuring compliance with regulations like HIPAA and GDPR. Additionally, the scalability and performance of systems must handle vast amounts of data efficiently and provide real-time processing capabilities. Ensuring accuracy and reliability in diagnostic tools and addressing biases in AI models are crucial for fair and effective healthcare solutions. Ethical and legal considerations also play a significant role, necessitating clear guidelines and accountability for AI-driven decisions. Gaining the trust of healthcare professionals and overcoming resistance to new technologies are essential for successful implementation.

Looking to the future, advancements in technology will drive the evolution of intelligent semantic healthcare. Innovations in AI, machine learning, and semantic web technologies, coupled with the integration of IoT and wearable devices, will enhance the capabilities of healthcare systems. Personalized medicine will benefit from semantic technologies by enabling more precise and individualized treatment plans, improving efficacy and patient outcomes. Enhanced decision support systems will offer advanced predictive analytics and proactive management, supporting better clinical decisions. Efforts toward global standards and improved data-sharing frameworks will promote interoperability and seamless integration. Ethical AI practices will ensure fairness, transparency, and accountability, fostering trust in technological solutions. Lastly, patient-centric care will be enhanced through greater access to health data and improved engagement, leading to more informed patient participation. As the field continues to evolve, addressing these challenges while leveraging future opportunities will be key to advancing healthcare delivery and improving patient care.

References

Abdel galil, W., 2023. Applications of artificial intelligence in the field of diagnostic medicine and future prospects. Int. J. Artif. Intell. Emerg. Technol 6 (2), 1–18. https://doi.org/10.21608/ijaiet.2024.275176.1005.

Abirami, S., Chitra, P., 2020. Energy-efficient edge based real-time healthcare support system. Adv. Comput. 117, 339–368. https://doi.org/10.1016/bs.adcom.2019.09.007.

Ahmed, Z., Mohamed, K., Zeeshan, S., Dong, X.Q., 2020. Artificial intelligence with multi-functional machine learning platform development for better healthcare and precision medicine. Database 2020. https://doi.org/10.1093/database/baaa010.

Amir-Aslani, A., Mangematin, V., 2010. The future of drug discovery and development: shifting emphasis towards personalized medicine. Technol. Forecast. Soc. Change 77 (2), 203–217. https://doi.org/10.1016/j.techfore.2009.09.005.

Bakare, S.S., Adeniyi, A.O., Akpuokwe, C.U., Emmanuella Eneh, N., 2024. Data privacy laws and compliance: a comparative review of the EU GDPR and USA regulations. Comput. Sci. IT. Res. J 5 (3), 528–543. https://doi.org/10.51594/csitrj.v5i3.859.

Bittner, T., Donnelly, M., Winter, S., 2005. Ontology and semantic interoperability large-scale 3D data integration: challenges and opportunities. CRC Press, United States, pp. 139–160. Available from: https://www.taylorfrancis.com/books/e/9781420036282.

Bose, R., Sugumaran, V., 2007. Semantic web technologies for enhancing intelligent DSS environments. Decis. Support. Glob. Enterp 221–238. https://doi.org/10.1007/978-0-387-48137-1_13.

Chowdhary, K., 2020. Natural language processing. Fundam. Artif. intell 603–649. https://doi.org/10.1007/978-81-322-3972-7_19.

Dash, S., Kumar Shakyawar, S., Sharma, M., Kaushik, S., 2019. Big data in healthcare: management, analysis and future prospects. J. Big Data 6 (1). https://doi.org/10.1186/s40537-019-0217-0.

Dissanayake, P.I., Colicchio, T.K., Cimino, J.J., 2020. Using clinical reasoning ontologies to make smarter clinical decision support systems: a systematic review and data synthesis. J. Am. Med. Inform. Assoc. 27 (1), 159–174. https://doi.org/10.1093/jamia/ocz169.

El Naqa, I., Kosorok, M.R., Jin, J., Mierzwa, M., Ten Haken, R.K., 2018. Prospects and challenges for clinical decision support in the era of big data. JCO. Clin. Cancer. Inform 2 (2), 1–12. https://doi.org/10.1200/cci.18.00002.

Gabriel, O.T., 2023. Data Privacy and Ethical Issues in Collecting Health Care Data Using Artificial Intelligence Among Health Workers: Center for Bioethics and Research.

González Sendino, R., Serrano, E., Bajo, J., Novais, P., 2024. A review of bias and fairness in artificial intelligence. Int. J. Interact. Multimed. Artif. Intell 9 (1), 5–17. https://doi.org/10.9781/ijimai.2023.11.001.

Huang, M., Han, H., Wang, H., Li, L., Zhang, Y., Bhatti, U.A., 2018. A clinical decision support framework for heterogeneous data sources. IEEE J. Biomed. Health Inform. 22 (6), 1824–1833. https://doi.org/10.1109/JBHI.2018.2846626.

Kumar, K.P., Kumar, V.R., Raghuveer, K., 2017. A survey on semantic Web technologies for the Internet of Things. 017 International Conference on Current Trends in Computer, Electrical, Electronics and Communication (CTCEEC). IEEE.

Locke, S., Bashall, A., Al-Adely, S., Moore, J., Wilson, A., Kitchen, G.B., 2021. Natural language processing in medicine: a review. Trends. Anaesth. Crit. Care 38, 4–9. https://doi.org/10.1016/j.tacc.2021.02.007.

Mirhaji, P., 2009. Public health surveillance meets translational informatics: a desiderata. JALA: J. Assoc. Lab. Autom. 14 (3), 157–170. https://doi.org/10.1016/j.jala.2009.02.007.

Ngafeeson, M., 2015. Understanding User Resistance to Information Technology in Healthcare: The Nature and Role of Perceived Threats.

Ohmann, C., Kuchinke, W., 2018. Future developments of medical informatics from the viewpoint of networked clinical research. Methods Inf. Med. 48 (01), 45–54. https://doi.org/10.3414/me9137.

Panahiazar, M., Taslimitehrani, V., Jadhav, A., Pathak, J., 2014. Empowering personalized medicine with big data and semantic web technology: promises, challenges, and use cases. In: Proceedings - 2014 IEEE International Conference on Big Data, IEEE Big Data, pp. 790–795. https://doi.org/10.1109/BigData.2014.7004307.

Panayides, A.S., Amini, A., Filipovic, N.D., Sharma, A., Tsaftaris, S.A., Young, A., Foran, D., Do, N., Golemati, S., Kurc, T., Huang, K., Nikita, K.S., Veasey, B.P., Zervakis, M., Saltz, J.H., Pattichis, C.S., 2020. AI in medical imaging informatics: current challenges and future directions. IEEE J. Biomed. Health Inform. 24 (7), 1837–1857. https://doi.org/10.1109/JBHI.2020.2991043.

Peng, C., Goswami, P., Bai, G., 2020. A literature review of current technologies on health data integration for patient-centered health management. Health Inf. J. 26 (3), 1926–1951. https://doi.org/10.1177/1460458219892387.

Riaz, M.S., 2024. Towards AI-Enabled Diagnostics From Noisy and Incomplete Data.

Sachdeva, S., Bhalla, S., 2012. Semantic interoperability in standardized electronic health record databases. J. Data. Inf. Qual 3 (1), 1–37. https://doi.org/10.1145/2166788.2166789.

Sarker, I.H., 2022. AI-based modeling: techniques, applications and research issues towards automation, intelligent and smart systems. SN Comput. Sci. 3 (2). https://doi.org/10.1007/s42979-022-01043-x.

Savova, G.K., Danciu, I., Alamudun, F., Miller, T., Lin, C., Bitterman, D.S., Tourassi, G., Warner, J.L., 2019. Use of natural language processing to extract clinical cancer phenotypes from electronic medical records. Cancer Res. 79 (21), 5463–5470. https://doi.org/10.1158/0008-5472.CAN-19-0579.

Savel, T.G., Foldy, S., 2012. The role of public health informatics in enhancing public health surveillance. Morb. Mortal. Wkly. Rep. 61.

Schönberger, D., 2019. Artificial intelligence in healthcare: a critical analysis of the legal and ethical implications. Int. J. Law Info Technol. 27 (2), 171–203. https://doi.org/10.1093/ijlit/eaz004.

Sheikhalishahi, S., Miotto, R., Dudley, J.T., Lavelli, A., Rinaldi, F., Osmani, V., 2019. Natural Language processing of clinical notes on chronic diseases: systematic review. JMIR Med. Inform. 7 (2), e12239. https://doi.org/10.2196/12239.

Sim, I., Tu, S.W., Carini, S., Lehmann, H.P., Pollock, B.H., Peleg, M., Wittkowski, K.M., 2014. The Ontology of Clinical Research (OCRe): an informatics foundation for the science of clinical research. J. Biomed. Inf. 52, 78–91. https://doi.org/10.1016/j.jbi.2013.11.002.

Sikdar, S., Guha, S., 2020. Advancements of healthcare technologies: Paradigm towards smart healthcare systems. Recent. Trends. Image. Signal. Process. Comput. Vis 113–132.

Thapa, C., Camtepe, S., 2021. Precision health data: requirements, challenges and existing techniques for data security and privacy. Comput. Biol. Med. 129, 104130. https://doi.org/10.1016/j.compbiomed.2020.104130.

Torab-Miandoab, A., Samad-Soltani, T., Jodati, A., Rezaei-Hachesu, P., 2023. Interoperability of heterogeneous health information systems: a systematic literature review. BMC Med. Inf. Decis. Making 23 (1). https://doi.org/10.1186/s12911-023-02115-5.

Unertl, K.M., Novak, L.L., Van Houten, C., Brooks, J.A., Smith, A.O., Harris, J.W., Avery, T., Simpson, C., Lorenzi, N.M., 2020. Organizational diagnostics: a systematic approach to identifying technology and workflow issues in clinical settings. JAMIA Open 3 (2), 269–280. https://doi.org/10.1093/jamiaopen/ooaa013.

Vidal, M.-E., Endris, K.M., Jozashoori, S., Karim, F., Palma, G., 2019. Semantic data integration of big biomedical data for supporting personalised medicine. Curr. Trends. Semant. Web. Technol. Theory. Practice 25–56.

Vorisek, C.N., Lehne, M., Klopfenstein, S.A.I., Mayer, P.J., Bartschke, A., Haese, T., Thun, S., 2022. Fast healthcare interoperability resources (FHIR) for interoperability in health research: systematic review. JMIR Med. Inform. 10 (7). https://doi.org/10.2196/35724.

Wagholikar, K.B., Sundararajan, V., Deshpande, A.W., 2012. Modeling paradigms for medical diagnostic decision support: a survey and future directions. J. Med. Syst. 36 (5), 3029–3049. https://doi.org/10.1007/s10916-011-9780-4.

Williams, S., Petrovich, E., 2023. Natural Language processing for unlocking insights from unstructured big data in the healthcare industry. Emerg. Trends. Mach. Intell. Big. Data 15 (10), 30–39.

Yeasmin, M.N., Al Amin, M., Joti, T.J., Aung, Z., Azim, M.A., 2024. Advances of AI in image-based computer-aided diagnosis: a review. Array, 100357.

Zafar, I., Anwar, S., kanwal, F., Yousaf, W., Un Nisa, F., Kausar, T., Ain, Q.ul, Unar, A.nullah, Kamal, M. A., Rashid, S., Ali Khan, K., Sharma, R., 2023. Reviewing methods of deep learning for intelligent healthcare systems in genomics and biomedicine. Biomed. Signal. Process. Control 86, 105263. https://doi.org/10.1016/j.bspc.2023.105263.

Zhang, S., Bamakan, S.M.H., Qu, Q., Li, S., 2018. Learning for personalized medicine: a comprehensive review from a deep learning perspective. IEEE Rev. Biomed. Eng. 12, 194–208. https://doi.org/10.1109/RBME.2018.2864254.

Index

'*Note:* Page numbers followed by "f" indicate figures and "t" indicate tables.'

D

E

I

T

U

V

W

X

Printed and bound by CPI Group (UK) Ltd, Croydon, CR0 4YY
21/09/2026
02220078-0005